AN INTRODUCTION TO THE THEORY OF CLIMATE

ATMOSPHERIC SCIENCES LIBRARY

A. S. MONIN
P. P. Shirshov Institute of Oceanology,
Academy of Sciences of the U.S.S.R.

AN INTRODUCTION TO THE THEORY OF CLIMATE

D. REIDEL PUBLISHING COMPANY

A MEMBER OF THE KLUWER ACADEMIC PUBLISHERS GROUP

DORDRECHT / BOSTON / LANCASTER / TOKYO

Library of Congress Cataloging-in-Publication Data

Monin, A. S. (Andreĭ Sergeevich), 1921–
An introduction to the theory of climate.

(Atmospheric sciences library)
Translation of: Vvedenie v teoriiû klimata.
Bibliography: p
Includes index.
1. Climatology. I. Title. II. Series.
QC981.M6613 1986 551.5 85-30115
ISBN 90-277-1935-7

Published by D. Reidel Publishing Company,
P.O. Box 17, 3300 AA Dordrecht, Holland.

Sold and distributed in the U.S.A. and Canada
by Kluwer Academic Publishers,
190 Old Derby Street, Hingham, MA 02043, U.S.A.

In all other countries, sold and distributed
by Kluwer Academic Publishers Group,
P.O. Box 322, 3300 AH Dordrecht, Holland.

Originally published by Gidrometeoizdat
under the title ВВЕДЕНИЕ В ТЕОРИЮ КЛИМАТА

Translated by Ron Hardin

Printed in The Netherlands

TABLE OF CONTENTS

PREFACE

During the last 20 years the study of, and the prediction of, changes in the climate of our planet have become an urgent social imperative, addressed to scientists the world over. The first principles on which to base such a study were formulated in 1974 in Stockholm, at the international GARP conference on the physical fundamentals of climate theory and climate modeling. In 1979 the World Meteorological Organization and the International Council of Scientific Unions decided to conduct a global program of climate research. This World Climate Program is designed mainly to investigate the variability of the climate on time scales ranging from a few weeks to a few decades and to create a scientific basis for the long-term forecasting of weather.

There is at present a definite need for a monograph which can serve as an introduction to the theory of climate. On a qualitative level (without the apparatus of theoretical physics and mathematics) such an introduction has already been presented, in Part I of a book on the history of climate by Yu. A. Shishkov and the author (Monin and Shishkov, 1979). Part II of that work gives factual data on climatic changes during the course of the Earth's history. The present book is designed to provide such an introduction on a quantitative level.

It could probably be claimed that a theory of climate does not yet actually exist. As an introduction to such a theory, let us present here the most important quantitative data on the main elements of the climate system: the atmosphere, the ocean, and the land. The biosphere will not be included, since biological topics should, in our opinion, be considered separately. We will also give considerable information about the theoretical methods currently used to describe these elements. As a rule, we will confine ourselves to formulations pertaining to the present climate of the Earth, without considering paleoclimates or climates of other planets.

The principal method of the future theory of climate would appear to be mathematical modeling, since modeling makes it possible to predict as well as to demonstrate. It should be noted, too, that mathematical models of climate are of more than just intrinsic interest. Since the climate is an important ecological factor influencing the world population, climate models are even now becoming a necessary element of the so-called global models being devised to enable quantitative predictions about the demographic and economic future of mankind.

Consequently, in this book special attention will be given to mathematical models of different elements of the climate system, such as the following:

radiation transfer in the atmosphere; the CO_2 cycle; aerosols; ozone; the circulation of the stratosphere; the oceanic thermocline; the upper layer of the ocean; deep equatorial countercurrents; seepage of moisture through the soil; underground, downhill, and river runoff; sea ice; and continental ice sheets. The last four chapters describe the hierarchy of mathematical models of the climate system, from the simplest of all to the most complicated. So many of these models have been constructed by different authors that even listing them all would be no simple matter, much less describing them. Thus, of the existing models, only those which are typical, or which seem to us to be the most important, will be considered, and we hereby apologize to all those whose models are not mentioned.

Climate is a global concept, and the problem of climatic variations is a global problem. Thus its solution requires a synthesis of diverse bits of knowledge garnered from different branches of the Earth sciences, physics, and mathematics. However, the systematic presentation of all this knowledge would call for a monograph of many volumes, which would no longer be merely an introduction.

The reader is assumed to have already at least a basic knowledge of geophysical hydrodynamics, dynamic meteorology, oceanic physics, descriptive climatology, hydrology, glaciology and oceanography, and probability theory (including the theory of random functions, mathematical statistics, and statistical hydrodynamics). It is also important to be acquainted with numerical methods of solving problems in geophysical hydrodynamics. In any case, information from these branches of science will be given here only very concisely.

The author considers it his pleasant duty to express his most sincere appreciation to G. S. Golitsyn for his help in selecting the material and for his valuable comments, to E. P. Borisenkov for his useful remarks, to N. I. Solntseva for her help in organizing the author's material into a manuscript, and to E. G. Agafonova and G. Yu. Aleksandrova for their great assistance in preparing the manuscript and the figures.

CHAPTER 1

THE CONCEPT OF CLIMATE

1. The Climate System

Weather commonly refers to the state of the lower layer of the atmosphere (including the human habitat) at some given geographical point and at some given time. It is described in terms of the meteorological elements directly perceived by humans, especially the air temperature, the speed and direction of the wind, the quantity and form of clouds, and the quantity and type of precipitation. Sometimes the atmospheric pressure, the air humidity, and the state of the Earth's surface are also cited (the presence of a snow cover, glazed frost, etc.), and in maritime regions the water temperature and the force of the waves are given. *Climate*, on the other hand, refers to the set of weather conditions typical of a given region, together with the frequency of these conditions and their seasonal variations (this is the definition of what E. E. Fedorov has called *complex climatology*).

When analyzing the formation of weather and climate, we sometimes find it advisable to broaden the everyday definitions of these concepts. First of all, it is obvious that the state of the lower layer of the atmosphere depends in many respects on the state of its upper layers. For instance, the atmospheric pressure at the Earth's surface is determined by the weight of the entire air column, the wind near the ground is determined mainly by the wind in the free atmosphere, clouds are distributed throughout the troposphere, and precipitation is typical of all cloud levels. Consequently, weather could be defined more aptly as the instantaneous state of the *entire thickness* of the atmosphere.

Secondly, future weather variations at a given geographical point are determined primarily by the weather at a given time in some surrounding region. This region will be greater, the longer the period of future time considered: for periods longer than a month, the weather in the whole hemisphere has to be taken into account (for example, for westerly transport at a typical velocity of 10 m/s = 36 km/h = 864 km/day, air in the middle latitudes can circumnavigate the globe in a month). Thus, it is advisable to consider the *global weather*, that is, the instantaneous state of the entire atmosphere. Thirdly, during long periods the atmosphere manages to interact appreciably with the ocean and with the upper layer of the land (that is, it exchanges with these substantial amounts of momentum, heat, moisture, and other substances), so that the evolution of the atmosphere is affected by the initial states of the land and sea under it, which, consequently, also have to be taken into account.

Therefore, any analysis of long-term variations in weather and climate must consider the whole atmosphere, the ocean, and the active layer of the land (which, for brevity, will in the following be called simply "the land"), as interacting parts of a single system, the *atmosphere–ocean–land system* or *climate system*. The initial state of the climate system will, *external conditions remaining unchanged*, completely determine its evolution in time.

Now let us compare the roles of the different parts of the atmosphere–ocean–land system. We note first that the mass of the atmosphere is 5.3×10^{15} tons, and that, assuming the specific heat capacity of air to be 1×10^3 J / (kg K), its total heat capacity is 5.32×10^{15} MJ/K. For the ocean we consider here only the upper layer, with its seasonal temperature fluctuations. This layer has a mean thickness of 240 m, a mass of 8.7×10^{16} tons, and a heat capacity of 36.45×10^{16} MJ/K. The active layer of the land has a mean thickness of 10 m, and, assuming a density of 2 tons/m^3, its mass is 3×10^{15} tons. This layer has a specific heat capacity of 0.8×10^3 J / (kg K) and a total heat capacity of 2.38×10^{15} MJ/K. Consequently, the A : O : L mass ratio is 1 : 16.4 : 0.55, and the heat-capacity ratio is 1 : 68.5 : 0.45. Thus, it is clear that, with regard to both mechanical and thermal effects, the ocean is the element of the atmosphere–ocean–land system with the most inertia.

For a quantitative description of states of the atmosphere–ocean–land system (of the global "weather"), we must first select complete sets of independent variables characterizing these states (various functions of these independent variables which may be of interest for one reason or another will be considered only secondarily). Since the atmosphere, ocean, and land are quite inhomogeneous in space, functions of the spatial coordinates (*fields*) have to be used if we are to describe their instantaneous states sufficiently fully. Consequently, an instantaneous state of the atmosphere–ocean–land system is determined mathematically as a multicomponent field (a complete set of independent quantities).

The following fields constitute a fundamental set of independent quantities describing the state of the atmosphere: the field of the three-dimensional wind-velocity vector $\mathbf{V}$, together with the fields of pressure p, temperature T, and water-vapor concentration, or specific humidity, q (in clouds the concentrations of liquid water q_w and ice q_i must also be taken into account, as well as, sometimes, such quantities as the size distribution of water droplets and ice crystals and the crystal shapes). In addition to the moisture content of the air, the concentrations of traces of other *thermodynamically active substances* affecting the radiative transfer of energy are also important; these include CO_2, O_3, and aerosols of various kinds. The principal features of these independent quantities are: cloudiness and precipitation of various forms; the air density ϱ; the entropy η; the potential vorticity; different forms of energy (the internal, potential, available potential, and kinetic energy of the mean circulation and of standing and moving vortices); fluxes of mass, momentum, entropy, various forms of energy (including radiative energy of various wavelengths), and thermodynamically active substances (primarily moisture); and, finally, specific weather phenomena.

The following can be taken as the main characteristics of the state of the ocean: the field of the three-dimensional current-velocity vector **V**, and the fields of the temperature T and salinity S (the values of T and S, together with the standard depth distribution of the pressure, are introduced into the equation of state in order to construct the field of the water density ϱ_w, after which the equation of hydrostatics is used to compute the pressure field, and then the heights of the isobaric surfaces, in particular the height of the ocean surface above the equilibrium level of the geoid — the sea level ζ). Also important are the CO_2 concentration (and the factors influencing it) and the state of the ocean surface (wave action and ice cover, including icebergs).

For a quantitative description of the states of the land, as an element of the atmosphere–ocean–land system, we have to determine its interaction with the atmosphere and the ocean. This includes the exchange with the atmosphere of the following quantities: (1) momentum (the topography of the land and its variability during the growth and melting of ice sheets must be known, as well as the roughness of the land surface, which depends primarily on the vegetation and on the presence or absence of a snow cover); (2) heat (the reflecting ability, or *albedo*, has to be known, as well as the emissivity for various wavelengths of sunlight and infrared radiation; these are also determined mainly by the vegetation and by the presence of a snow cover; we must also know the temperature field, including the effective surface temperature, which determines the self-radiation and the contact heat exchange); (3) moisture (the moisture field has to be known, including the moisture content of the soil, snow, ice, and permafrost and, in particular, the effective air humidity at the ground); (4) carbon dioxide (taking into account volcanic exhalations, combustion processes, decay of organic matter, and the effect of vegetation); and (5) aerosols (taking into account, in particular, volcanism, dust storms, and industrial emissions). Finally, the runoff from rivers and glaciers has to be determined as well (the hydrological characteristics of drainage basins and the ice-flow fields in glaciers have to be known).

The atmosphere–ocean–land system is *global*, and its states typically possess spatial inhomogeneities with scales ranging from global (horizontal extents of the order of 10^4 km, vertical extents of the order of 10 km, and volumes of the order of 10^9 km^3) down to very small, of the order of the most minute turbulent inhomogeneities in the atmosphere and ocean, and on land of the order of plant foliage and structural irregularities and pores in the soil. Assuming a minimum scale of about 1 mm, the atmosphere–ocean–land system contains approximately 10^{27} homogeneous minivolumes. If the state of each of these is described by, say, ten quantities, then the atmosphere–ocean–land system can be said to have 10^{28} degrees of freedom. Since the individual description of all these degrees of freedom is obviously impossible in practice, it is advisable to divide inhomogeneities of the fields being considered into *large-scale inhomogeneities*, which can be described individually, and *small-scale inhomogeneities*, describable only statistically. Consequently, a description of states of the atmosphere–ocean–land system will have to include individual characteristics of large-scale inhomogeneities and statistical characteristics of small-scale inhomogeneities.

Small-scale inhomogeneities in the atmosphere and ocean are produced by high-frequency hydrodynamic processes with periods ranging from fractions of a second to minutes (acoustic waves, turbulence, and gravity waves on a water surface) and from minutes to hours (internal gravity waves, convection, inertial oscillations). Their statistical regime depends on the individual state of the large-scale inhomogeneities in the atmosphere–ocean–land system, and thus it can, in principle, be *parametrized*. This means that statistical characteristics describing the effect of small-scale inhomogeneities on large-scale inhomogeneities (or characteristics which are interesting for other reasons) can be expressed in terms of the individual characteristics of the large-scale inhomogeneities.

The states of the atmosphere–ocean–land system vary with time. These variations have, firstly, components with very definite periods: *diurnal oscillations*, due to daily changes in the *insolation* (that is, in the irradiation by solar heat) caused by the Earth's rotation; *tidal oscillations*, due to gravitational effects of the Moon and Sun on the rotating Earth (the main ones being the semidiurnal and diurnal lunar tides M_2 and O_1, with periods of about 12 h 25 min and 24 h 50 min, and the semidiurnal and diurnal solar tides S_2 and P_1, with periods of half a day and one day); and *seasonal oscillations* (including all monsoon effects), due to changes in the insolation as the Earth revolves about the Sun (because of the inclination of the equator to the ecliptic and the eccentricity of the Earth's orbit).

Secondly, the time variations of the state of the atmosphere–ocean–land system have the following very intense irregular components: *synoptic oscillations*, caused by Rossby waves (having in the atmosphere scales of the order of 10^3 km and periods of several days, and having in the ocean scales of the order of 10^2 km and periods of several weeks); *global oscillations*, with periods ranging from weeks to months (including the approximately biweekly cycle of oscillations of the zonal atmospheric circulation in the middle latitudes, as well as self-oscillations in the atmosphere–ocean system having periods of the order of months); and *interannual oscillations*, which have their greatest intensity in the period range from 2 to 5 years (including the 26-month rhythm of oscillations of the zonal flow in the equatorial stratosphere, the quasi-biennial El Niño phenomenon in the eastern equatorial Pacific, the 3.5-year self-oscillations of the northern branch of the Gulf Stream, the shifts of thermal anomalies along gigantic oceanic gyres, and the processes produced by interannual variations of the albedo and radiation budget of the Earth).

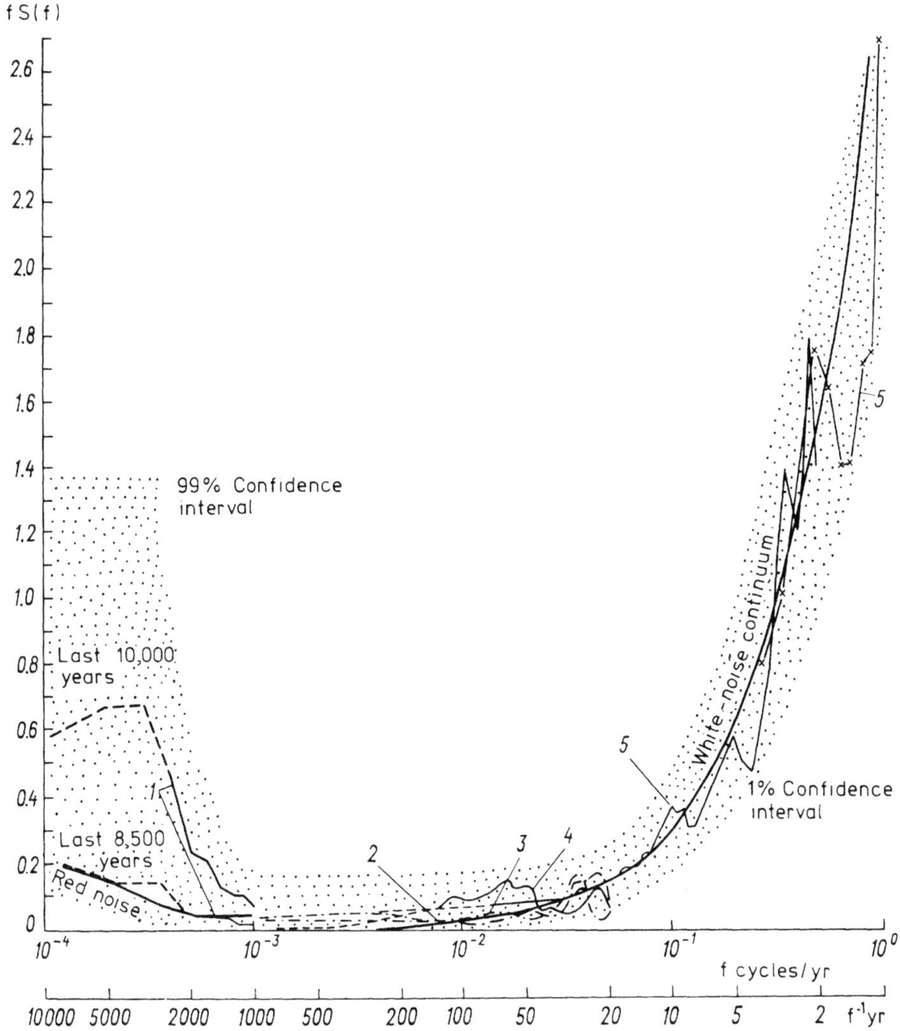

Fig. 1.1. Spectrum of air-temperature oscillations in North Atlantic sector of world, according to Kutzbach and Bryson. The ordinate is the amplitude squared $S(f)$ times the frequency f; the abscissae are the frequencies f on a log scale and the corresponding periods. (1) Central England, botanical data; (2) Central England, historical evidence; (3) Iceland, historical evidence; (4) Greenland, $\delta\,^{18}O$ data; (5) Central England, instrumental data.

Factual data indicate that long-period oscillations in the atmosphere–ocean–land system, with periods of years (*intrasecular oscillations*) and hundreds of years (*intersecular oscillations*), are on the average considerably less intense than the short-period oscillations mentioned above (although they may still be of great interest, as, for instance, in the case of the "Little Ice Age" of the 17th–19th centuries and the period of warming during the first half of the 20th century).

Long-period oscillations with periods of thousands of years (transitions from glacial conditions to nonglacial conditions) or tens of thousands of years (alternations of glacial periods and interglacials), on the other hand, are very intense. As an example, Fig. 1.1 shows the energy spectrum $f\,S(f)$ of the air-temperature oscillations in the North Atlantic region for periods f^{-1} from 2 to 10,000 years (according to Kutzbach and Bryson, 1974), which has a broad minimum in the period interval from 20 to 1000 years.

2. The Definition of Climate

For such purposes as organizing agriculture, the construction of dwellings, transport, and many other kinds of human activity, and also to understand the biogeography of flora and fauna, we don't have to know the individual chronological sequence of states of the atmosphere–ocean–land system. Rather we must have *statistics* of these states, that is, their limits of variation and their frequencies of occurrence over a long time interval. To ensure *stability* of the statistical characteristics describing states of the atmosphere–ocean–land system, that is, to make sure that they are not influenced by the specific choice of the long time interval, the length of this interval should be selected from a period range in which the oscillation spectra of the system characteristics (such as those shown in Fig. 1.1) have a minimum. In practice, it is convenient to consider some short-period part of this interval, that is, a period region several decades long. This gives us the following definition: *a statistical ensemble of states of the atmosphere–ocean–land system during a time period several decades long is called a climate.*

Here a statistical ensemble is taken to be a set $\mathfrak{A}$ of elements with a specified probability measure $P(A)$ which indicates for each measurable subset $A \subset \mathfrak{A}$ its probability $P\{a \in A\}$.

Since the states of the atmosphere–ocean–land system are multicomponent fields, a climate is defined mathematically as a *multicomponent random field*. A complete description of it would have to specify all the finite-dimensional probability distributions for values of the field components in all possible finite sets of points in space. However, such a complete description is, of course, unrealizable in practice. Frequently, only the first and second moments are considered, that is, the climatic mean values, variances, and correlations of the dynamically independent quantities and certain functions of these. To describe deviations of the one-dimensional probability distributions from the Gaussians, on the other hand, the third and fourth moments are sometimes also considered. Such a description of the climate obviously remains within the framework of the single-point and two-point probability distributions.

Since the atmosphere–ocean–land system is global, climate is by definition a global concept. Local climates, that is, single-point probability distributions at fixed points in space, are manifestations of the global climate. However, the totality of local climates at all points in space would still not represent the global climate, since it would not contain multipoint probability distributions (in particular, for instance, two-point correlations).

The climate is determined by three groups of factors: (1) *external astronomical factors*, such as the irradiance of the Sun, the position and motion of the Earth in the solar system, the inclination (tilt) of its axis of rotation to the orbital plane, and its rotation rate. These factors govern the effects on the Earth of the other bodies in the solar system: firstly the insolation, and secondly the gravitational effects of the Sun, Moon, and other planets. The latter effects produce the tides, as well as oscillations of the characteristics of the orbital motion and rotation of the Earth, and thus oscillations of the distribution of insolation over the outer boundary of the atmosphere;

(2) *external* (relative to the atmosphere–ocean–land system) *geophysical factors*, such as the size and mass of the Earth, its axial rotation, intrinsic gravitational and magnetic fields, and internal heat sources (as well as the associated geothermal heat flux and volcanism);

(3) *internal geophysical factors* (properties of the atmosphere–ocean–land system), such as the mass and composition of the atmosphere (including its permanent components, as well as its variable thermodynamically active substances), the features of the underlying surface (including the geographical distribution of continents and oceans and the topography of the land surface), the bottom topography, mass, and composition of the ocean, and the structure of the active layer of the land.

It is not known whether the climate is determined uniquely by all these factors or whether for given fixed values of all the climatogenic factors different climates are possible, due to slight differences in the initial states of the atmosphere–ocean–land system. These doubts arise mainly because of the climates of the Pleistocene, when all of the climatogenic factors apparently remained unchanged, but the atmosphere–ocean–land system was nevertheless sometimes in a glacial regime and sometimes in a nonglacial regime.

Some dynamic systems, known as *ergodic systems*, or *transitive systems*, have the property that, during the course of time, the trajectory representing their evolution in the corresponding phase space (the points of which represent possible states of the system) goes around all the points of this space. Thus the statistics of the states traversed by the system during an infinitely long time will be determined by the probability measure specified for the entire phase space. For nonergodic, or *intransitive*, dynamic systems, on the other hand, phase trajectories emanating from different initial points (origins) in general go around different sets of points in the phase space. Thus the statistics of the states traversed during an infinitely long time may in this case be different if the initial states are different. E. Lorenz suggested that the climate system is *almost intransitive*, that is, its phase space breaks up into a number of sets $\mathcal{A}_i$ with certain conditional probability measures $P_i(A)$,

$A \subset \mathcal{A}_i$, and the phase trajectories can, for a long but finite time, remain in each of these sets (representing the corresponding climate $P_i(A)$, $A \subset \mathcal{A}_i$) and occasionally pass from one of these sets to another.

Lorenz cites a mathematical example of an almost intransitive system: an idealized nonstationary trimodal rolling convection in a fluid, in which the dimensionless stream function ψ and the temperature deviation from the linear vertical profile ϑ in the xz plane have the form:

$$\psi = X\sqrt{2}\sin\frac{k_1 x}{H}\sin\frac{\pi z}{H};$$
$$\vartheta = Y\sqrt{2}\cos\frac{k_1 x}{H}\sin\frac{\pi z}{H} - Z\sin\frac{2\pi z}{H}, \tag{2.1}$$

while the amplitudes X, Y, Z are given as functions of the time t by the equations:

$$\begin{aligned} \mathrm{d}X/\mathrm{d}t &= -\sigma X + \sigma Y; \\ \mathrm{d}Y/\mathrm{d}t &= rX - Y - XZ; \\ \mathrm{d}Z/\mathrm{d}t &= -bZ + XY, \end{aligned} \tag{2.2}$$

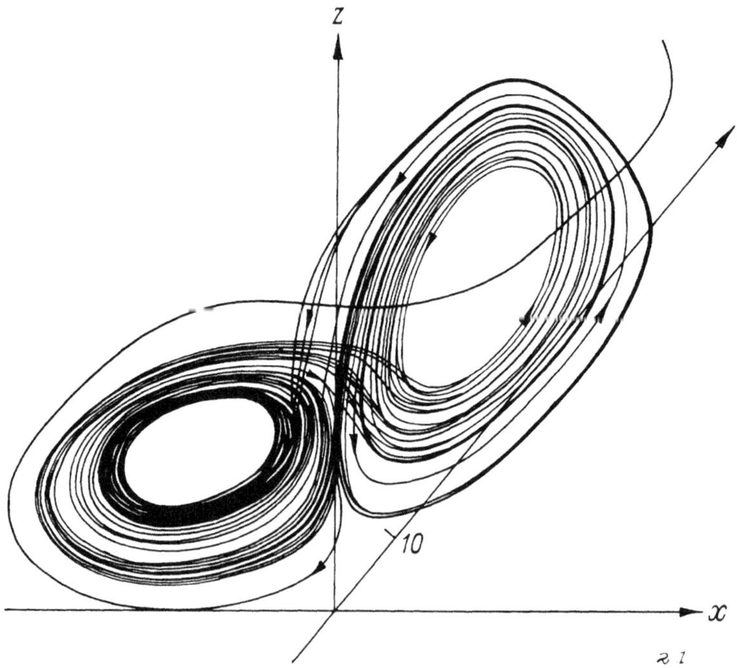

Fig. 2.1. Typical trajectory of system (2.2).

where σ, r, and b are numerical constants (σ is the Prandtl number, and r is the ratio of the Rayleigh number to its critical value, at which rolling convection begins).

For high Prandtl numbers $\sigma > b + 1$ and Rayleigh numbers $r > r_{cr} = \sigma (\sigma+b+3)(\sigma - b - 1)^{-1}$ in XYZ phase space, systems (2.2) have three fixed unstable points: $O = (0, 0, 0)$, $C = (\sqrt{br - b}, \sqrt{br - b}, r - 1$ and $C' = (-\sqrt{br - b}, -\sqrt{br - b}, r - 1)$. Lorenz integrated equations (2.2) numerically for $b = 8/3$, $\sigma = 10$ (corresponding to $r_{cr} \approx 24, 74$), and $r = 28$. Each trajectory in the phase space was found to lie, sooner or later, in the vicinity of one of the points C or C', at which time it makes around it several spiraling loops, after which it moves away, reaches the vicinity of the second point, and circles around it similarly. The sequence of these transitions from point to point is irregular in form and sensitive to changes in the initial data (see the example of such a trajectory in Fig. 2.1). Lorenz suspected that the trajectories here approximate some limit set in phase space: a so-called *strange attractor*, which in this case is a two-dimensional surface of infinitely many sheets, the intersection of which with some straight line is a Cantor set of points, that is, a nowhere dense closed set without isolated points.

By the way, the main evidence indicating that climate may not be uniquely determined for fixed values of the climatogenic factors, namely the alternation of glacial and nonglacial climates during the Pleistocene, is apparently false after all (although this result does not rule out the possibility that the climate system may be intransitive). Factual data indicate that the climatic oscillations in the Pleistocene were produced by oscillations of the values of the climatogenic factors: they have the periods characteristic of the insolation oscillations due to astronomical oscillations of the elements of the Earth's orbit and the inclination of the equator to the ecliptic (see §5).

CHAPTER 2

ASTRONOMICAL FACTORS

3. Insolation

The energy of all the processes in the atmosphere–ocean–land system can be traced back to the incoming flux of solar heat and to possible heat sources inside the Earth. At the Earth's surface the geothermal heat flux is, on the average, 5.022×10^{-2} W/m^2. Since this is only 1/ 30,000 of the solar heat flux, in most problems of the theory of climate the geothermal source of heat need not be taken into account: *the primary energy source for the climate system is insolation*.

SOLAR CONSTANT

The flux of solar heat at the Earth's mean distance from the Sun r_0 is known as the *solar constant* I_0:

$$I_0 = S_\odot/4\pi r_0^2, \tag{3.1}$$

where $S_\odot$ is the radiated power (irradiance) of the Sun, for which measurements of I_0 have given a value of about $S_\odot \approx 4 \times 10^{20}$ MW. More precisely, on the basis of numerous measurements, the most probable value of I_0 has been placed between 1368 and 1377 W/m^2 (the maximum spread of values, from 1322 to 1428 W/m^2, lies completely within the range of measurement error and is purely random, without any regularity in time; hence the term "solar constant"). In subsequent calculations we will use the value $I_0 = 1356$ W/m^2, for which the mean diurnal flux of solar heat, as an average per unit area of the Earth's surface, is estimated as $\pi a^2/4\pi a^2\, I_0 = 30$ MJ/m^2 (where a is the Earth's radius).

INSTANTANEOUS INSOLATION

The flux of solar heat incident upon a unit horizontal area of the outer limit of atmosphere over a given point on the Earth at a fixed time (the instantaneous insolation) is

$$I_1 = I_0 \left(\frac{r_0}{r}\right)^2 \cos \zeta;$$

$$\cos \zeta = \sin \varphi \sin \delta + \cos \varphi \cos \delta \cos \psi, \tag{3.2}$$

where r is the distance of the Earth from the Sun at this time, and ζ is the Sun's zenith angle at the given point on the Earth at the given time (φ is the geographical latitude of the place; δ is the geocentric declination of the Sun, that is, the angle between the Sun's direction and the plane of the Earth's equator; $\psi = 2\,\pi t/T_{\odot}$ is the hour angle, that is, the angle between the plane of the meridian of the given place and the plane passing through the Earth's rotation axis and the Sun, reckoned to the west of the meridian; t is time; $T_{\odot} = 86{,}400$ s is the length of the solar day). Formula (3.2) is suitable for the daylight hours (that is, for $\cos\zeta \geqslant 0$, which is equivalent to $\cos\psi \geqslant -\text{tg}\,\varphi\,\text{tg}\,\delta$; in polar regions during seasons when $\text{tg}\,\delta \geqslant \text{ctg}\,\varphi$ this inequality is satisfied for any ψ, that is, the sun never sets), while at night $I_1 = 0$. The values of r and δ vary with the time of year. The dependence of r on t is given by *Kepler's equation*,

$$E - e\sin E = \frac{2\pi\,(t - t_0)}{T_0}; \qquad \cos E = \frac{r_0 - r}{r_0 e}, \tag{3.3}$$

where E is the so-called *eccentric anomaly*, e is the eccentricity of the Earth's orbit, t_0 is the time of the Earth's passage through perihelion, and T_0 is the length of the *tropical year* (that is, the time between successive arrivals of the Sun at the vernal equinox; for a positive *precession of the equinoxes* this time is shorter than the *sidereal year*; for a positive [that is, in the direction of the Earth's motion] rotation of the *line of apsides*, joining the perihelion and aphelion, the time between successive passages of the Earth through perihelion, the *anomalistic year*, will be longer than the sidereal year). Instead of the geocentric declination of the Sun δ, it is more convenient to use its geocentric longitude λ (reckoned eastward from the point of vernal equinox), calculated from the formula

$$\sin\delta = \sin\varepsilon\sin\lambda, \tag{3.4}$$

where ε is the angle of inclination of the equator to the ecliptic (the *obliquity* of the ecliptic). The dependence of λ on time is given by *Lacaille's formula*:

$$\text{tg}\,\frac{v}{2} = \sqrt{\frac{1+e}{1-e}}\,\text{tg}\,\frac{E}{2}; \quad v = \lambda - P \tag{3.5}$$

together with Kepler's equation (3.3). Here P is the longitude of the perihelion, and v is the so-called *true anomaly*. The length of the solar day $T_{\odot}$ enters into formula (3.2) for the instantaneous insolation via the hour angle ψ. It is defined by the formula

$$1/T_{\odot} = 1/T - 1/T_0, \tag{3.6}$$

where, as previously, T_0 is the length of the tropical year, while T is the period of the Earth's axial rotation proper (the *sidereal day*).

DAILY INSOLATION

It is advisable to consider the diurnal oscillations of the insolation separately from the annual variations of its daily sums. The daily insolation is found by integrating expression (3.2) with respect to time over the daylight hours

($\cos \psi \geqslant - \operatorname{tg} \varphi \operatorname{tg} \delta$; note, however, that for the polar regions on days when the Sun does not set, that is, for $\operatorname{tg} \delta \geqslant \operatorname{ctg} \varphi$, the integration is over the whole day).

Assuming for this integration that r and δ are quasi-constant, we obtain the following expression for the mean daily insolation I:

$$I = \frac{1}{T_\odot} \int_{-t_0}^{t_0} I_1 \, dt = \frac{I_0}{\pi} \left(\frac{r_0}{r} \right)^2 (\psi_0 \sin \varphi \sin \delta + \\ + \cos \varphi \cos \delta \sin \psi_0), \tag{3.7}$$

where $\psi_0 = 2\pi t_0 / T_\odot = \arccos(-\operatorname{tg} \varphi \operatorname{tg} \delta)$ is the hour angle of the Sun at sunset (on polar days when the Sun does not set $\psi_0 = \pi$, and on polar nights when it never rises $\psi_0 = 0$). The annual variations of this quantity at each fixed latitude φ are found with the aid of formulas (3.3)–(3.5). The first to carry out such a calculation was M. Milankovitch (1939). Figure 3.1 shows the results of the calculation.

The daily insolation is obviously independent of the longitude, so that its distribution always possesses *latitudinal zonality*. This zonality is especially marked during the winter of the corresponding hemisphere, while during the transition seasons it can be observed simultaneously in both hemispheres. Since in the polar part of the winter hemisphere the Sun does not rise above the horizon, the insolation there (between 66.6° and the pole at the winter solstice) will be zero; in Fig. 3.1 the regions of polar night are shaded. At the equator at the winter solstice of the Northern Hemisphere the daily insolation is 35.49 MJ/m^2. At the equinoxes the insolation is a maximum at the equator (at the vernal equinox of the Northern Hemisphere the insolation is 37.37 MJ/m^2 in a day) and zero at the poles. At the summer solstice of the Northern Hemisphere the daily insolation reaches a maximum (45.10 MJ/m^2) at the North Pole, and at this time it is only 33.18 MJ/m^2 at the equator. However, the insolation does not vary monotonically with the latitude at this time: between 40 and 45°N there is a secondary diurnal maximum (41.65 MJ/m^2) and between 60 and 65°N there is a secondary minimum (41.00 MJ/m^2). But since the annual total insolation in the polar regions is much less than the annual total at the equator (see below), the latitudinal zonality of the insolation shows up clearly on the yearly curves. Consequently, despite the deviation from latitudinal zonality in the distribution of reflected and absorbed radiation, due to the horizontal nonuniformity of the Earth's surface and atmosphere (especially the variable distribution of clouds, which has the greatest effect on the Earth's reflectivity), the insolation produces an approximate latitudinal zonality of the climate.

Figure 3.1 clearly demonstrates the seasonal oscillations of the insolation caused by the obliquity ε of the Earth's equator to the ecliptic and by the eccentricity e of the Earth's orbit. For $\varepsilon < 45°$ the polar circles (bounding the polar regions, within which there are days when the Sun does not rise or does

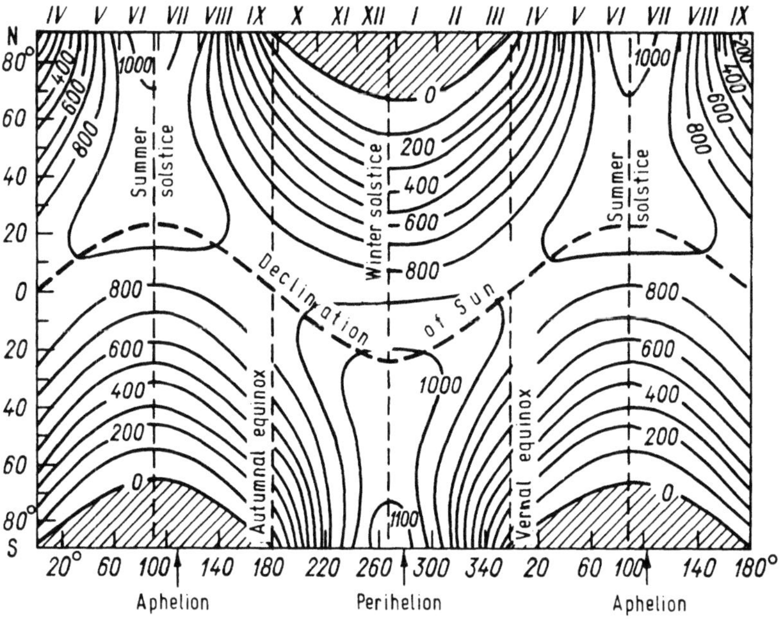

Fig. 3.1. Daily insolation at upper limit of atmosphere in cal/(cm^2 d) for I_0 = 1356 W/m^2, as function of latitude and time of year. 1 cal/cm^2 = 0.0419 MJ/m^2.

not set) lie outside of the tropical zone (within which there are days when the Sun reaches the zenith), and because e is small the insolation at a given latitude goes through one maximum and one minimum during the course of a year (if we count the polar night as one minimum).

If $\varepsilon = 0$ the seasonal oscillations of the insolation at all latitudes would be in phase, with a true anomaly v. If $e = 0$ the seasonal oscillations in different hemispheres would differ only in phase (by π radians). If $P = 0$, that is, if the perihelion and aphelion coincided with the equinox points, then the seasonal oscillations in one hemisphere could be derived from the oscillations in the other hemisphere by varying the direction of the time, reckoned from one of the equinoxes (for instance, in one hemisphere the summer half of the year would proceed from perihelion to aphelion, and in the other hemisphere from aphelion to perihelion). Actually, none of these three quantities ε, e, and P is zero, phase shifts relative to the longitude or declination of the Sun appear in the seasonal oscillations of the insolation, and there is asymmetry in the insolation of the Northern and Southern Hemispheres. For instance, in the Northern Hemisphere in summer the Earth is near aphelion ($r \approx 1.52 \times 10^{13}$ cm), and in winter near perihelion ($r \approx 1.47 \times 10^{13}$ cm), so that the summer insolation is less in the Northern Hemisphere than in the Southern Hemisphere, while the winter insolation is greater. In the annual sum, the asymmetry of the insolation of the hemispheres disappears.

SEASONAL AND ANNUAL INSOLATION

In order to find the total insolation for a season or for a year, we have to integrate expression (3.7) over the appropriate time interval. Using the following relation implied by (3.3) and (3.5):

$$\left(\frac{r_0}{r}\right)^2 \mathrm{d}t = \frac{T_0\,\mathrm{d}\lambda}{2\pi\sqrt{1-e^2}}, \tag{3.8}$$

passing to integration over the Sun's longitude λ, eliminating δ with the aid of formula (3.4), and assuming I_0, T_0, e, and ε to be constant during this integration, we obtain

$$W = \frac{I_0\,T_0 \sin\varphi \sin\varepsilon}{2\pi^2\sqrt{1-e^2}} \int_{\lambda_1}^{\lambda_2} (\psi_0 - \mathrm{tg}\,\psi_0) \sin\lambda\,\mathrm{d}\lambda;$$

$$\cos\psi_0 = -\frac{\mathrm{tg}\,\varphi \sin\varepsilon \sin\lambda}{\sqrt{1-\sin^2\varepsilon\,\sin^2\lambda}}. \tag{3.9}$$

For those values of φ and λ for which the above expression for $\cos\psi_0$ is less than -1 (polar day) it must be assumed that $\psi_0 = \pi$, whereas if it is greater than $+1$ (polar night) then $\psi_0 = 0$. Since for the substitution $(\varphi, \lambda) \to (-\varphi, \lambda + \pi)$ the value of (3.9) does not change, in the following we can restrict ourselves just to the Northern Hemisphere. In it the points $\lambda = 0, \pi/2, \pi$, and $3\pi/2$ divide the year into the astronomical spring, summer, autumn, and winter. In the northern polar zone $(\varphi = \pi/2 - \varepsilon)$ the interval $(\lambda_1, \pi - \lambda_1)$, where $\sin\lambda_1 = \cos\varphi/\sin\varepsilon$ corresponds to endless day, and $(\pi + \lambda_1, 2\pi - \lambda_1)$ to endless night (point λ_1 corresponds to spring, $\pi - \lambda_1$ to summer, $\pi + \lambda_1$ to autumn, and $2\pi - \lambda_1$ to winter). During the rest of the year the Sun rises and sets daily there.

Substitution of the variable $\lambda \to (\pi - \lambda)$ in (3.9) shows that the values of W for spring and summer are identical; similarly, the values of W for autumn and winter are identical. Thus we can lump together the seasons with the same sums of received heat and consider a summer half year (spring + summer) with a total insolation W_s and a winter half year (autumn + winter) with a total insolation W_w:

$$W_{s,w} = \frac{I_0 T_0}{2\pi\sqrt{1-e^2}}\left[S(\varphi, \varepsilon) \pm \sin\varphi \sin\varepsilon\right];$$

$$S(\varphi, \varepsilon) = \frac{\sin\varphi \sin\varepsilon}{2\pi} \int_0^{2\pi} (\psi_0 - \mathrm{tg}\,\psi_0) \sin\lambda\,\mathrm{d}\lambda. \tag{3.10}$$

On the other hand, the total annual insolation W_0 is given by the sum

$$W_0 = W_s + W_w = \frac{I_0 T_0 S(\varphi, \varepsilon)}{\pi\sqrt{1-e^2}}. \tag{3.11}$$

Function $S(\varphi, \varepsilon)$, describing the distribution of the annual insolation along the meridian, is expressed in terms of elliptic functions. At the equator $S(0, \varepsilon) = (\pi/2)\, E(\sin \varepsilon)$, where $E(k)$ is a complete elliptic integral of the second kind. At the pole $S(\pi/2, \varepsilon) = \sin \varepsilon$. The ratio of the annual insolation at the equator and at the pole $2E(\sin \varepsilon)/\pi \sin \varepsilon$ decreases with an increase in ε; for the Earth ($\varepsilon = 23°27'$) this ratio is about 2.41. The distribution of the annual insolation W_0 along the latitudes φ is given in Table 3.1. The instantaneous insolation of the entire Earth is $I_0(r_0/r)^2 \pi a^2$. The daily insolation $I_0(r_0/r)^2 \pi a^2 T_\odot$, as an average on a unit area of thc Earth's surface $4\pi a^2$, is $\frac{1}{4} I_0 (\frac{r_0}{r})^2 T_\odot$ (actually it is distributed just over the part of the Earth's surface outside of the region of polar night). Now, by integrating this expression over a year, the time being expressed in days, that is, over $dt/T_\odot$, taking (3.8) into account we obtain the mean annual insolation over the Earth's surface $I_0 T_0 / 4\sqrt{1 - e^2}$.

TABLE 3.1

Annual values of insolation W_0

$\varphi°$	W_0 MJ/m^2	$\varphi°$	W_0 MJ/m^2
0	1089	50	746
10	1077	60	620
20	1031	70	515
30	960	80	469
40	863	90	452

The seasonal nature of the insolation, that is, the difference between the amounts of insolation during the summer and winter half years,

$$W_s - W_w = \frac{I_0 T_0 \sin \varphi \sin \varepsilon}{\pi \sqrt{1 - e^2}}, \qquad (3.12)$$

is due to the difference in the intensity of irradiation of a given latitude zone, as well as to the difference between the durations T_s and T_w of these half years. Actually, using (3.8) and Kepler's ellipse equation

$$\frac{r}{r_0} = \frac{1 - e^2}{1 - e \cos (\lambda - P)} \qquad (3.13)$$

and carrying out the calculations with an accuracy to the first power of e, we obtain

$$T_s = \int_{\lambda = 0}^{\lambda = \pi} dt = \frac{T_0}{2\pi \sqrt{1 - e^2}} \int_0^\pi \left(\frac{r}{r_0}\right)^2 d\lambda \approx \frac{T_0}{2\pi} \int_0^\pi [1 + 2e \cos \times$$

$$\times (\lambda - P)]\, d\lambda = \frac{T_0}{2}\left(1 + \frac{4}{\pi} e \sin P\right), \qquad (3.14)$$

and, similarly,

$$T_w \approx \frac{T_0}{2}\left(1 - \frac{4}{\pi} e \sin P\right). \qquad (3.14')$$

INSOLATION DURING CALORIC HALF YEARS

Sometimes it is convenient to consider half years of equal duration. For instance, Milankovitch considered the amounts of insolation Q_s and Q_w during the summer and winter *caloric half years*, defined for each fixed latitude φ as the half years of duration $T_0/2$ for which the daily insolation at latitude φ on any day of the summer half year is greater than on any day of the winter half year. The geocentric longitudes of the Sun λ_1 and λ_2 at the beginnings of the winter and summer caloric half years must satisfy the condition $\int_{\lambda_1}^{\lambda_2} dt = T_0/2$, which, by analogy with (3.14), can be put into the form

$$\lambda_2 - \lambda_1 + 2e\,[\sin(\lambda_2 - P) - \sin(\lambda_1 - P)] = \pi, \qquad (3.15)$$

and indicates that λ_1 and $\lambda_1' = \pi - \lambda_2$ are small quantities of the order of e, related by the expression $\lambda_1 + \lambda_1' \approx 4e \sin P$. Further, in view of formula (3.9), we get

$$Q_{s,w} = W_{s,w} \mp Q;$$

$$Q = \frac{I_0 T_0 \sin\varphi \sin\varepsilon}{2\pi^2\sqrt{1 - e^2}}\left(\int_0^{\lambda_1} + \int_{\pi-\lambda_1}^{\pi}\right)(\psi_0 - \operatorname{tg}\psi_0)\sin\lambda\, d\lambda. \qquad (3.16)$$

Because λ_1 and $\pi - \lambda_2$ are small, formula (3.4) shows that the corresponding declinations of the Sun δ_1 and δ_2 are also small. Since $\cos\psi_0 = -\operatorname{tg}\varphi\ \operatorname{tg}\delta$, the corresponding hour angles ψ_0 will be close to $\pi/2$, their sines close to unity, and their tangents close to $1/\cos\psi_0 \approx \operatorname{ctg}\varphi/\sin\delta = -\operatorname{ctg}\varphi/\sin\varepsilon\sin\lambda$. Thus the integrand in (3.16) can be replaced by $(\pi/2 + \operatorname{ctg}\varphi/\sin\varepsilon\sin\lambda) \times \sin\lambda \approx \operatorname{ctg}\varphi/\sin\varepsilon$. Then the sum of the integrals in (3.16) will be approximately equal to

$$(\lambda_1 + \lambda_1')\frac{\operatorname{ctg}\varphi}{\sin\varepsilon} \approx 4e\sin P\,\frac{\operatorname{ctg}\varphi}{\sin\varepsilon},$$

and, with the aid of (3.10), we obtain from (3.16) the following formulas of Milankovitch:

$$Q_{s,w} \approx \frac{I_0 T_0}{2\pi}\left[S(\varphi,\varepsilon) \pm \sin\varphi\sin\varepsilon \mp \frac{4}{\pi} e \sin P\cos\varphi\right]. \qquad (3.17)$$

It should be noted that these quantities are functions of the obliquity ε and the elements of the Earth's orbit r_0 (via I_0 and T_0), e, and P.

4. The Astronomical Theory of Climatic Fluctuations

The obliquity of the ecliptic ε and the elements of the Earth's orbit vary in time, because of *perturbations* of the Earth's motion produced by the gravitational attraction of the other planets in the solar system. Consequently, the distributions of insolation, and quantities (3.17) in particular, vary with time, which may well lead to climate variations.

In the theory of perturbations of planetary orbits, Laplace demonstrated that, accurate to the first power of the eccentricity e, the semimajor axis r_0 of an elliptical orbit is not perturbed. Accordingly, to this same accuracy, we can say that I_0 and the year length T_0 (which is according to Kepler's third law proportional to $r_0^{3/2}$) do not vary in time, and that thus the mean annual insolation over the Earth's surface $I_0T_0/4\sqrt{1-e^2}$ is constant as well. Variations of the obliquity ε and the elements of the orbit e and P can thus only cause redistributions of the total annual insolation over the seasons of the year and the geographical latitudes. In particular, it is evident from (3.12) that an increase in ε will increase the thermal differences between seasons, that is, it will make the climate more continental; from (3.14) and (3.14′) it follows that an increase in $e \sin P$ makes the differences in the duration of the astronomical seasons greater. It can be assumed, for example, that warm winters and cool summers promote the growth of glaciers. Therefore, in order to evaluate possible fluctuations of the climate, Milankovitch suggested calculating the periodic fluctuations of the distributions of the insolation during the summer and winter caloric half years (3.17). Accurate to small first-order variations $\delta\varepsilon$ of the obliquity and of the orbital elements $e \sin P$, the amounts of insolation during the summer and winter caloric half years (Q_{Ns} and Q_{Nw} in the Northern Hemisphere and Q_{Ss} and Q_{Sw} in the Southern Hemisphere) are given by the formulas

$$\begin{aligned} Q_{Ns} &= W_s + R_s\delta\varepsilon - me\sin P; \\ Q_{Sw} &= W_w + R_w\delta\varepsilon - me\sin P; \\ Q_{Nw} &= W_w + R_w\delta\varepsilon + me\sin P; \\ Q_{Ss} &= W_s + R_s\delta\varepsilon + me\sin P, \end{aligned} \tag{4.1}$$

where

$$\begin{aligned} R_s &= \frac{I_0T_0}{2\pi}\left[\left(\frac{\partial S}{\partial \varepsilon}\right)_0 + \sin\varphi\cos\varepsilon_0\right]; \\ R_w &= \frac{I_0T_0}{2\pi}\left[\left(\frac{\partial S}{\partial \varepsilon}\right)_0 - \sin\varphi\cos\varepsilon_0\right]; \\ m &= \frac{2I_0T_0\cos\varphi}{\pi^2\sqrt{1-e^2}}, \end{aligned} \tag{4.2}$$

the values of $W_{s,w}$ and $R_{s,w}$ being chosen for a fixed value (today's value) of the obliquity $\varepsilon = \varepsilon_0$ and, like m, being assumed to be independent of time.

Milankovitch and subsequent investigators represented the oscillations of the insolation (4.1) in terms of equivalent latitudes φ_* $(t; \varphi)$, determined from the relations Q $(\varphi_*, 0) = Q$ (φ, t). Taking into account that the largest oscillations between glacial and nonglacial conditions during the Pleistocene took place in the middle latitudes around $\varphi = 65°$N, they computed only the oscillations of the equivalent latitude $\varphi_*(t; 65°\text{N})$. However, much information was thereby lost, so here we will calculate the insolation anomalies

$$\begin{aligned} \delta Q_{\text{Ns}} &= R_s\delta\varepsilon - m\delta\,(e \sin P); \\ \delta Q_{\text{Sw}} &= R_w\delta\varepsilon - m\delta\,(e \sin P); \\ \delta Q_{\text{Nw}} &= R_w\delta\varepsilon + m\delta\,(e \sin P); \\ \delta Q_{\text{Ss}} &= R_s\delta\varepsilon + m\delta\,(e \sin P) \end{aligned} \tag{4.3}$$

completely. R_s, R_w, and m in (4.3) are functions only of the latitude (and do not change when φ is replaced by $-\varphi$), while $\delta\varepsilon$ and δ $(e \sin P)$ are functions only of time (and do not change when the Sun's longitude λ is replaced by $\lambda + \pi$). Coefficient R_s, the factor of $\delta\varepsilon$, increases from the equator to the high latitudes (while R_w increases from the equator to the middle latitudes); coefficient m, the factor of $\delta(e \sin P)$, increases from the poles to the equator.

When calculating oscillations $\delta\varepsilon$ (and, in general, when describing the precessions of the Earth's rotation axis in the space of the fixed stars), we will assume the Earth to be an ellipsoid of revolution with principal moments of inertia $A = B < C$ and we will use a geocentric coordinate system $X_1X_2X_3$ with its origin 0 at the Earth's center; the X_1X_2 plane coincides with the ecliptic at some fixed time, the X_1 axis is directed toward the vernal equinox at that same time, and the X_3 axis points toward the north ecliptic pole. The directions of the principal inertial axes of the Earth x_i (see Fig. 4.1) are defined by the *Euler angles*: the angle of inclination of the equator to the ecliptic $\vartheta < \pi/2$ (reckoned from X_3 to x_3); the precession angle ψ between the X_1 axis and the line of nodes, that is, the line of intersection between the equator and the ecliptic (reckoned on the ecliptic from the vernal equinox eastward to the descending node); and the angle φ in the equatorial plane between the x_1 axis and the line of nodes (reckoned eastward from the descending node).

The rotation of the Earth is then made up of rotations at a rate $d\varphi/dt$ about the x_3 axis, at a rate $d\psi/dt$ about the X_3 axis, and a rate $d\vartheta/dt$ about the line of nodes. Projected onto axes x_i, these rotations have the form:

$$\begin{aligned} \omega_1 &= -\frac{d\vartheta}{dt}\cos\varphi - \frac{d\psi}{dt}\sin\vartheta\sin\varphi; \\ \omega_2 &= \frac{d\vartheta}{dt}\sin\varphi - \frac{d\psi}{dt}\sin\vartheta\cos\varphi; \\ \omega_3 &= \frac{d\psi}{dt}\cos\vartheta + \frac{d\varphi}{dt} \end{aligned} \tag{4.4}$$

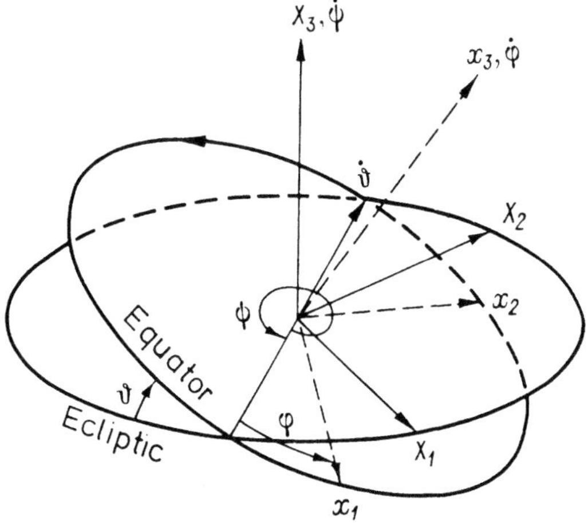

Fig. 4.1. X_1 X_2 X_3 and x_1 x_2 x_3 reference systems and Euler angles.

(these relations are called the Euler kinematic equations).

The equations of the dynamics of the Earth's rotation, in terms of the Euler angles ϑ, ψ, φ, expressing the law of conservation of angular momentum, are derived with the aid of the Lagrangian of the rotating Earth:

$$\mathscr{L} = \frac{1}{2}\ [A\,(\omega_1^2 + \omega_2^2) + C\omega_3^2] + U, \tag{4.5}$$

where U is the attraction potential of the other celestial bodies, and quantities ω_i are given by formulas (4.4). The first two Lagrange equations obtained with the aid of (4.5) have the form:

$$\begin{aligned} \frac{d\psi}{dt} &= \frac{1}{C\omega_3 \sin\vartheta}\frac{\partial U}{\partial\vartheta} + \frac{A}{C\omega_3}\left[\cos\vartheta\left(\frac{d\psi}{dt}\right)^2 - \frac{1}{\sin\vartheta}\frac{d^2\vartheta}{dt^2}\right]; \\ \frac{d\vartheta}{dt} &= -\frac{1}{C\omega_3 \sin\vartheta}\frac{\partial U}{\partial\psi} + \\ &\quad + \frac{A}{C\omega_3}\left[\sin\vartheta\frac{d^2\psi}{dt^2} + 2\cos\vartheta\frac{d\psi}{dt}\frac{d\vartheta}{dt}\right] \end{aligned} \tag{4.6}$$

(the third equation has the form $\omega_3 = \text{const}$). The terms with square brackets are small in this case, and when they are dropped we get the so-called *Poisson equations*. Potential U is the sum of the attraction potentials of the other celestial bodies:

$$\begin{aligned} U_k = G\,m_k \int \frac{\varrho dV}{R_k} &\approx G\,\frac{m_0 m_k}{r_k} + G\,\frac{m_k}{r_k^3} \times \\ &\times \left[\frac{2A + C}{2} - \frac{3}{2}\,\frac{A\,(x_{1k}^2 + x_{2k}^2) + Cx_{3k}^2}{r_k^2}\right] \end{aligned} \tag{4.7}$$

$$\text{const} - \frac{3}{2} G (C - A) \frac{m_k}{r_k^3} \cos^2 \gamma_k;$$

$$\cos \gamma_k = \frac{x_{3k}}{r_k} = \cos \vartheta \cos \vartheta_k + \sin \vartheta \sin \vartheta_k \sin (\psi_k - \psi),$$

where G is the gravitational constant, dV is a volume element inside the Earth, R_k is the distance from it to the kth celestial body, m_k is the mass of this body (m_0 is the mass of the Earth), (x_{1k}, x_{2k}, x_{3k}) are the coordinates of this body in the rotating $x_1x_2x_3$ reference system, and $(r_k, \vartheta_k, \psi_k)$ are the spherical coordinates of this body in the $X_1X_2X_3$ reference system. If (4.7) is used, the first Poisson equation becomes

$$\frac{d\psi}{dt} = -\frac{3G}{\omega_3}\frac{C - A}{C} \sum_k \frac{m_k}{r_k^3} \cos \gamma_k [\text{ctg}\, \vartheta \sin \vartheta_k \sin (\psi_k - \psi) - \cos \vartheta_k] \approx -\frac{3G}{2\omega_3}\frac{C - A}{C}\left(\frac{m_{☽}}{r_{☽}^3} + \frac{m_{\odot}}{r_{\odot}^3}\right)\cos \vartheta, \quad (4.8)$$

it being taken into account in the second of these equalities that the Moon and Sun make the main contribution to the precession and that for them $\vartheta_k \approx \pi/2$, $\sin \vartheta_k \approx 1$, and multiplier $\sin^2 (\psi_k - \psi)$ can be replaced by its mean value 1/2. Now, for the values $G \approx 6.673 \times 10^{-8}$ cm^3/(g s^2), $\omega_3 \approx 7.292 \times 10^{-5}$ s^{-1}, $\frac{C-A}{C} \approx 1/305.8$, $m_{☽} \approx m_0/81.53$, $m_{\odot} \approx 332{,}290\ m_0$, $m_0 \approx 5.975 \times 10^{27}$ g, $r_{☽} \approx 3.844 \times 10^{10}$ cm, $r_{\odot} \approx 1.49504 \times 10^{13}$ cm, and $\vartheta = \varepsilon_0 \approx 23° 27' 08.26''$, we obtain $d\psi/dt \approx -50''$/yr. Astronomical measurements give $d\psi/dt \approx -50.2564''$/yr, corresponding to a precession period $2\pi(d\psi/dt)^{-1} \approx 25{,}735$ tropical years. As will be seen below, the precession period plays a significant role in the astronomical fluctuations of the insolation (4.3). The second Poisson equation (4.6) describes the *nutation* of the Earth's rotation axis, but this will not be discussed here.

An exact integration of equations (4.6) and a calculation of the perturbations of the elements of the Earth's orbit requires that the locations of all the planets be known as functions of time. As a first approximation the planets can be assumed to have Keplerian motion along elliptical orbits, the location of each planet being specified by six elements: the position of the orbital plane is given by its angle to the ecliptic (its *inclination* i) and by the *longitude of the ascending node* Ω; the orientation of the orbit in this plane is given by the *longitude of the perihelion* $P = \Omega + \omega$ (where ω is the perihelion longitude on the orbit reckoned from the ascending node); the shape of the orbit is given by its *semimajor axis* r_0 (from which the *mean motion* of the planet is determined, that is, the frequency of its revolution around the Sun $\omega_k \sim (1 + m_k/m_{\odot})^{1/2} r_{ok}^{-3/2}$) and by the *eccentricity* e; the position of the planet on its orbit is given by the time t_0 of its *perihelion passage* (which gives the *true anomaly* v as a function of time, that is, the longitude of the planet on the orbit reckoned from the perihelion, or the so-called *argument of the latitude* $u = \omega + v$, that is, the longitude on the orbit reckoned from the node).

In closer approximations, the *perturbations* of the motion of each planet caused by the attraction of the other planets have to be taken into account. Due to these perturbations, the orbital elements vary in time. Elements Ω and P are subject to *secular* perturbations, which increase with time: the lines of nodes rotate in a sense opposite to the direction of motion of the planets, while the major axes of the orbits, the *lines of apsides*, rotate predominantly in the direction of planetary motion. The other elements of the orbits are apparently subject only to nonincreasing perturbations.

According to the theorem of V.I. Arnold, for most initial conditions the motion of n interacting planets is always close to *Lagrangian* motion along ellipses (with mean motions ω_k) having the following characteristics: small, nearly constant inclinations i_k and eccentricities e_k; nearly constant semimajor axes r_{0k}; lines of nodes and lines of apsides which so rotate that the *Laplace vectors* $\mathbf{e}_k$ (of length $m_k^{1/2} r_{ok}^{1/4} e_k$) directed along the major axes of the ellipses can be represented as sums $\mathbf{e}_k = \Sigma_{l=1}^{n} \mathbf{e}_{kl}$ of vectors $\mathbf{e}_{kl}$, which rotate uniformly with low frequencies μ_l (the same for all planets), and the vectors $\mathbf{i}_k$ (of length $m_k^{1/2} r_{0k}^{1/4} \mathbf{i}_k$) directed along the lines of nodes can be represented as sums $\mathbf{i}_k = \Sigma_{l=1}^{n-1} \mathbf{i}_{kl}$ of vectors $\mathbf{i}_{kl}$, which rotate uniformly with low frequencies v_l. Therefore, the Lagrangian motion of n planets is described by $3n - 1$ frequencies ω_n, μ_n, v_k (which are, in general, incommensurable) and can thus be called *conditionally periodic*: the functions of time characterizing it can be represented as superpositions of harmonics with various frequencies $N_k \omega_k + N'_k \mu_k + N''_k v_k$, where $N_k, N'_k, N''_k = 0, \pm 1, \pm 2, \ldots$ In particular,

$$\begin{aligned} \varepsilon &= h^* + \Sigma A_i \cos (a_i t + b_i); \\ e \sin P &= \Sigma C_i \sin (c_i t + d_i); \\ e &= F_0 + \Sigma F_i \cos (f_i t + g_i). \end{aligned} \tag{4.9}$$

These trigonometric series, proposed by Laplace and Lagrange, have since been calculated by a number of persons, for instance, by Stockwell in 1873 and by Michkovitch (1931), whose solutions were later used by Milankovitch (1939) and Sharaf and Budnikova (1968), and even more recently by Bretagnon and by Berger (1975, 1977). According to the calculations, in the series (4.9) for ε the largest amplitudes A_i = 0.828 89; 0.167 77; 0.139 76; 0.069 65; 0.055 97° have harmonics with periods $2\pi/a_i$, equal respectively to 40,900, 52,500, 39,500, 29,500, and 28,600 years (a period of about 41,000 years is the most representative). In the series for $e \sin P$ the largest amplitudes $10^7 C_i$ = 183,400; 163,395; −148,347; 104,435 have harmonics with periods $2\pi/c_i$ = 23,700, 22,400, 18,900, and 19,100 years (precession periods of about

23,000 and 19,000 years are the most representative). In the series for e the largest amplitudes $10^7 F_i = -88{,}000, -78{,}402, 61{,}898, 55{,}147$, and $-50{,}068$ have harmonics with periods $2\pi/f_i = 94{,}600, 121{,}600, 99{,}400, 129{,}800$, and 1,922,800 years (periods of about 100,000 years are the most representative). According to the calculations of Sharaf and Budnikova, during the last 30 million years the values of e have ranged from 0.0007 to 0.0658 (present value 0.016751), the values of ε from 22.068 to 24.568°, and the values of $e \sin P$ from -0.07 to $+0.03$ (present value 0.016454).

The indicated values of ε and $e \sin P$ were used at the Institute of Oceanology of the U.S.S.R. Academy of Sciences to calculate the amounts of insolation (4.1) during the period -1 million years $\leqslant t \leqslant +1$ million years, with time intervals of 5000 years and latitude intervals of 10°. Table 4.1 gives the mean amounts of insolation $\bar{Q}_{NS}(\varphi)$, $\bar{Q}_{SW}(\varphi)$, $\bar{Q}_{NW}(\varphi)$, and $\bar{Q}_{SS}(\varphi)$ for this period.

TABLE 4.1
Mean amounts of insolation (MJ/m²) according to (4.1)

$\varphi°$	$\bar{Q}_{NS}$	$\bar{Q}_{SW}$	$\bar{Q}_{NW}$	$\bar{Q}_{SS}$
90	5434	0	0	5434
80	5432	75	159	5508
70	5542	443	608	5708
60	5962	1253	1480	6201
50	6419	2254	2538	6731
40	6754	3264	3634	7127
30	6930	4211	4634	7353
20	6922	5066	5522	7379
10	6729	5782	6260	7207
0	6344	6344	6834	6834

Figure 4.2 shows graphs of the insolation anomalies (4.3) for a period -1 million years $\leqslant t \leqslant +50{,}000$ years. The left half of each graph gives the anomaly distribution over the entire Earth for the summer half year of the Northern Hemisphere δQ_{NS} and δQ_{SW}, while the right half of each graph gives the distribution for the summer half year of the Southern Hemisphere δQ_{NW} and δQ_{SS}. Regions of negative anomaly are shaded.

These graphs show that the insolation anomalies during caloric half years reach a maximum, as a rule, in summer polar regions, where they may be as high as ±6 kcal/cm² (251 MJ/m²). In general, the anomalies decrease toward the winter poles, although they are still not small (for instance, an annual anomaly of 4 kcal/cm² (167.6 MJ/m²), for the latent heat of fusion of ice 335 kJ/kg, can cause the melting of an ice sheet 2.5 km deep in 5000 years). The strongest component of the anomalies, observed in the high and middle latitudes of the summer hemispheres, has a quasi-periodic nature with a period of about 40,000 years (that is, it is produced by fluctuations of ε); this

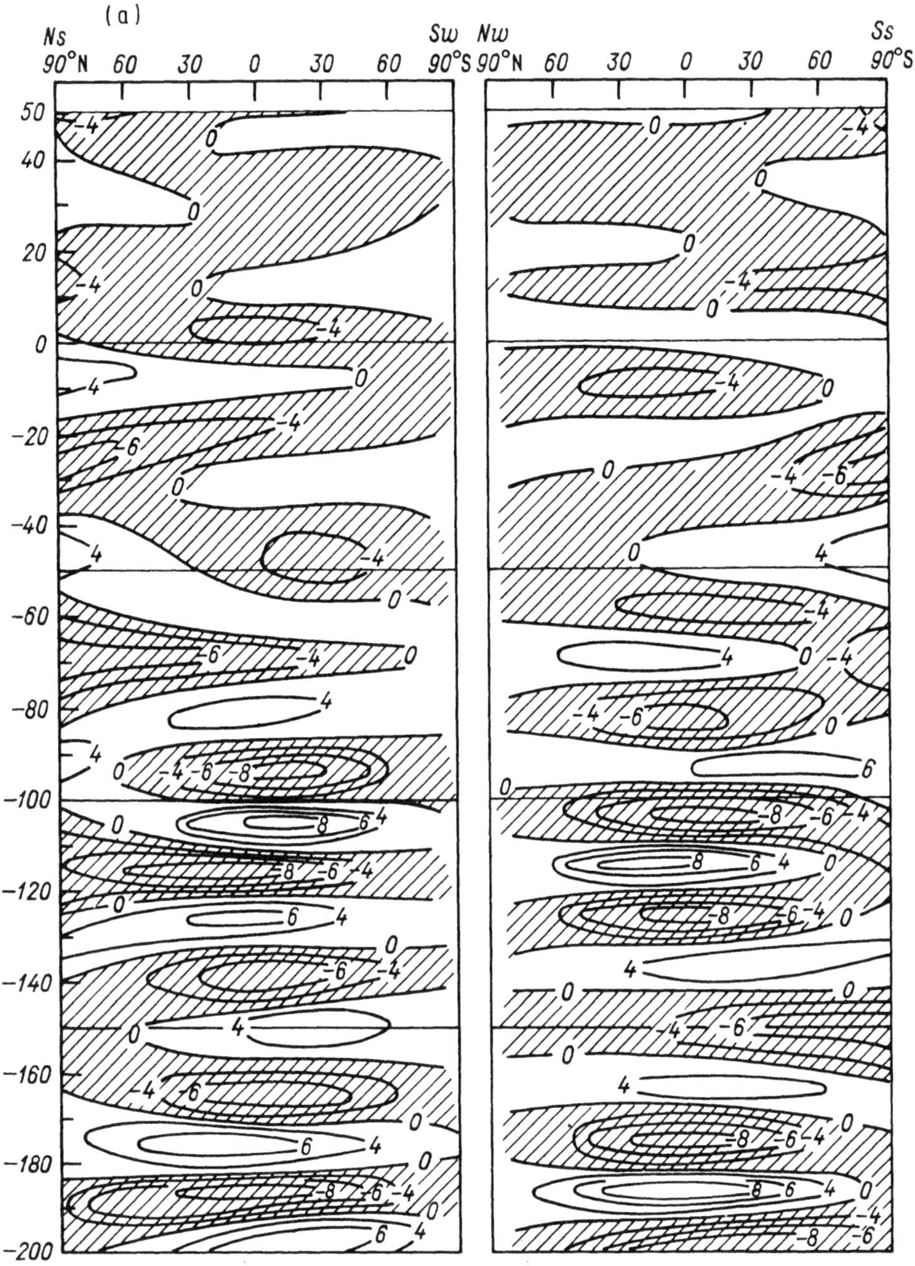

Fig. 4.2(a). Anomalies of insolation (kcal/cm²) during caloric half years. 1 kcal/cm² = 41.9 MJ/m². −200,000 yr ≤ t ≤ + 50,000 yr.

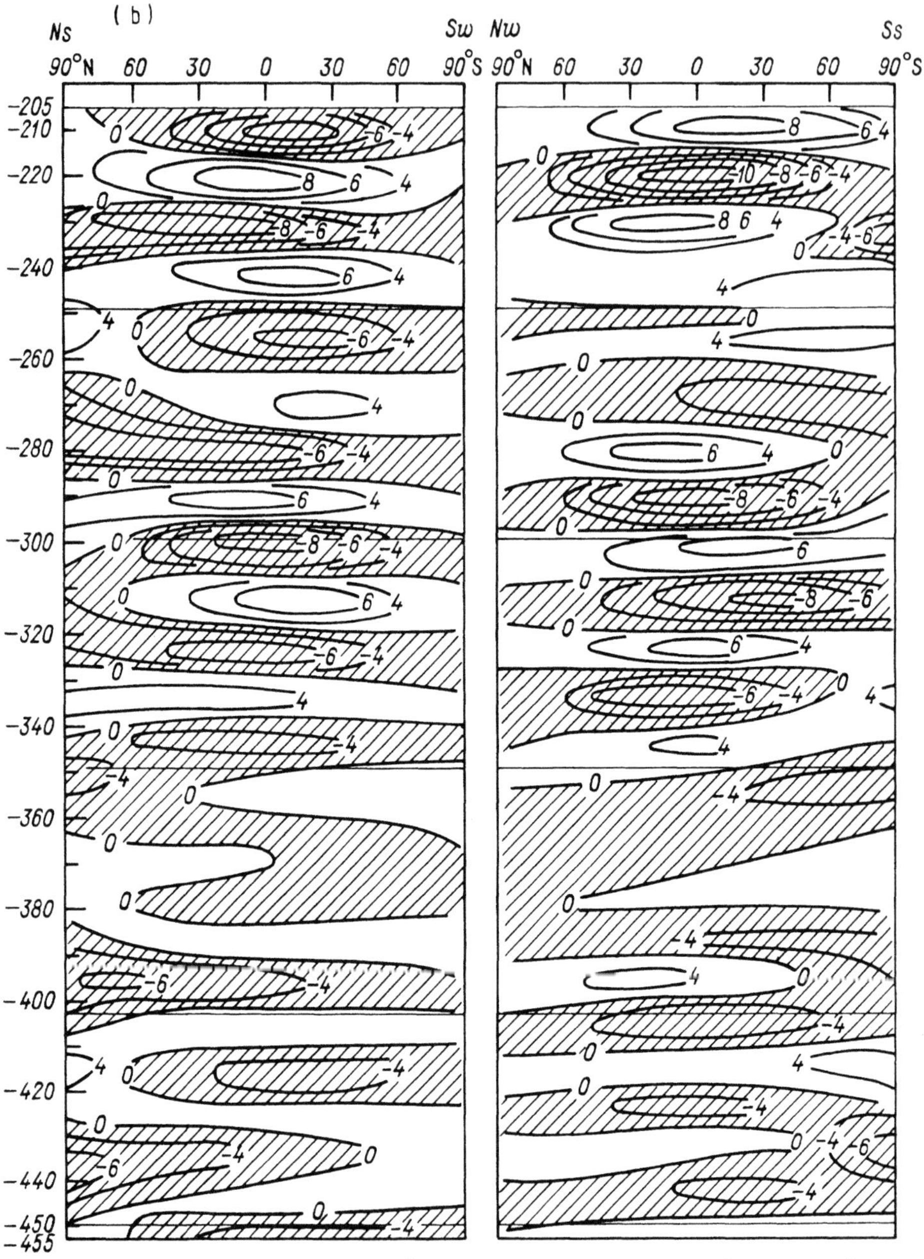

Fig. 4.2(b). Anomalies of insolation (kcal/cm^2) during caloric half years. 1 kcal/cm^2 = 41.9 MJ/m^2. $-455{,}000$ yr $\leqslant t \leqslant -205{,}000$ yr.

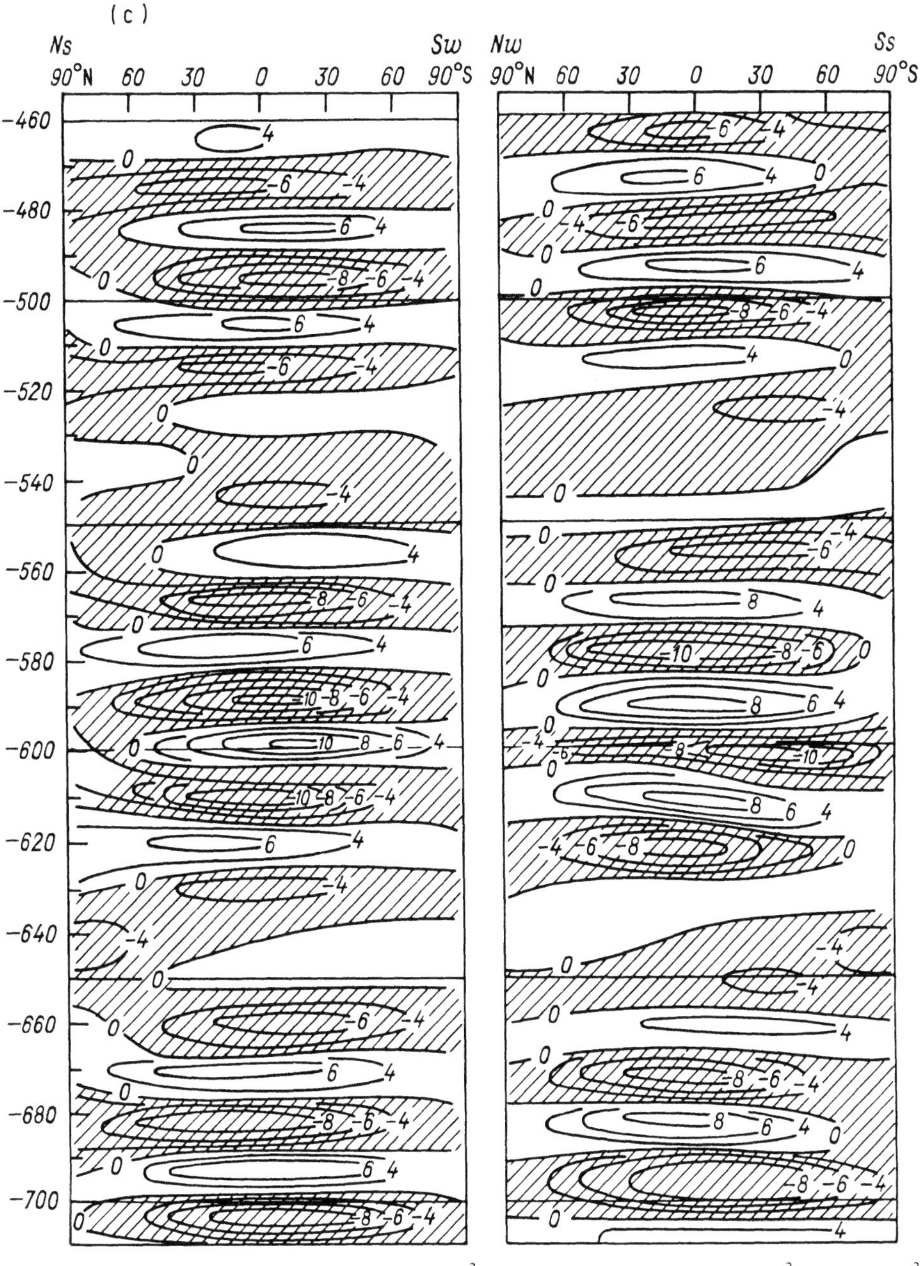

Fig. 4.2(c). Anomalies of insolation (kcal/cm^2) during caloric half years. 1 kcal/cm^2 = 41.9 MJ/m^2. –680,000 yr ≤ t ≤ – 460,000 yr.

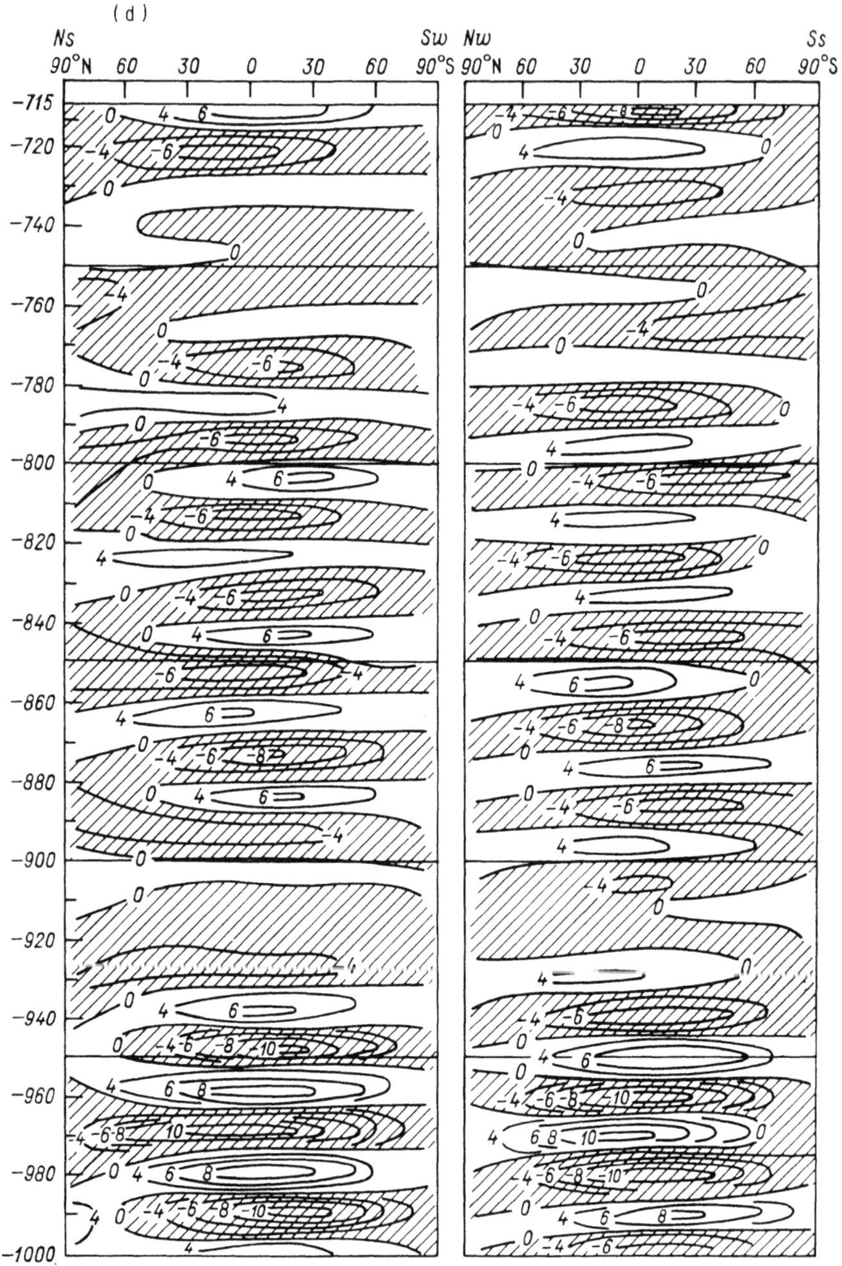

Fig. 4.2(d). Anomalies of insolation (kcal/cm^2) during caloric half years. 1 kcal/cm^2 = 41.9 MJ/m^2. −1 million yr ⩽ t ⩽ − 715,000 yr.

component has the same sign in the summer half years of both the Northern and Southern Hemispheres (mirror symmetry of the left and right halves of the graphs). A somewhat less intense component of the anomalies has during a given season the same sign over the entire Earth, while within a given hemisphere it changes sign from the winter half year to the summer half year (asymmetry of the left and right halves of the graphs) and regions of the same sign cover long time intervals, of the order of 60,000 to 150,000 years.

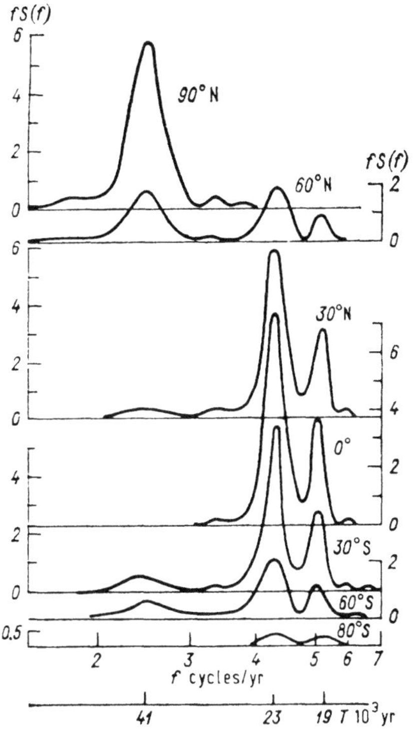

Fig. 4.3. Energy spectra of insolation fluctuations $\delta Q_{NS}(\varphi, t)$ and $\delta Q_{SW}(\varphi, t)$ at various latitudes.

If we base our analysis of climatic fluctuations during the Pleistocene on the scheme of A. Penck and E. Brückner (1909) for the Alps, then the insolation minima in the high and middle latitudes of the summer hemispheres in Fig. 4.2 can be dated, respectively, to periods 25,000, 70,000, and 115,000 years ago (Würm III, II, I); 190,000 and 230,000 years ago (Riss II, I); 435,000 and 475,000 years ago (Mindel II, I); 550,000 – 590,000 years ago (Günz); and 685,000, 855,000, and 970,000 years ago (Danube); the names of the stages of glaciation of the Pleistocene are given in parentheses. At present the Earth is in a typical interglacial period. The next insolation minimum, which spans the 20,000 years to come, will have a peak after 10,000 years and a configuration perfectly analogous to the previous minimum of Würm III (although this future minimum will be somewhat less marked). Consequently, from this point of view the astronomical theory predicts for the next few millenia a cooling of the climate, to say the least.

Figure 4.3 gives the energy time spectra $fS(f)$ of insolation fluctuations $\delta Q_{Ns}(\varphi, t)$ and $\delta Q_{Sw}(\varphi, t)$ at various latitudes φ. The spectra are seen to have a peak at a period of 41,000 years and a double peak at 23,000 and 19,000 years (the second peak being somewhat lower). The peak at 41,000 years is a maximum at the summer pole ($\varphi = 90°$), decreases toward the equator (becoming almost undetectable at latitudes of 20 to 0°), and then increases to a small height at 40–60°, after which it rapidly disappears as the winter pole is approached. The double peak at 23,000 and 19,000 years is a maximum in the equatorial zone $20° \geq \varphi \geq -20°$ and goes to zero as the poles are approached. At latitude $\varphi = 60°$N all three peaks are seen to be about equal in height.

Hays, Imbrie, and Shackleton (1976) found decisive proof that astronomical oscillations of the insolation (4.3) manifested themselves as climatic

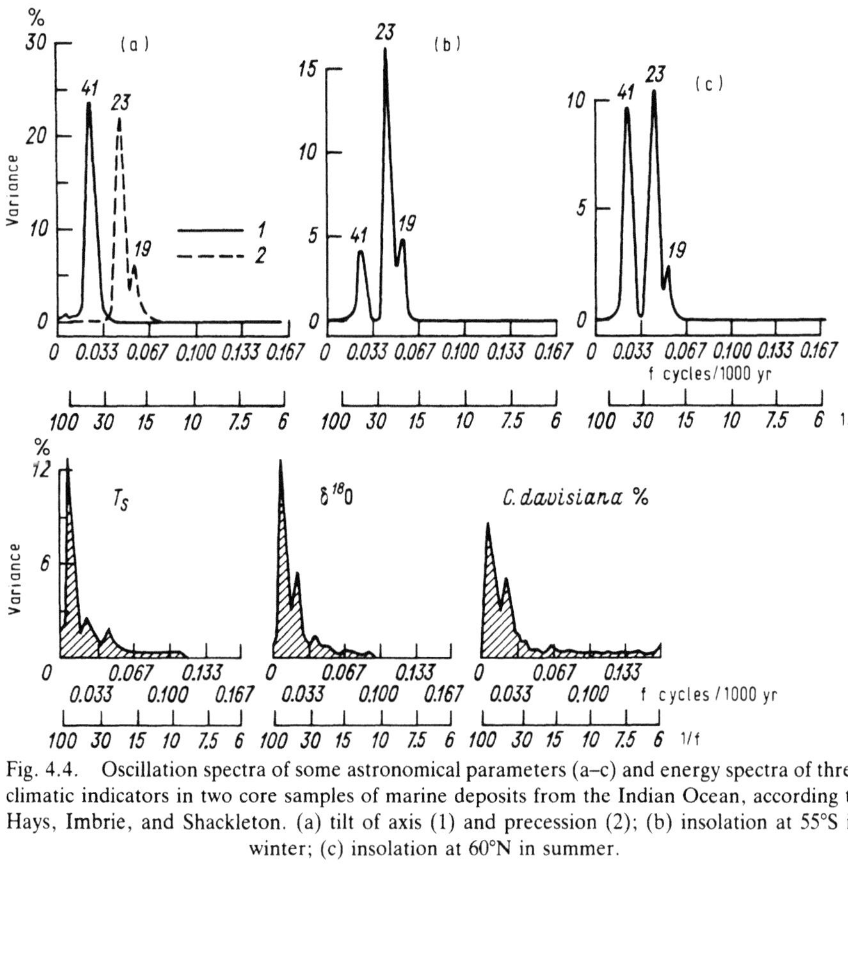

Fig. 4.4. Oscillation spectra of some astronomical parameters (a–c) and energy spectra of three climatic indicators in two core samples of marine deposits from the Indian Ocean, according to Hays, Imbrie, and Shackleton. (a) tilt of axis (1) and precession (2); (b) insolation at 55°S in winter; (c) insolation at 60°N in summer.

fluctuations during the Pleistocene. These investigators reconstructed the absolute chronology (over the last 468,000 years) of three climatic indicators in two core samples of marine deposits from the central part of the Indian Ocean at 43–46°S. The indicators were: the relative content of the heavy oxygen isotope ^{18}O in shells of planktonic foraminifera (reflecting changes in the isotopic composition of seawater due to the accumulation and melting of continental ice sheets); the species composition of communities of radiolaria (reflecting fluctuations of the temperature T_s of the upper ocean layer; and the relative content of the radiolaria species *Cycladophora davisiana* (reflecting fluctuations of the salinity of the water). The spectra calculated from these six time series (Fig. 4.4) have maxima with periods of about 100,000, 41,000, 23,000, and 19,000 years, as predicted by the astronomical theory. The period of about 100,000 years, corresponding to oscillations of the eccentricity e of the Earth's orbit, is poorly defined in the spectra of insolation fluctuations (Fig. 4.3), but it shows up clearly in the spectra of Fig. 4.4, as well as in the spectra of many other climatic series. This is because of the nonlinear response of the climate to fluctuations of the insolation, giving a combination frequency $\omega_{100} \approx \omega_{19} - \omega_{23}$; another combination frequency $\omega_{10} \approx \omega_{19} + \omega_{23}$ can be discerned in, for example, an ^{18}O spectrum for a core of Greenland ice obtained at Camp Century.

Insolation fluctuations (4.3) have always occurred, but a strong reaction of the climate to these, with an alternation of glacial and nonglacial regimes, apparently took place only during epochs with a cold climatic background, determined mainly by the distribution of continents and oceans. The variations of this distribution during the Cenozoic (closing of the Tethys Ocean, opening of the Atlantic and Indian Oceans, collision of India with Asia to form the Himalayas, separation of South America and Australia from Antarctica with formation of the Antarctic Circumpolar Current, etc.) caused a cooling of the climatic background to the Pleistocene level, whereby continental ice sheets were able to form at the insolation minima.

CHAPTER 3

RADIATION FACTORS

5. The Radiative Influx of Heat

The radiation field consists of an aggregate of electromagnetic waves having various lengths, propagation directions, amplitudes, and polarizations. At each fixed point waves with a fixed length and direction of propagation can be described in terms of the mean squares $I_l = \overline{|\xi_l|^2}$ and $I_r = \overline{|\xi_r|^2}$ of the amplitudes ξ_l and ξ_r of the fluctuations $E_l(t)$ and $E_r(t)$ of two mutually perpendicular components of vector **E**. Here **E** is the strength of the electric field in a plane perpendicular to the direction of wave propagation. Alternatively, the sum of these, the *intensity* $I = I_l + I_r$, and the difference $Q = I_l - I_r$ can be used instead of I_l and I_r. T_o complete the description of the waves, we introduce the amplitude and phase correlations $U = \overline{2\xi_l\,\xi_r \cos\delta}$ and $V = \overline{2\xi_l\,\xi_r \sin\delta}$, where δ is the phase difference between fluctuations $E_l(t)$ and $E_r(t)$. Quantities I, Q, U, and V are known as *Stokes parameters*. For an elliptically polarized wave, with constant ξ_l/ξ_r and δ, we have $Q = I \cos 2\varphi \cos 2\chi$, $U = I \cos 2\varphi \sin 2\chi$, and $V = I \sin 2\varphi$, where χ is the angle between direction l and the major axis of the ellipse made by the end of vector $\mathbf{E}(t)$, while φ is the degree of ellipticity, that is, tg φ is the ratio of the ellipse axes. Angle φ is positive for right-handed polarization and negative for left-handed polarization, and for waves of natural unpolarized radiation $Q = U = V = 0$. When scattering occurs, the transition from vector $\mathbf{I}(\mathbf{n}') = \{I_l, I_r, U, V\}$ of the incident wave with direction $\mathbf{n}'$ to vector $\mathbf{I}(\mathbf{n})$ of the scattered wave with direction $\mathbf{n}$ is effected with the aid of the so-called *angular matrix* $\Gamma(\mathbf{n}, \mathbf{n}')$.

Parameters Q, U, and V of the radiation field in the Earth's atmosphere and ocean are usually small (compared with I). Therefore, when calculating the radiative influx of heat, we can assume approximately that this radiation is unpolarized and describable in terms of a single parameter, the *spectral intensity* $I_\lambda(\mathbf{x}, \mathbf{n}, t)$. The spectral intensity is defined as the amount δF of energy of electromagnetic waves with lengths in a unit inverval $\delta\lambda$ centered at λ and with directions in a unit solid angle $\delta\Omega(\mathbf{n})$ with an axis $\mathbf{n}$ passing (during a unit time interval δt centered at time t) through a unit area $\delta\sigma$ centered at point $\mathbf{x}$, this area being perpendicular to direction $\mathbf{n}$ (where $\mathbf{n}$ is the unit vector). Here the wavelength λ is designated by the same symbol as the geographical longitude, but this should not cause any confusion.

The vector integral

$$\mathbf{F}_\lambda\,(\mathbf{x}, t) = \int I_\lambda\,(\mathbf{x}, \mathbf{n}, t)\;\mathbf{n}\;d\Omega\,(\mathbf{n}), \tag{5.1}$$

taken over all directions $\mathbf{n}$, is the *spectral flux of radiative energy*. By integrating this quantity over all wavelengths λ, we obtain the integral flux of radiant energy $F(\mathbf{x}, t)$. The rate of radiative influx of heat to a unit mass of air or water, which appears on the right-hand side of the equation of entropy evolution, is given by the formula

$$\varepsilon_r = -\frac{1}{\varrho}\,\mathrm{div}\,\mathbf{F} \approx -\frac{1}{\varrho}\,\frac{\partial F_z}{\partial z} = -\frac{1}{\varrho}\,\frac{\partial}{\partial z}(F^{\uparrow} - F^{\downarrow}), \qquad (5.2)$$

where the approximate equality signifies the approximation of a *locally plane-parallel medium* with its transverse direction along the vertical. This approximation is valid, since the horizontal divergence of field F is much less than the vertical divergence, and it is suitable for directions of radiation propagation making angles of more than 5° to the horizontal. Quantities $F^{\uparrow}$ and $F^{\downarrow}$ in (5.2) are the vertical components of, respectively, the *upward and downward fluxes* of radiative energy, obtained by integrating (5.1) over values of $\mathbf{n}$, respectively, from the upper hemisphere ($\mathbf{n}^{\uparrow}$) and from the lower hemisphere ($\mathbf{n}^{\downarrow}$). In the atmosphere $F^{\uparrow}$ and $F^{\downarrow}$ are about 350 W/m^2, and $\varrho\varepsilon_r$ is about 35 W/(m^2 km). The minus sign in (5.2) indicates that, as the energy flux increases, in the direction of its propagation the medium loses energy, that is, its self-radiation exceeds the absorption. In the atmosphere and ocean, assumed approximately to be locally plane-parallel media, it is convenient to reckon directions $\mathbf{n}$ from the vertical, that is, to describe them in terms of the zenith angle ϑ and the azimuth ψ. The spectral intensity of the radiation will then be a function of λ, z, ϑ, and ψ (it depends on the horizontal coordinates and time only parametrically, via the characteristics of the medium). This function satisfies the *equation of radiation transfer*:

$$\cos\vartheta\,\frac{\partial I_\lambda(z, \vartheta, \psi)}{\varrho(z)\,\partial z}$$
$$= \alpha_\lambda(z)\,[B_\lambda(z) - I_\lambda(z, \vartheta, \psi)] + \sigma_\lambda(z) \times$$
$$\times \left[\int_0^{2\pi} \mathrm{d}\psi' \int_0^{\pi} \gamma_\lambda(z; \vartheta, \psi; \vartheta', \psi')\, I_\lambda(z, \vartheta', \psi') \sin\vartheta'\, \mathrm{d}\vartheta' - \right.$$
$$\left. - I_\lambda(z, \vartheta, \psi)\right], \qquad (5.3)$$

where α_λ and σ_λ are the spectral coefficients of absorption and scattering per unit mass of the medium (measured in cm^2/g, whereas coefficients $\varrho\alpha_\lambda$ and $\varrho\alpha_\lambda$, calculated per unit volume of the medium, have units of reciprocal length); γ_λ is the spectral scattering function (the conditional probability of a direction $\mathbf{n} = (\vartheta, \psi)$ of the scattered radiation for a fixed direction $\mathbf{n}' = (\vartheta', \psi')$ of the incident radiation), usually a function only of the cosine

$$\cos\theta = \mathbf{n}\cdot\mathbf{n}' = \cos\vartheta\cos\vartheta' + \sin\vartheta\sin\vartheta'\cos(\psi - \psi')$$

of the scattering angle θ and normalized by the condition $\int \gamma_\lambda(\cos\theta)\,\mathrm{d}\Omega(\mathbf{n}) = 1$; and B_λ is the ratio of the spectral coefficients of radiation and absorption. In this section, in accordance with our assumption of a locally plane-parallel medium, we will reckon heights z from the actual surface of the Earth at the given geographical point, rather than from the surface of the geoid. Coefficients α_λ and σ_λ and scattering function γ_λ are given by the formulas

$$\alpha_\lambda = \sum_k c_k \alpha_{\lambda k}; \qquad \sigma_\lambda = \sum_l c_l \sigma_{\lambda l}; \qquad \sigma_\lambda \gamma_\lambda = \sum_l c_l \sigma_{\lambda l} \gamma_{\lambda l}, \tag{5.4}$$

where c_k and $\alpha_{\lambda k}$ are the concentrations of the various absorbing substances and their spectral absorption coefficients (per unit mass of absorbing substance), and c_l, $\sigma_{\lambda l}$, and $\gamma_{\lambda l}$ are the concentrations of the various scattering substances, their spectral scattering coefficients (per unit mass of scattering substance), and their scattering functions. The lower 70-km layer of the atmosphere is in a state of *local thermodynamic equilibrium*, that is, each infinitesimal volume of air absorbs and emits radiation like an ideally black cavity at the same temperature, which is in thermodynamic equilibrium. Consequently, in particular, *Kirchhoff's law* is obeyed, that is, the ratio B_λ of the spectral coefficients of radiation and absorption does not depend on the nature of the absorbing substances and is a universal function of the wavelength and temperature, a *Planck function*

$$B_\lambda = \frac{2c^2h}{\lambda^5}\left(e^{ch/\lambda kT} - 1\right)^{-1}, \tag{5.5}$$

where c is the velocity of light, h is Planck's constant, and k is the Boltzmann constant. Note that B_λ has a maximum at a wavelength $\lambda_m = ch/xkT$, where x is the root of the equation $x = -\ln(1 - x/5)$; the integral of B_λ over all λ is $(\sigma/\pi)\ T^4$, where $\sigma = 2\pi^5k^4/15c^2h^3$ is the Stefan–Boltzmann constant (the integral radiation flux of a perfect black body in the hemisphere is thus $B(T) = \sigma T^4$).

The radiation observed in the atmosphere can be divided into *short-wave radiation* and *long-wave radiation*. Short-wave radiation comprises the direct, reflected, and scattered solar radiation in the wavelength interval from 0.17 to 4 μm (in which is concentrated 99% of the radiation energy of a perfect black body for the temperature of the solar surface); of this, 8% of the energy is in the ultraviolet (0.1–0.39 μm), 56% of the energy is in the visible range (0.39–0.76 μm), with a maximum for blue light λ_m = 0.4738 μm, and 36% of the energy is in the near infrared (0.76–4 μm). Less than 0.4% of the solar radiation is in the range $\lambda > 5$ μm. Long-wave radiation comprises the infrared self-radiation of the atmosphere, the Earth's surface, and clouds in the interval from 4 to 100 μm. Note that, for a temperature of 300 K, 99% of the radiation energy of a perfect black body is concentrated in the 3–80 μm range with a maximum around 10 μm, whereas for a temperature of 200 K 99% of the energy is between 4 and 120 μm with a maximum around 15 μm. Less than 0.4% of the terrestrial radiation is at wavelengths $\lambda < 5$ μm.

In the atmosphere the main absorbers of short-wave radiation are water vapor and ozone (as well as oxygen), and to a slight degree CO_2, aerosol, and cloud particles. Water vapor has weak bands at $\lambda = 0.543$ to 0.847 μm and several absorption bands in the near infrared for $\lambda > 0.7$ m; these bands are around 0.72; 0.81 ($\varrho_{\sigma\tau}$); 0.93 (Φ); 1.13 (ψ); 1.42; 1.89; 2.01 – 2.05(Ω): 2.25 – 3.0(χ); 3.17 μm (the designations of some of the bands are given in parentheses). Ozone has strong absorption bands in the ultraviolet region (a Hartley band at 0.22–0.29 μm and a band at 0.31–0.36 μm, where about 1.5% of the energy of the solar radiation is absorbed), with a volume absorption coefficient of 135 cm^{-1} at $\lambda = 0.255$ μm, as well as a weak Chappuis band in the visible range (0.44–0.75 μm) with a maximum around $\lambda = 0.602$ μm (oxygen absorbs ultraviolet radiation in the 0.13–0.24 μm band and has two narrow absorption lines in the visible range at $\lambda = 0.69$ and 0.76 μm). Carbon dioxide has weak absorption bands in the near infrared at $\lambda = 1.4$, 1.6, 2.0, 2.7, and 4.3 μm (which are usually negligible). The absorption by the aerosol is quite variable in time and space; on the average, it is comparable to the absorption by water vapor. The so-called *absorption index* $\varrho\delta_\lambda\lambda/4\pi$ for the tropospheric aerosol generally has a value of the order of 0.01. At high relative humidities the aerosol particles become water-coated and their optical properties approximate those of water droplets. It should be noted that there is practically no short-wave emission in the atmosphere at low temperatures.

Short-wave radiation is scattered in the atmosphere by microinhomogeneities of the air density (molecular or Rayleigh scattering), by aerosol particles, and by cloud particles. Scattering by inhomogeneities or particles smaller than the wavelength of the radiation is called Rayleigh scattering. It is characterized by the volume scattering factor $\varrho\sigma_\lambda = \frac{128}{3}\pi^5 N\beta^2\lambda^{-4}$, where $\beta = (n^2 - 1)/4\pi N$ is the so-called polarizing capacity. In the latter expression n is the refractive index of the medium and N is the number of particles per unit volume. A scattered beam is partially plane-polarized in a plane perpendicular to the scattering plane, and the ratio of the intensities parallel to and perpendicular to the scattering plane is equal to $\cos^2\theta$, where θ is the scattering angle. For air at standard pressure for $\lambda = 0.3$ to 1 μm the scattering factor $\varrho\sigma_\lambda = 1.525 \times 10^{-6}$ to 1.09×10^{-8} cm^{-1}. The phase function for Rayleigh scattering does not depend on the wavelength and has the form $\gamma(\cos\theta) = \frac{3}{16\pi}(1 + \cos^2\theta)$. The scattering factor for scattering by large particles of a dielectric substance (such as water) has the form

$$\varrho\sigma_\lambda = \pi r^2 N f\left(\frac{2\pi r}{\lambda}\right),$$

where r is the particle radius, and $f(x)$ is a function which for $0 < x < 6.1$ increases to 4 and then decreases, fluctuating in value, to a final limit of 2. The phase functions for scattering by large particles are greatly elongated in the forward direction. For a cloud with a narrow drop-size distribution in the interval $\lambda = 0.6$ to 3.8 μm, we have $\sigma_\lambda \approx 2500$ cm^2/g (it drops to 1500 cm^2/g for $\lambda = 2.8$ to 3.4 μm), while the mean elongation of the scattering function (the ratio of the probabilities that $\cos\theta > 0$ or $\cos\theta < 0$) is of the order of several tens.

Long-wave radiation in the atmosphere is absorbed (and emitted) by water vapor, carbon dioxide, ozone, aerosol particles, and cloud particles. The absorption by water vapor is greatest in a wavelength band from 5 to 7.5 μm with a maximum at $\lambda = 6.27$ μm, where the absorption coefficient, smoothed over the lines, reaches 200 cm^2/g. The 8.5 – 12 μm interval is the *atmospheric window*, within which water vapor almost does not absorb radiation (smoothed absorption coefficient of the order of 0.1 cm^2/g). The region from 12 to 100 μm contains the rotational bands of water vapor (in the 11–12 μm interval there is also some absorption by the water-vapor dimer). Carbon dioxide has weak absorption bands around $\lambda = 4.8$ and 5.2 μm, but its highest absorption is in the 12.9–17.1 μm band, centered near 15 μm, where $\alpha_\lambda = 134$ cm^2/g and the effect of CO_2 turns out to be greater than that of water vapor. Ozone has strong but narrow absorption bands around 4.75 μm, in the vicinity of the atmospheric window (in the interval 9.35–11 μm, centered at $\lambda = 9.569$ μm), and around 14.1 μm. The absorption of long-wave radiation by the aerosol is quite variable (at high relative humidities it may be comparable to the absorption by water vapor); it can play an especially important role in the atmospheric window, where the absorption index $\varrho\alpha_\lambda\lambda/4\pi$ has been estimated to be about 0.2–0.5. The absorption by cloud particles is quite great; for instance, if the drop-size distribution is narrow, coefficient α_λ increases from 1190 to 3530 cm^2/g in the interval $\lambda = 5$ to 14 μm and then drops, fluctuating in value, to 1580 cm^2/g at $\lambda = 40$μm.

Long-wave radiation is scattered by the aerosol and by cloud particles. The role of scattering is particularly great in regions where multiple scattering makes the mean free paths of the photons many times longer (up to $h\,\overline{\sec\vartheta}$, on the average, where h is the thickness of the cloud, ϑ is the zenith angle of the direction of wave propagation, and $\overline{\sec\vartheta}$ is the "diffusivity parameter", with values of the order of several units). Thus, the total absorption of long-wave radiation by a cloud is nearly 100%, and in the long-wave part of the spectrum a sufficiently dense cloud acts like a black body. Such a cloud is sometimes assumed approximately to emit into the surrounding space just like a black body at the corresponding boundary temperature (this approximation is, however, inapplicable in the vicinity of cloud boundaries, where the values of ε_r are especially high, leading to heating of the lower boundary and cooling of the upper boundary and influencing the cloud evolution considerably).

In the ocean only the short-wave radiation entering from the atmosphere is propagated (the long-wave radiation is absorbed by the thin surface layer of water). This short-wave radiation is absorbed by the water, by substances dissolved in it (mainly the organic "yellow substance"), and by suspended matter (primarily phytoplankton pigments). It is scattered both molecularly (by inhomogeneities of the water density, of the orientation of water molecules, and of the concentration of dissolved substances) and by suspended particles (terrigenous particles with sizes $r < 1$ μm and biogenous particles 1 – 20 μm in size). The absorption bands of water are mostly associated with bands of a single molecule (water vapor). However, because of intermolecular interactions these bands are shifted toward higher λ and blurred, forming a continuous spectrum, in which as λ increases from 0.25 to 0.46 μm the value of $\varrho\alpha_\lambda$ drops from 0.19 m^{-1} to a blue-green minimum of 0.002 m^{-1}, and then increases to 2.62 m^{-1} at $\lambda = 0.75$ m, after which it drops to 2.02 m^{-1} at $\lambda = 0.8$ μm. Absorption by the yellow substance increases toward the short-wave region and also increases with depth; absorption by suspended matter has a small maximum at $\lambda = 0.41$ to 0.43 μm (and in the red lines of chlorophyll at $\lambda = 0.67$ to 0.68 μm), and it decreases with depth. The molecular scattering at inhomogeneities of the water density and of the orientation of water molecules is in visible light characterized by the quantity

$$\varrho\sigma_\lambda\gamma_\lambda = A_\lambda\left(1 + \frac{1 - \Delta_\lambda}{1 + \Delta_\lambda}\cos^2\theta\right),$$

where A_λ is approximately proportional to λ^{-4} and is at $\lambda = 0.436$ μm equal to 2.08×10^{-4} m^{-1}, while Δ_λ, the depolarization coefficient, is equal to 0.088 at $\lambda = 0.436$ μm. The scattering at inhomogeneities of the concentration of dissolved substances has a Rayleigh phase function. The total volume coefficient of molecular scattering decreases from 4×10^{-2} to 3.6×10^{-4} m^{-1} as λ increases from 0.25 to 0.8 μm. The scattering by suspended matter is very great, is variable ($\varrho\sigma_\lambda$ varies from 0.02 to 3.0 m^{-1} and drops only slightly as λ increases), and its phase functions are greatly elongated forward (50 times more than atmospheric scattering functions).

The foregoing information (together with some details which were not given here) is in principle sufficient for a numerical integration of the spectral equations of radiation transfer (5.3), a summation over all the spectral lines, and a calculation of the radiative fluxes and influxes of heat. However, additional complications arise because real absorption lines have a finite spectral width and a definite form. In the troposphere these effects are due mainly to molecular collisions, giving a so-called Lorenz shape to the lines and making their width proportional to the pressure (after the radiation has traversed a path with substantially varying pressures, in the *Curtis-Godson approximation* the lines are assumed to have a Lorenz shape with a width corresponding to the mean pressure over the path). The mean absorption coefficient over the entire line width decreases with an increase in the mass of the absorbing substance traversed (*Forbes effect*), since after complete absorption at the center of the line the maximum values of the absorption coefficient are left out of the averaging. The simplest means of parametrizing all these "line-width effects" is to assume a power-law dependence on the pressure and temperature of the mean absorption coefficient in the given line of form $(p/p_0)^n\ (T/T_0)^l$, with exponents n and l which depend in general on the absorbing substance, the part of the spectrum, and the absorbing mass traversed. If the dependence on the mass is ignored, while the dependence on λ is reduced just to the difference between the short-wave and long-wave radiation, then these correction factors can be made to pertain to the densities of the corresponding absorbing substances, by introducing their "effective masses":

$$m_k(z) = \int_0^z (p/p_0)^{n_k} (T/T_0)^{l_k} \varrho_k(z)\, dz, \tag{5.6}$$

rather than to the absorption coefficients. Then the absorption coefficients can be assumed to be independent of the height.

In the case of long-wave radiation (taking scattering into account only via the diffusivity parameter $\overline{\sec \vartheta}$) we confine ourselves to an approximate calculation with the aid of so-called *integral transmission functions*, defined as the following functionals:

$$\begin{aligned} &D[m_1(z)-m_1(z'), \ldots, m_n(z)-m_n(z')] \\ &= 2\pi \int \alpha_\lambda \frac{B_\lambda [T(z')]}{B[T(z')]} \int_0^{\pi/2} \exp \Big\{ -\sec\vartheta \sum_{k=1}^{n} \alpha_{\lambda k} [m_k(z) - \\ &- m_k(z')] \Big\} \cos\vartheta \sin\vartheta\, d\vartheta. \end{aligned} \tag{5.7}$$

From equation (5.3), with $\sigma_\lambda = 0$ and with boundary conditions $I_\lambda(0, \vartheta, \varphi) = B_\lambda[T(0)]$ for $0 < \vartheta < \pi/2$ and $I_\lambda(\infty, \vartheta, \varphi) = 0$ for $\pi/2 < \vartheta < \pi$, we can derive the formulas

$$\begin{aligned} F_L^{\uparrow}(z) &= B[T(0)] D[m_1(z), \ldots, m_n(z)] + \\ &+ \int_0^z dz'\, B[T(z')] \frac{d}{dz'} D[m_1(z) - \\ &\quad m_1(z'), \ldots, m_n(z) - m_n(z')]; \\ F_L^{\downarrow}(z) &= \int_z^\infty dz'\, B[T(z')] \frac{d}{dz'} D[m_1(z') - \\ &\quad - m_1(z), \ldots, m_n(z') - m_n(z)], \end{aligned} \tag{5.8}$$

where

$$\frac{d}{dz'} = \sum_{k=1}^{n} \frac{dm_k(z')}{dz'} \frac{\partial}{\partial m_k(z')}.$$

Let m_1, m_2, m_3 be the masses of water vapor, carbon dioxide, and cloud particles (and in (5.6) let us set $n_l = 1/2$, $n_2 = 4/5$, $n_3 = l_1 = l_2 = l_3 = 0$). E. M. Feigel'son then approximates the transmission function as follows:

$$D\,(m_1, m_2, m_3) = e^{-\alpha\beta_3 m_3}\Bigl[\sum_{k=1}^{3} a_k^{(1)}\, e^{-\alpha_k^{(1)}}\beta_1 m_1 + $$
$$+ \Bigl(\sum_{k=1}^{3} a_k^{(2)} e^{-\alpha_k^{(2)}\beta_1 m_1}\Bigr)\Bigl(\sum_{k=1}^{3} a_k^{(3)} e^{-\alpha_k^{(3)}\beta_2 m_2}\Bigr)\Bigr], \qquad (5.9)$$

where $a_k^{(n)}$ and $\alpha_k^{(n)}$ are especially selected parameters and α is the mean absorption coefficient of cloud droplets (for a narrow droplet-size distribution it can be assumed that $\alpha \approx 1700$ cm^2/g), while $\beta_1 = \beta_2 = 5/3$ and $\beta_3 \sim 10$ are diffusivity parameters. A more thorough approach would also take into account the absorbing mass of the aerosol m_4 in the layers outside the clouds. A rougher approximation, on the other hand, would describe the clouds as radiating black bodies; then the transmission function $D(m_1, m_2, 0)$ is used in the layers outside the clouds and formulas of the type of (5.8) are derived for these layers.

For short-wave radiation we have to set $B_\lambda \equiv 0$ in equation (5.3) and then solve it just for the daytime. In the calculation of the radiation field in the atmosphere, the boundary conditions for equation (5.3) have the form

$$I_\lambda\,(\infty, \mathbf{n}^\downarrow) = I_{0\lambda}\delta(\mathbf{n}^\downarrow + \mathbf{n}_\odot^\uparrow); \qquad (5.10)$$

$$I_\lambda\,(0, \mathbf{n}^\uparrow) = \int_0^{2\pi} d\psi' \int_{\pi/2}^{\pi} r_\lambda(\mathbf{n}^\uparrow, \vartheta', \psi')\, I_\lambda\,(0, \vartheta', \psi') \cos\vartheta' \sin\vartheta'\, d\vartheta', \qquad (5.11)$$

where $I_{0\lambda}$ is the spectral solar constant, and $r_\lambda(\mathbf{n}^\uparrow, \mathbf{n}^\downarrow)$ is the spectral coefficient of directional reflection from the underlying surface. The surface of the land can often be considered orthotropic, that is, it is such that r_λ does not depend on $\mathbf{n}^\uparrow$ and $\mathbf{n}^\downarrow$ (this is apparently also the case for parts of the ocean surface which are covered with foam); then $r_\lambda = A_\lambda/\pi$, where A_λ is the spectral albedo, so that, for instance, the reflection of direct solar radiation gives

$$I_\lambda\,(0, \mathbf{n}^\uparrow) = \frac{A_\lambda}{\pi} I_\lambda\,(0, -\mathbf{n}_\odot^\uparrow) \cos\vartheta_\odot.$$

The reflection from a wavy ocean surface in regions without foam was calculated by Yu. A. R. Mullamaa, neglecting diffraction at the edges of elementary areas, shading of some area elements by others, and multiple reflections. In this case the reflection can be found with the aid of Fresnel's formulas, according to which an amount of radiation $I_\lambda\,(0, \mathbf{n}^\downarrow)$ is reflected from an elementary area with an outward normal $\mathbf{n}_1$ in the direction $\mathbf{n}^\uparrow = \mathbf{n}^\downarrow + 2\mathbf{n}_1 \cos i_1$, where $\cos i_1 = -(\mathbf{n}^\downarrow \cdot \mathbf{n}_1)$ is the cosine of the angle of incidence i_1. If the incident radiation is not polarized, then the intensity of the reflected radiation $I_\lambda(0, \mathbf{n}^\uparrow)$ is obtained by multiplying $I_\lambda(0, \mathbf{n}^\downarrow)$ by the expression

$$f_\lambda = \frac{1}{2}\Bigl[\frac{\sin^2(i_1 - i_2)}{\sin^2(i_1 + i_2)} + \frac{\mathrm{tg}^2(i_1 - i_2)}{\mathrm{tg}^2(i_1 + i_2)}\Bigr],$$

where i_2 is the refraction angle ($\sin i_2 = \sin i_1/n_\lambda$, where n_λ is the refractive index of water relative to air for radiation with a wavelength λ). In practice, when calculating the reflection from a wavy ocean surface, the values of $I_\lambda(0, \mathbf{n}^\uparrow)$ for fixed $\mathbf{n}^\uparrow$ have to be averaged over the probability distribution of the directions $\mathbf{n}_1$ of the reflecting areas (which is a characteristic of the wave action) and, as in formula (5.11), integrated over the directions $\mathbf{n}^\downarrow$ of the incident radiation. Note that polarization effects, which have so far been neglected here, may influence the reflection processes. Finally, it should be mentioned that apart from the reflected radiation a small contribution to $I_\lambda(0, \mathbf{n}^\uparrow)$ may be made by the radiation scattered by the water thickness; an accurate calculation of this, however, is very difficult.

It is difficult to compute the field of short-wave radiation in the atmosphere, because of the impossibility of taking multiple scattering into account, especially in clouds. At the same time, it is important to calculate this field, since the direct heating of the atmosphere by short-wave radiation turns out to be sizable: during the daytime in summer it is comparable to the cooling via the emission of long-wave radiation (clouds redistribute this heat in space, increasing it in the layer above the clouds and reducing it below the clouds, the total increase in heating being some tens of percents). In one method of calculating the radiation field precisely, an additional term $1/[c_0\varrho\,(\partial I_\lambda/\partial t)]$ is introduced on the left-hand side of equation (5.3), where c_0 is the velocity of light. With the aid of this unsteady-state equation, the paths of a great number of randomly selected photons can be distinguished, after which an averaging over all the photons is carried out (Monte Carlo method) or analytical solutions asymptotic in t are sought.

In practice it is convenient to calculate separately the radiation fluxes in the cloud layers and in the layers above, between, and below the clouds. This requires continuity of the radiation intensities at the cloud boundaries and a consideration of the angular distributions there of the transmitted light (the *law of limb darkening*, to be found separately) and the reflected light (approximately isotropic distribution). In visible light a satisfactory solution of equation (5.3) can apparently be obtained with the aid of the *two-flux approximation*, that is, by an approximate derivation from (5.3) of the equations for the upward and downward radiation fluxes $F^\uparrow(z)$ and $F^\downarrow(z)$. In the near infrared region another simplification is possible, namely to neglect in (5.3) the integral term outside the clouds, and for clouds to use the separately found photon-path distributions to determine the probabilities of reflection and transmission of radiation and the mean free paths $l = h\,\overline{\sec\vartheta}$. For all regions (inside and outside the clouds) the integral transmission function of the absorbing substances is used.

For ocean regions without foam the penetration of short-wave radiation into the seawater through a wavy surface is calculated approximately using Fresnel's formulas, according to which the radiation $I_\lambda(0, \mathbf{n}^\downarrow)$ passes through an area element with an outward normal $\mathbf{n}_1$ in the direction

$$\mathbf{n}_2^{\downarrow} = \frac{\sin i_2}{\sin i_1}\mathbf{n}^{\downarrow} - \frac{\sin (i_1 - i_2)}{\sin i_1}\mathbf{n}_1$$

and is attenuated by a factor of $(1 - f_\lambda)^{-1}$. When calculating the field of short-wave radiation in the ocean, we have to average the values of $I_\lambda(0, \mathbf{n}_2^{\downarrow})$ with fixed $\mathbf{n}_2^{\downarrow}$ over the probability distribution for the directions $\mathbf{n}_1$ of refracting areas and then integrate over the directions $\mathbf{n}^{\downarrow}$ of the incident radiation. The integration of equation (5.3) (for $B_\lambda \equiv 0$) with this boundary condition can be carried out using the Monte Carlo method or else with the aid of some approximate analytical method (for instance, the two-flux, small-angle, or transport approximation).

Radiative heat exchange plays a very important role in the formation of the temperature fields in the atmosphere and ocean. Let us demonstrate this role using the atmosphere as an example. To do this, we compute the thermal state of the atmosphere which would exist (its so-called *radiation equilibrium*) if of all the actual heat-exchange mechanisms only radiative heat exchange were operating in it. Such a calculation was first carried out by Emden (1913) in a simplified form, whereby the atmosphere was assumed to be transparent to short-wave radiation and not to scatter long-wave radiation, but rather to absorb it like a *gray body* (the absorption coefficient α is independent of the wavelength). Then, from (5.3) we obtain the following equations for the fluxes of radiative energy:

$$\frac{\partial F^{\uparrow}}{\partial m} = \alpha\beta(\sigma T^4 - F^{\uparrow}); \qquad \frac{\partial F^{\downarrow}}{\partial m} = -\alpha\beta(\sigma T^4 - F^{\downarrow}), \tag{5.12}$$

where $m = \int_0^z \varrho_a \, dz$ is the mass of absorbing substances, and $\beta = 5/3$ is the diffusivity parameter. The boundary conditions at the Earth's surface ($z = 0$, $m = 0$) and at the upper limit of the atmosphere ($z = \infty$, $m = M$) are taken to be

$$F^{\uparrow}(0) - F^{\downarrow}(0) = (1 - A)\, I_0 \cos \vartheta_\odot; \qquad F^{\downarrow}(\mathrm{M}) = 0. \tag{5.13}$$

Since for radiation equilibrium it must be true that $F^{\uparrow} - F^{\downarrow} = \text{const}$, the air temperature T and the temperature of the Earth's surface $T_s = [\sigma^{-1}F^{\uparrow}(0)]^{1/4}$ can easily be found from (5.12)–(5.13) to be

$$T = \left\{\frac{1-A}{2\sigma} I_0 \cos \vartheta_\odot \,[1 + \alpha\beta\,(M - m)]\right\}^{1/4};$$

$$T_s = \left\{\frac{1-A}{2\sigma} I_0 \cos \vartheta_\odot \,(2 + \alpha\beta M)\right\}^{1/4}. \tag{5.14}$$

For $\alpha M = 1$ and $(1 - A)\, I_0 \cos \vartheta_\odot = 209.4$ W/m^2, we obtain $T_s = 283$ K and $T(0) = 261$ K, that is, at the Earth's surface there is a temperature jump of 22 K; in the lower troposphere the air temperature drops rapidly with height (for such an unstable stratification, convection must develop, that is, the radiation equilibrium turns out to be convectively unstable), while the upper troposphere and the stratosphere are in this case found to be isothermal. On the whole, both the vertical and the latitudinal temperature contrasts in the atmosphere for radiation equilibrium are considerably greater than the actual contrasts. This means that the contributions of the other mechanisms of heat transfer (mainly convective heat exchange, that is, the transfer of sensible and latent heat via large-scale and small-scale air movements) in the atmosphere are appreciable.

Kibel' (1943) calculated the thermal equilibrium in the atmosphere, allowing for the absorption (in the gray-body approximation) of short-wave radiation as well as long-wave radiation, and also introducing convective (turbulent) heat exchange. This provided, for an appropriate choice of parameters, a realistic temperature distribution for the troposphere of the middle latitudes and isothermy in the stratosphere. Manabe and Strickler (1964) carried out a very detailed calculation of radiation equilibrium in a cloudy or cloudless atmosphere, with summation over all the spectral lines, as well as of thermal equilibrium with convective adjustment in layers with unstable stratification ($\partial T/\partial z > 10°C/km$ or $6.5°C/km$). Some of their results are shown in Fig. 5.1 (for $I_0 = 139.4\ W/m^2$; $\cos \vartheta_\odot = 0.25$; $A = 0.102$; $|\partial T/\partial z|_{cr} = 6.5°C/km$ and for the vertical distributions of absorbing substances and clouds typical of April at a latitude of 35°N).

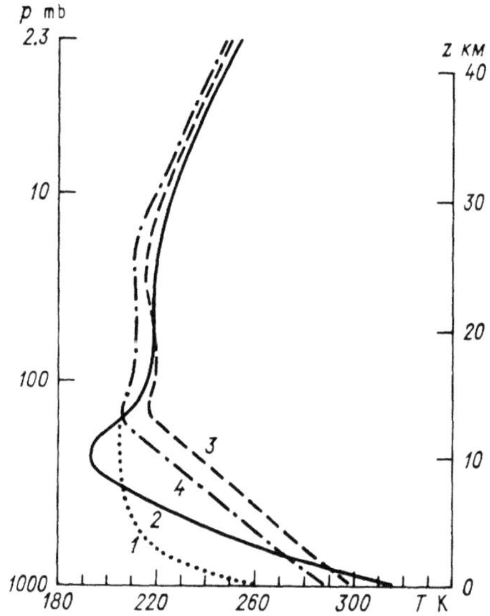

Fig. 5.1. Vertical temperature profiles. (1) for radiation equilibrium according to Emden's theory; (2) for radiation equilibrium in a cloudless atmosphere; (3) for thermal equilibrium in a cloudless atmosphere; (4) for thermal equilibrium in a cloudy atmosphere (2–4 according to Manabe and Strickler).

For radiation equilibrium some very high temperature gradients in the troposphere were obtained in this case (but there was practically no temperature jump at the Earth's surface), with a temperature at the Earth's surface which was too high and a temperature in the stratosphere that was too low; temperature inversions were obtained at the tropopause and above 27 km. The latitudinal distributions of the temperature, and especially of the height of the tropopause, showed a much worse fit with reality, which was to be expected, since in this calculation the transfer of heat by large-scale currents was not taken into account at all.

6. Carbon Dioxide

The main climatic function of carbon dioxide is to produce the greenhouse effect in the Earth's atmosphere. Although it allows through the downward flux of short-wave solar radiation, carbon dioxide absorbs the upward flux of long-wave radiation from the Earth's surface and thus contributes to the heating of the lower atmosphere. The greenhouse effect can be estimated roughly with the aid of the following formula of Eddington, which is related to Emden's formulas (5.14):

$$T_s = T_0 \left(1 + \frac{3}{4} \tau \right)^{1/4} , \tag{6.1}$$

where τ is the effective optical thickness of all the components of the atmosphere producing the greenhouse effect, and T_0 and T_s are the temperatures of the surface air (the air near the ground) in the absence of and in the presence of the greenhouse effect. For the atmosphere at present $T_0 = 255$ K, $T_s = 288$ K, and $\tau = 0.84$, the contribution of CO_2 to τ being about 0.24. If CO_2 were absent (that is, if $\tau \approx 0.6$), then the surface temperature would be $T_s \approx 280$ K, that is, the air at the ground would be 8 K less than the present value. Thus it is clear that the greenhouse effect of CO_2 has a very substantial influence on the climate. The data to be presented below indicate that the CO_2 concentration in the atmosphere is at present on the rise (due to the burning of fossil fuels and the cutting down of tropical forests). Extrapolation of this process predicts a doubling of the CO_2 concentration during the period from 2000 to 2100 and a possible further rise by a factor of 2 to 4 (until the reserves of fossil fuels give out). The corresponding increase δT_s in the greenhouse effect can be estimated using the following formula, ensuing from (6.1) for small increments $\delta\tau$ of the optical thickness:

$$\frac{\delta T_s}{T_s} = \frac{1}{4} \left[1 - \left(\frac{T_0}{T_s} \right)^4 \right] \frac{\delta \tau}{\tau} . \tag{6.2}$$

Here it just has to be taken into account that in view of the Forbes effect the optical thickness of CO_2 varies not in proportion to the mass, but rather considerably more gradually; according to laboratory measurements, a doubling of the current CO_2 concentration will increase the optical thickness by an amount $\delta\tau = 0.024$, which according to formula (6.2) is equivalent to $\delta T_s = 0.7$K.

Much more detailed estimates of the effect of doubling the CO_2 concentration were obtained by Manabe and Wetherald (1975), with the aid of a three-dimensional (nine-level, with an interval of about 500 km along the horizontal) numerical model of the general circulation of the atmosphere, taking into account the positive feedbacks of the greenhouse effect with the water-vapor content in the atmosphere and with the increase in the albedo of the Earth's surface due to the decrease of the snow cover, but simplified in the following respects: (1) not only was the CO_2 concentration fixed, but the ozone concentration and the cloud amount were fixed as well (the cloud cover being specified as a function of latitude and height on the basis of recent mean annual factual data), so that the negative feedback between the greenhouse effect and the cloud amount was not taken into account; (2) heat transport by ocean currents was not taken into account at all; (3) no seasonal fluctuations were taken into account, the calculation being carried out for fixed (mean annual) external conditions (with two variants of the initial conditions) for about 800 days and the mean state of the atmosphere being estimated for the last 100 days (mean for the two variants of the initial conditions); (4) the calculation pertained to a limited region ($8.3° \leqslant \theta \leqslant 90°$ with smoothing at the boundaries, $-60° \leqslant \lambda \leqslant 60°$ with a cyclical continuity for the longitude), with an idealized distribution of continents and oceans (ocean in the region $23.5° \leqslant \theta \leqslant 90°$, $0° \leqslant \lambda \leqslant 60°$).

Two numerical experiments were carried out: one for the present CO_2 concentration of 300 ppm and one for a doubled concentration of 600 ppm. Figure 6.1 shows the effect of a doubling of the CO_2 concentration, in terms of the difference of the mean zonal temperature distributions in these two experiments. Due to the enhancement of the greenhouse effect, the troposphere and lower stratosphere (up to a height of 18–20 km) are seen to become 2 – 3 K warmer (the lower half of the troposphere in the polar region warms up by as much as 6 – 9 K, the mean warming of the surface air being about 3 K), whereas the upper stratosphere becomes appreciably cooler. It should be noted that there is at the same time an increase in the air humidity and the amount of precipitation (by 7%). However, if the foregoing simplifications of the model are dispensed with, these results may be quantitatively quite different. Therefore, more reliable estimates are obtained with models of an interacting ocean-atmosphere system with a real geometry, taking seasonal fluctuations into account and calculating the cloud amount.

Data for 1977 gave a mean value of 333 ppm for the volume CO_2 concentration in the atmosphere. However, this value fluctuates appreciably in both time and space, since the sources and sinks of CO_2 are distributed irregularly throughout the atmosphere and they tend to vary with time. Green plants are

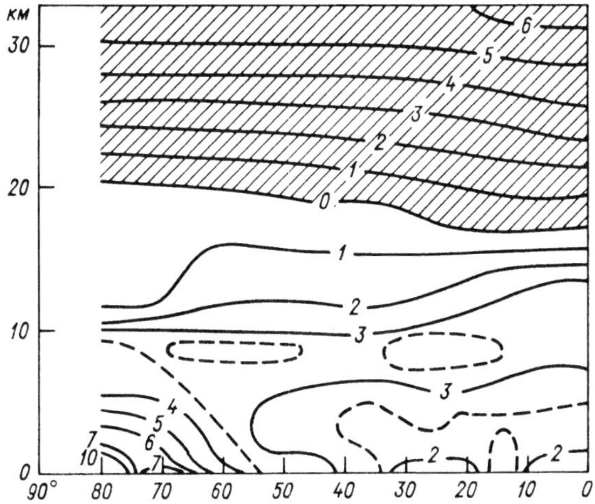

Fig. 6.1. Increment of mean zonal temperature (kelvins) due to doubling of present CO_2 concentration, according to numerical experiment of Manabe and Wetherald (1975); region of negative variations is shaded.

both a source and a sink for CO_2: on the one hand, they are continually breathing, that is, they absorb oxygen and give off CO_2; on the other hand, in sunlight plants are intensely photosynthesizing, and thus absorbing CO_2 and giving off oxygen. As a consequence, for example, the CO_2 content in the air between corn plants can vary by a factor of two during a 24-hour period, from a maximum at night to a minimum in the daytime. To be sure, only a small distance away from the vegetation these fluctuations of the CO_2 concentration are smoothed out considerably, but at the same time the seasonal fluctuations of this concentration in the middle and high latitudes, with a maximum in late winter and a minimum in late summer, prove to be quite pronounced. As an example, we cite the data in Fig. 6.2, obtained at Mauna Loa Observatory (Hawaii, 19.5°N, 155.6°W), where from 1958 on regular measurements of the CO_2 concentration in the air were carried out. The dots on the graph indicate the mean monthly values. Inspection of the figure shows the seasonal fluctuations to be very regular (with a gradual rise in the CO_2 concentration over a seven-month period from summer to winter and a rapid drop over a five-month period from winter to summer), with an amplitude of about 5 ppm.

Further northward the amplitude of the seasonal variations in the CO_2 concentration increases (reaching, for example, 15 ppm on Long Island, New York). The maximum amplitude is in the Arctic, where the seasonal effects of the vegetation are supplemented by a summertime absorption of CO_2 by the waters of the ice-free arctic seas, which have been transported there to cool by warm currents and are undersaturated with carbon dioxide (at the same time, CO_2 is carried away from the Arctic by cold currents and is released into the air in the tropics as the water warms up). Further toward the equator the

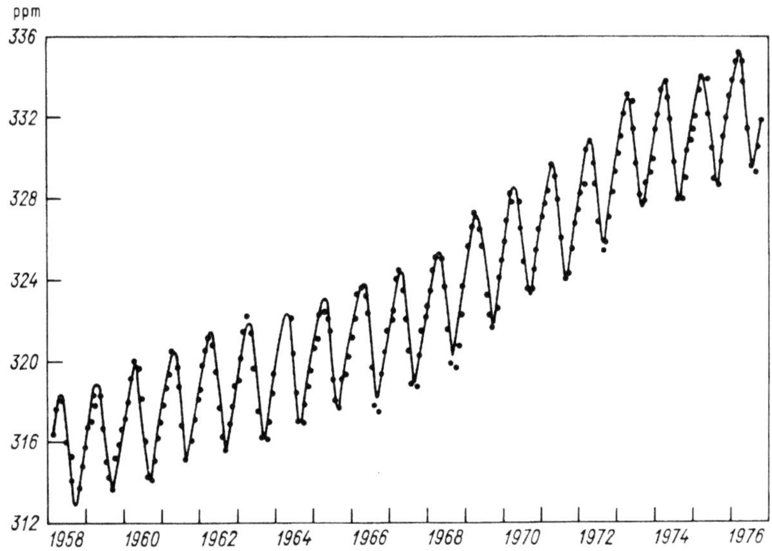

Fig. 6.2. Variations of CO_2 concentration in air at Mauna Loa Observatory (Hawaii) from 1958 to 1976.

amplitude of the seasonal CO_2 fluctuations decreases. In the Southern Hemisphere, where both the total seasonal fluctuations of the vegetation (mainly of land forests) and the interlatitudinal (meridional) water exchange in the oceans are weaker, the seasonal CO_2 fluctuations turn out to be small. The overall picture of the seasonal fluctuations of the CO_2 concentration in the atmosphere at different latitudes, according to Bolin and Keeling (1963), is shown in Fig. 6.3. These investigators found the main natural source of CO_2 for the atmosphere to be the tropical regions of the ocean, and the main industrial source to be the middle latitudes. During the course of a year 2×10^{10} tons of CO_2 are transported from the tropics to the polar regions of the Northern Hemisphere, whereas north of 45°N the vegetation in summer absorbs about 1.5×10^{10} tons of CO_2. At the same time, of course, onto this mean zonal pattern are superimposed a nonzonal (longitudinal) variations, created by differences between the continents and oceans, and irregular fluctuations, produced by synoptic processes in the atmosphere and ocean. Consequently, the regular compilation of synoptic charts of the CO_2 content in the atmosphere in the future would be useful.

In addition to the seasonal fluctuations, Fig. 6.2 reveals the trend of the CO_2 concentration to increase through the years, from 315 ppm in 1958 to 333 ppm in 1976, that is, a rise of 18 ppm in 18 years. This increase slowed down somewhat from 1963 to 1967, speeded up from 1968 to 1973, and then slowed down again. A similar pattern was observed at the South Pole, where the CO_2 concentration was slightly lower, the increase being from 314 ppm in 1958 to 331 ppm in 1976. There were slight fluctuations, with a period of approxi-

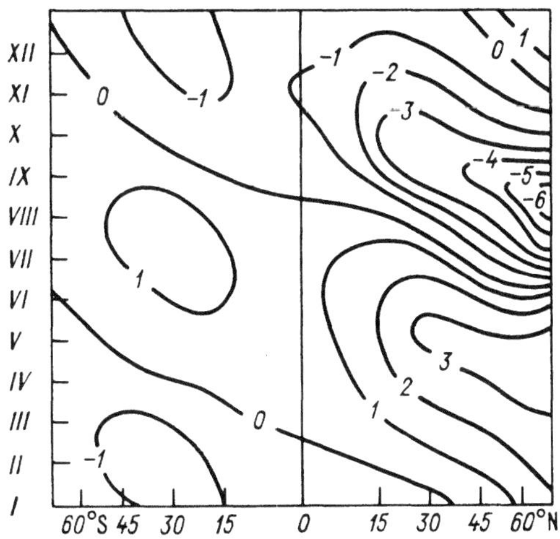

Fig. 6.3. Seasonal fluctuations of CO_2 concentration (ppm) in the atmosphere at different latitudes, according to Bolin and Keeling (1963).

mately four years, superimposed onto this trend. With regard to the prediction of future climatic variations, it is of great interest to ascertain the reasons for this trend and to extrapolate it into the time to come.

In order to explain the variations of the CO_2 content in the atmosphere over the years, we have to evaluate the sources and sinks of atmospheric carbon dioxide. The first step in this evaluation is to compute the amounts of CO_2 in the atmosphere and in any reservoirs associated with it which contain CO_2 in either the free or bound state (the ocean, the biosphere, soil and water on land, fossil fuels and other carbonaceous fuels). A useful measurement unit for this is trillions of tons, or teratons, of carbon (1 Tton = 10^{12} metric tons = 10^{18} g).

For the present atmosphere the volume concentrations of CO_2 in ppm given by the formula $10^6\, p_{CO_2}/p$, where p_{CO_2} and p are the partial pressure of CO_2 and the air pressure, can be converted into trillions of tons of carbon (Ttons C) with the aid of the relation 1000 ppm $CO_2 \approx 2.134$ Ttons C. The latter relation ensues from the formula for the mass of carbon $M_C = (\mu_C/\mu_{CO_2})\, q_{CO_2}\, M$, where $M \approx 5.157 = 10^3$ Ttons is the mass of the atmosphere, $\mu_C \approx 12$ and $\mu_{CO_2} \approx 44$ are the relative molecular masses of carbon and carbon dioxide, and $q_{CO_2} = (\mu_{CO_2}\mu)\,(P_{CO_2}/P)$ is the specific concentration of CO_2 ($\mu \approx 29$ being the relative molecular mass of air). For a volume concentration of 333 ppm, this relation gives a CO_2 content in the atmosphere in 1976 equal to 0.711 Ttons C (of which about 85%, or 0.604 Ttons C, was in the troposphere, the other 0.107 Ttons C being in the stratosphere). The preindustrial level of atmospheric CO_2 (in 1860) can be set tentatively at 290 ppm, or 0.619 Ttons C, giving an increment over 116 years of 0.092 Ttons C, or 14.8% of the preindustrial level.

TABLE 6.1
Contents of various forms of carbon in waters of the World Ocean

	Form	Amount, Ttons C
1	Inorganic forms in solution	38.2
1a	in upper 100-m layer	0.864
2	Dissolved organic matter	1.8
2a	in upper 100-m layer	3×10^{-2}
3	Dead organic suspended matter (detritus)	2.7×10^{-2}
4	Living organisms	1.4×10^{-3}
4a	Zooplankton	8.6×10^{-4}
4b	Zoobenthos	3.6×10^{-4}
4c	Nekton	8.8×10^{-5}
4d	Phytoplankton	5.6×10^{-5}
4e	Phytobenthos	9×10^{-6}

Table 6.1 shows the CO_2 content (or, more precisely, the contents of various forms of carbon) in the waters of the World Ocean. Rows 2 and 3 of the table were taken from the summary of Romankevich (1977), while rows 4a–4e are based on data of the Institute of Oceanology of the U.S.S.R. Academy of Sciences. It is clear from the table that the bulk of the carbon in the waters of the World Ocean is contained in inorganic forms in solution; this amount is equivalent to 140 Ttons CO_2, or 54 times the CO_2 content in the atmosphere. The carbon in organic matter is 1/20 of the amount contained in inorganic matter; the bulk of the C_{org} is in solution, and only about 0.1% of the total C_{org} is contained in living organisms (mostly in zooplankton and, to a somewhat less extent, in zoobenthos).

The inorganic forms of CO_2 dissolved in ocean water (the so-called carbonate system) are: (1) free dissolved $CO_{2\,dis}$, (2) its chemical combination with water (undissociated carbonic acid H_2CO_3), (3) the bicarbonate ion HCO_3^-, and (4) the carbonate ion $CO_3^=$. Forms 1 and 2, related by the chemical equilibrium $CO_2 + H_2O \rightleftarrows H_2CO_3$, are practically indistinguishable. Carbonic acid dissociates in two stages, described by the equilibrium equations:

$$CO_2 + H_2O \rightleftarrows H^+ + HCO_3^-; \qquad K'_1 = \frac{a_H\,[HCO_3^-]}{a_{CO_2} a_{H_2O}}; \tag{6.3}$$

$$HCO_3^- \rightleftarrows H^+ + CO_3^=; \qquad K'_2 = \frac{a_H\,[CO_3^=]}{[HCO_3^-]}, \tag{6.4}$$

where K'_1 and K'_2 are the first and second dissociation constants for carbonic acid, and square brackets indicate the concentrations; a_H is the *activity* of the hydrogen ions, defined for any substances in solution by the formula

$$a = \exp\frac{\mu - \mu_0}{RT} = \gamma m,$$

where μ and μ_0 are the chemical potentials of the particles, that is, their partial molar free energies in the state being considered and in some standard state, $R = 8.31439$ J/(mole K) is the molar gas constant, m is the concentration, and γ is the practical activity coefficient, which is close to unity for hydrogen ions in seawater (direct measurements are made not of a_H, but rather of the pH value $\approx -\lg a_H$, which is equal to 7 in a neutral solution at 25°C; in seawater, which due to the release of hydroxyl OH^- during the hydrolysis of bicarbonates and carbonates has a slightly alkaline reaction, the pH usually varies from 7.5 to 8.4, decreasing with a rise in temperature or pressure); a_{CO_2} is the activity of molecular CO_2 and undissociated carbonic acid; $a_{H_2O} = 1 \times 10^{-4}$ Cl‰ to 9.69×10^{-4} Cl‰ is the activity of H_2O molecules in seawater (equal to the ratio of the pressures of saturated water vapor over seawater and over fresh water). Coefficients K_1' and K_2', which are functions of the temperature, salinity, and pressure, have been tabulated by oceanographers. In order to calculate the elements of the carbonate system in seawater, measurements are made of the pH and the total (titration) alkalinity Alk, defined as the number of millimoles of hydrochloric acid (1 mole of HCl corresponds to 36.47 g) required to neutralize 1 liter of seawater to a pH value of 5.4 – 5.5. In addition to the carbonate alkalinity

$$A_K = [HCO_3^-] + 2\,[CO_3^=], \tag{6.5}$$

which is, on the average, 2.4 mmoles/l (or about 7% of the salinity), a small contribution to Alk (about 4%) is made by the borate alkalinity, related primarily to the equilibrium

$$H_3BO_3 \rightleftarrows H^+ + H_2BO_3^-; \qquad K'_B = \frac{a_H[H_2BO_3^-]}{[H_3BO_3]} \tag{6.6}$$

and defined as

$$A_B = \frac{2.2 \times 10^{-5}\text{Cl‰}}{1 + a_H/K'_B}, \tag{6.7}$$

where the numerator contains the sum of the molar concentrations of boric acid and borate ions, expressed in terms of the chlorinity Cl (related to the salinity by the expression $S = 0.03 + 1.805$ Cl). The other terms in

$$\text{Alk} = A_K + [H_2BO_3^-] + [OH^-] - [H^+] + [HS^-] - [NH_4^+]$$

generally do not exceed 0.01 mmole/l. Once pH and A_K have been determined and K_1' and K_2' are known, it is easy to calculate $[HCO_3^-]$, $[CO_3^=]$, and a_{CO_2} from (6.3) – (6.5). The sum $[CO_2]$ of the concentrations of molecular CO_2 and undissociated carbonic acid is found from a_{CO_2} by multiplying it by α_s/α_0, where α_s and α_0 are the coefficients of CO_2 saturation in seawater and fresh water, respectively (that is, the ratios of concentration $[CO_2]$ to the partial pressure $p^*_{CO_2}$ of the gas over the solution, which according to Henry's law are for very dilute solutions independent of the concentrations). Then the total carbon dioxide is calculated as

$$[\Sigma\ CO_2]=[CO_2] + [HCO_3^-] + [CO_3^=] \tag{6.8}$$

(this can, by the way, also in principle be measured directly, for instance, by means of gas chromatography).

Finally, by dividing $[CO_2]$ by α_s, we obtain the equilibrium partial pressure of carbon dioxide over the solution $p^*_{CO_2}$. If it is less than the actual atmospheric value of p_{CO_2}, then carbon dioxide will pass from the atmosphere to the solution. For $p^*_{CO_2} > p_{CO_2}$, on the other hand, carbon dioxide will be released by the ocean into the atmosphere. Factual data indicate that the sign of the difference $p^*_{CO_2} - p_{CO_2}$ varies from one part of the ocean to another, its values being of the order of 10^{-5} atm (or 1 Pa). The variations $\delta(\Sigma CO_2)$ of the total carbon dioxide in the water for variations $\delta(p_{CO_2})$ of the atmospheric carbon dioxide are given by the formula

$$\frac{\delta\ (\Sigma\ CO_2)}{\Sigma\ CO_2} = B\ \frac{\delta\ (p_{CO_2})}{p_{CO_2}}, \tag{6.9}$$

where the buffering factor B in the upper layer of the present-day ocean varies from 0.090 at a temperature of 17.7°C to 0.075 at 3.5°C (due to the buffer effect, which is typical of weak acid solutions in general, large amounts of a strong acid or strong base have to be added in order to vary the pH of the solution slightly). Consequently, in view of the buffer effect, a 10% increase in the atmospheric carbon dioxide will raise the total carbon dioxide in the upper ocean layer ΣCO_2 by only 1%. Since at present ΣCO_2 of the upper layer of the ocean is approximately equal to ΣCO_2 of the atmosphere, the upper layer of the seawater will take on only 10% of the increments in the atmospheric CO_2.

Chemical equilibria (6.3)–(6.4) depend appreciably on the CO_2 content in the solution and the pH value associated with it. For instance, the addition of CO_2 to the solution, via an increase of the CO_2 content of the atmosphere, the respiration of organisms, or the oxidation of an organic substance, increases the concentration of hydrogen ions H^+ (reduces the pH) and promotes the conversion of carbonates to bicarbonates; however, this reaction, which is written from right to left, involves the deposition of calcium carbonate, which reduces the pH and causes bicarbonates to change into dissolved carbon dioxide. On the other hand, a reduction of the dissolved CO_2 due to a drop in the atmospheric CO_2 content or due to absorption during plant photosynthesis reduces the concentration of H^+ ions (increases the pH) and causes bicarbonates to change into carbonates, whereas the dissolution of calcium carbonate, which raises the pH (via the hydrolysis of Ca^{++}), promotes the conversion of dissolved CO_2 into bicarbonates. Figure 6.4 shows the percentages of $[CO_2]$, $[HCO_3^-]$, and $[CO_3^=]$ for various values of the pH. Inspection of the figure shows that, at the pH values of 7–8.5 typical of seawater, the

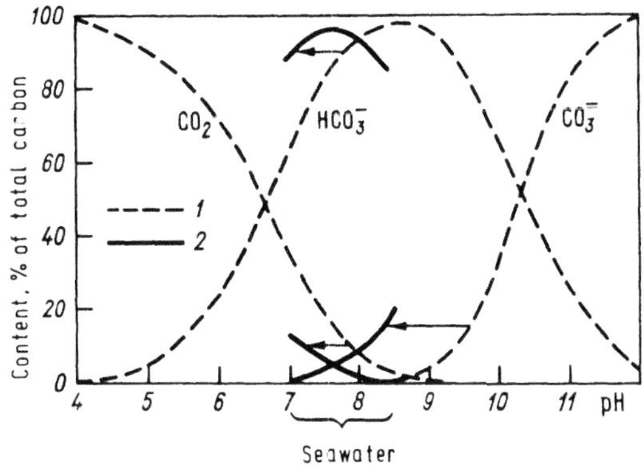

Fig. 6.4. Carbonate system for pressure $p=1$ atm (or 10^5 Pa) and temperature 0°C, as a function of pH. (1) fresh water; (2) seawater.

main form of carbon dioxide present is the bicarbonate HCO_3^-: for pH=7 it amounts to 80%, the rest being CO_2; for pH=8.5 the bicarbonate fraction is also 80%, the rest being $CO_3^=$. With a rise in temperature, the curves in Fig. 6.4 shift to the right (but ΣCO_2 decreases), whereas with a rise in salinity or pressure they shift to the left.

The fate of the CO_2 in the ocean depends considerably on the precipitation or dissolution of calcium carbonate there, that is, on shifts in the chemical equilibrium:

$$Ca^{++} + CO_3^= \rightleftarrows CaCO_3; \qquad K_{sp} = a_{Ca++} a_{CO_3^=}, \tag{6.10}$$

where the quantity K_{sp} (defined for saturation of a solution in equilibrium with a solid salt) is called the *solubility product*. It has different values for the two crystalline forms of calcium carbonate: calcite and (denser) aragonite. At a pressure of 1 atm (or 10^5 Pa) and a temperature of 25°C, for calcite $K_{sp} = 10^{-8.34}$, whereas for aragonite $K_{sp} = 10^{-8}$, that is, it is about twice as great. Often the apparent solubility product $K'_{sp} = [Ca^{++}]_{sat} [CO_3^=]_{sat}$ is used instead of K_{sp} (for calcite in seawater with a chlorinity of 19‰, K'_{sp} varies from 6.72×10^{-7} to 3.82×10^{-7} with a temperature rise from 0 to 30°C), and the calcium carbonate saturation $r = [Ca^{++}] [CO_3^=] \cdot (K'_{sp})^{-1}$ is calculated. In the ocean a "compensation depth" z_c can be discerned, above which the water is supersaturated with calcium carbonate (and it readily precipitates out in the form of shells of marine organisms) and below which it is undersaturated (and carbonate deposits become dissolved). In the Atlantic, and apparently in the Indian Ocean as well, $z_c \approx 4500$ m for calcite and 1000 m for aragonite; in the North Pacific $z_c \approx 500$ and 300 m, and in the equatorial and southern parts of the Pacific $z_c = 3500$–4000 m for calcite. The increasing absorption of atmospheric CO_2 by the ocean will make the compensation depth greater and will cause lower-lying deposits to dissolve (provided they are suitably mixed); according to some computations, the calcite available in the ocean deposits is sufficient to absorb the CO_2 which would be produced by burning of all existing reserves of fossil fuel.

The problem of the CO_2 content (that is, the content of carbon in its various forms) on land is every bit as difficult as for the ocean. Table 6.2 gives the amounts of carbon present on land. The data in rows 1 and 1a, cited by Bolin (1976), are based on the 1971 FAO-UNESCO *Soil Map of the World* and on other sources. The substances in row 1a are, according to FAO terminology, district and gelic histosols, which practically do not participate at all in the exchange with the atmosphere. The approximately 2 Ttons of C in the humus, on the other hand, are concentrated primarily in the upper meter of the soil layer, in amounts ranging from 4 to 60 kg/m^2 (18 kg/m^2, on the average), and this carbon gradually oxidizes, releasing CO_2 into the atmosphere at about the same rate as the accumulation of decaying organic matter.

TABLE 6.2
Content of carbon (in various forms) on land

	Form	Amount, Ttons C
1	Nonliving organic matter in soil (humus)	2.9
1a	including peat-forming substances	0.86
2	Carbon dioxide in fresh-water bodies	0.45
3	Vegetation	0.828
3a	including forests	0.744
4	Combustible minerals (fossil fuels)	5.037 (0.757)
4a	Coal	4.306 (0.615)
4b	Oil	0.246 (0.072)
4c	Shale	0.324 (0.041)
4d	Natural gas	0.161 (0.029)
5	Carbonaceous sedimentary rocks	5×10^4

The estimated biomass 0.828 Ttons C of the vegetation on land (spread over an area of 1.49×10^8 km^2, giving a mean biomass density of 5.56 kg C/m^2) was taken from Whittaker and Likens (1975). A breakdown is given in Table 6.3. These data indicate that forests account for about 90% of the vegetation biomass, and that tropical jungles account for about 40%. It should be noted that cultivated land constitutes less than 1% of the total vegetation.

The data in rows 4–4d of Table 6.2 were taken from Perry and Landsberg (1977). The first numbers are the estimated reserves, and the numbers in parentheses are the reserves which have been confirmed to be exploitable using present-day technology. The latter are comparatively small, their carbon content being comparable to the amounts of carbon in the atmosphere and biosphere. The estimated carbon content of carbonaceous sedimentary rock (row 5 of Table 6.2) was obtained by assuming that during the fusion of the Earth's crust with a mass of 5×10^7 Ttons the mantle material simultaneously gave off 3% as volcanic gases, 10% of which was CO_2, and that this then went into carbonaceous sediments. Since this amount is thousands or tens of thousands of times greater than the CO_2 contents in the above-mentioned reservoirs, it is clear that the latter must have varied considerably during the course of the Earth's history, because of relatively slight disturbances of the equilibrium between the volcanic release of CO_2 and the accumulation of carbonaceous deposits.

TABLE 6.3
Biomass of land vegetation

	Type of vegetation	Biomass, Ttons C	Density, kg C/m^2
1	Forests	0.744	15.33
1a	Tropical jungles	0.3442	20.25
1b	Seasonal tropical forests	0.1181	15.75
1c	Subtropical evergreen forests	0.0788	15.75
1d	Summergreen forests of middle latitudes	0.0945	13.5
1e	Boreal forests	0.108	9.0
2	Savannah	0.027	1.8
3	Shrubland (chaparral, etc.)	0.023	2.7
4	Swamps	0.0135	6.75
5	Meadows	0.0065	0.72
6	Cultivated land	0.0063	0.45
7	Semideserts	0.0056	0.31
8	Tundra and alpine meadows	0.0022	0.27
9	Deserts	0.0002	0.01

Table 6.4 gives the estimated flows of CO_2 between the indicated reservoirs (in gigatons of carbon per year, where 1 Gton = 10^9 metric tons = 10^{-3} Tton). A plus sign denotes an influx of atmospheric CO_2, and a minus sign denotes a CO_2 loss. Inspection of the table indicates that the CO_2 flows can be divided into three groups: (1) the most intensive process (one to three orders of magnitude more intensive than the others) is CO_2 absorption during plant photosynthesis (rows 1–1b of the table), so that it should be approximately balanced out by the oxidation of organic matter by micro-organisms, animals, and fires (row 2); (2) anthropogenic release of CO_2 and its accumulation in the atmosphere (rows 3–5d), which is an order of magnitude less intensive than the previous process and which should apparently be balanced out by CO_2 absorption by the oceans and the biosphere (row 6); (3) geological processes of CO_2 release and absorption (rows 7–9), which are two orders of magnitude weaker than the foregoing processes.

In rows 1 and 1b of Table 6.4 photosynthesis refers to the *net* primary production of the plant biomass, that is, the total, or *gross*, production minus the amount used by the plants for respiration. Table 6.5 gives detailed data of Whittaker and Likens for plant photosynthesis on land. It is seen that about

TABLE 6.4

CO_2 flows between reservoirs

	Flow mechanism	Amount, Gtons C/yr	Source
1	Plant photosynthesis (pure production)	−77.61	Whittaker and Likens (1975)
1a	Plant photosynthesis on land	−52.86	" "
1aa	including photosynthesis in forests	−33.25	" "
1b	Photosynthesis of plants in the ocean	−24.75	" "
2	Oxidation by microorganisms and animals, forest and steppe fires	+78?	
3	Mean rate of increase of atmospheric CO_2 from 1958 to 1976	+2.134	
3a	Present rate of increase of atmospheric CO_2	+2.3	
4	Industrial emission of CO_2	+5	
5	CO_2 release due to destruction of forests		
5a	Estimate based on FAO forestry data	+1	Bolin (1976)
5b	Confidence intervals	+(2 – 18) +(4 – 8)	Woodwell *et al.* (1977)
5c	Estimate based on increase in cultivated land from 1950 to 1970	+1.3	Revelle and Munk (1977)
5d	Estimated burning of wood and oxidation of humus	+1.9	Wong (1978)
6	Absorption of CO_2 by ocean and biosphere	−4?	
7	Release of CO_2 by volcanoes	$+4\times 10^{-2}$	Bowen (1966)
8	Mean rate of accumulation of carbonaceous sediments during Earth's history	-10^{-2}	
8a	Present rate of storage of carbon in bottom sediments	-8.5×10^{-2}	Romankevich (1977)
9	Weathering of carbonate and silicate rocks	-10^{-2}	

60% of the primary production is from forests, and that half of these forests are tropical jungles; about 12% of the primary production is from savannah, and 8% is from cultivated land. Since on cultivated land neither the density of the biomass nor the density of the primary production is increased, these lands are important only in that they provide the *qualities* of the primary production necessary for humans and in that they speed up the turnover of CO_2 among the corresponding reservoirs. It is noteworthy, too, that swamps have by far the highest density of primary production.

Inspection of lines 1a and 1b of Table 6.4 shows that the primary production of the plant biomass (expressed as the amount of carbon) in the ocean is only about half the production on land. The bulk of it (18.7 Gtons C/yr) takes place in the open ocean; 4.3 Gtons C/yr are produced on the continental

TABLE 6.5
Photosynthesis of plants on land

	Type of vegetation	Production, Gtons C/yr	Production density, kg C/(m^2 yr)
1	Forests	33.25	
1a	Tropical jungles	16.83	0.99
1b	Seasonal tropical forests	5.40	0.72
1c	Subtropical evergreen forests	2.92	0.585
1d	Summergreen forests of middle latitudes	3.78	0.54
1e	Boreal forests	4.32	0.36
2	Savannah	6.08	0.405
3	Shrubland (chaparral, etc.)	2.68	0.315
4	Cultivated land	4.10	0.293
5	Swamps	2.70	1.35
6	Meadows	2.43	0.27
7	Semideserts	0.74	0.041
8	Tundra and alpine meadows	0.50	0.063
9	Lakes	0.36	0.18
10	Deserts	0.02	0.001

shelf, 1 Gton C/yr in estuaries, 0.7 Gton C/yr in reefs and shallow-water phytobenthos, and only 0.1 Gton C/yr in zones of upwelling. Almost all this production involves phytoplankton. Table 6.6 shows the distribution of the biological productivity of the ocean among the various types of organisms (according to data of the Institute of Oceanology of the U.S.S.R. Academy of Sciences).

TABLE 6.6
Biological productivity of the ocean

	Type of organism	Production (wet weight), Gtons/yr	Gtons C/yr
1	Phytoplankton	550	22.5
2	Zooplankton	53	2.12
3	Zoobenthos	3	0.11
4	Nekton	0.2	0.017
5	Phytobenthos	0.2	0.009

The above-mentioned increase in atmospheric CO_2, which is indicated in rows 3 and 3a of Table 6.4, can apparently be attributed to the anthropogenic factors cited in rows 4 and 5. The industrial production of CO_2 during the 113-year period from 1860 to 1973 was estimated to be 128 Gtons C, 98% of which was from the burning of fossil fuels and 2% from the manufacture of cement. This production is accelerating rapidly, and the present rate (about 5 Gtons C/yr) is already considerably higher than the mean rate over the indicated period. Different investigators give different estimates of the liberation of CO_2 associated with the destruction of forests (rows 5a–5d of Table 6.4). The most detailed calculations are those of Revelle and Munk (1977), who found that from 1860 to 1970 the world area of cultivated land rose from 500 to 1370 million hectares, 30% at the expense of forests; this is equivalent to a plant biomass of 50 Gtons C of carbon dioxide.

If to the above amount we now add the effect of oxidation of, say, 15% of the soil humus, or about 30 tons C/hectare, and then subtract the biomass of the cultivated plants, the total release of carbon dioxide amounts to 72 Gtons C, or 0.63 Gtons C/yr, averaged over the 110 years in question. However, this process is also accelerating rapidly. During the two decades 1950–1970 the world area of cultivated land increased by 310 million hectares (88 in the USSR, 74 in Africa south of the Sahara, 50 in Latin America, 84 in Asia, 7 in China, and 6 million hectares in Australia and New Zealand), which according to the above-indicated calculation norms was equivalent to 26 Gtons C of carbon dioxide, or about 1.3 Gtons C/yr, on the average. A similar result was obtained by Wong (1978), who estimated the rate of CO_2 liberation from the burning of wood to be 1.6 Gtons C/yr and that from the oxidation of soil humus to be 0.3 Gtons C/yr.

The contributions of the burning of fossil fuels and the destruction of forests to the observed increase of atmospheric CO_2 can be estimated from the ratios of radiocarbon ^{14}C and stable heavy carbon ^{13}C to the principal stable carbon isotope ^{12}C in tree rings. Radiocarbon, with a halflife of 5730 ± 40 years, is produced in the atmosphere (from 1937 to 1967 at a rate of 2.2 ± 0.4 atoms per cm^2 of the Earth's surface per second) as a result of the bombardment of nitrogen ^{14}N nuclei by cosmic rays (but in 1961–1962 also as a result of hydrogen-bomb tests, which doubled the ^{14}C content of the atmosphere), and then spreads throughout the atmosphere, hydrosphere, and biosphere. Since ^{14}C is not present in fossil fuels, industrial burning of fuel from 1890 to 1950 resulted in a reduction of the ^{14}C content of the atmosphere by $(2 \pm 0.3)\%$. This effect is known as the "industrial effect," or (after its discoverer) the "Suess effect". A certain contribution to the reduction of the ^{14}C concentration in the atmosphere can also be traced to the "de Vries effect", a lessening of the ^{14}C production in the atmosphere with an increase in solar activity, due to a screening of galactic cosmic rays by the solar wind. A similar role, by the way, may be played by an enhancement of the geomagnetic field, or by a redistribution of ^{14}C among the reservoirs of the atmosphere-ocean-land system during climatic fluctuations.

The stable heavy isotope ^{13}C is present in the atmosphere as well as in fossil fuels and the biosphere, but in the latter its ratio to ^{12}C is somewhat lower than in the atmosphere (lower by an amount $\delta\ ^{13}C = 18‰$, where $\delta\ ^{13}C$ is the relative deviation of the value of $^{13}C/^{12}C$ from the so-called PDB value in a sample of "P–D" belemnite taken as a standard). Thus, the burning of fossil fuels and the destruction of forests should reduce the value of $\delta\ ^{13}C$ in the atmosphere. Stuiver (1978) has estimated this reduction to be 1.5‰ for the period from 1850 to 1950. Using isotope data, he obtained a value of 120 Gtons C for the influx of CO_2 to the atmosphere from the continental biosphere; together with the influx of 26 Gtons C from 1960 to 1970, this gives a total yield by the biosphere of 146 Gtons C for 1850–1970, or twice the previously cited estimate of 72 Gtons C by Revelle and Munk. However, the "Stuiver effect", namely the reduction of $\delta\ ^{13}C$ in tree rings for 1850–1950, may also have other causes, for instance, differences in the heights of the trees studied, relative to the surrounding vegetation. Note, too, that this effect does not exhibit the exponential increase with time which is typical of the burning of fossil fuels and the destruction of forests, so that its interpretation should apparently be reconsidered.

The geological processes of liberation and absorption of atmospheric CO_2 listed in rows 7–9 of Table 6.4 are of four types: (1) CO_2 release by volcanos; (2) weathering of carbonaceous and siliceous rocks on land, attended by absorption of CO_2 from the atmosphere, with a subsequent transport of bicarbonate ions via rivers to the ocean and then a deposition of carbonates on the bottom with a release of CO_2 into solution; (3) direct deposition of carbonaceous sediments. The mean rate of accumulation of carbonaceous sediments is found by dividing the accumulated sediment mass 5×10^4 Ttons C (row 5 of Table 6.2) by the age of the Earth 4.5×10^9 yr. The present rate of accumulation of carbonaceous sediments, according to Romankevich (1977), is 8.5×10^{-2} Gtons C/yr (of a total of 1–3 Gtons C/yr reaching the bottom with the suspended matter; the remaining 91–97% of the carbon in the deposited suspended matter becomes mineralized and dissolves; the dissolving of calcium carbonate below the "compensation depth" also enters in here). The absorption of atmospheric CO_2 during the weathering of carbonaceous and siliceous rocks takes place at rates with this same order of magnitude.

Models known as *reservoir models* are widely used for the quantitative description and prediction of variations in the CO_2 content of the atmosphere and of the reservoirs associated with it. In these models each reservoir (designated as k) is characterized by the amount Q_k of CO_2, or carbon in other forms, contained in it. The accuracy of such a model is, of course, somewhat limited, since it does not allow for inhomogeneities of the spatial distribution of CO_2 in the reservoirs. More accurate, but at the same time much more complicated mathematically, are the three-dimensional models of the climate system described in §3; these include the CO_2 field or the fields of the various forms of carbon.

Linear models of reservoirs are described by the equation

$$\frac{dQ_k}{dt} = \sum_{l \neq k} a_{lk} Q_l - Q_k \sum_{l \neq k} a_{kl} + f_k, \tag{6.11}$$

where a_{kl} is the relative rate of transfer of CO_2 from the kth to the lth reservoir, and f_k is the external source of CO_2 for the kth reservoir. In the absence of external sources, the total amount Q of CO_2 in all the reservoirs does not vary with time. The matrix $\mathcal{A}$ of the coefficients of system of equations (6.11) has one simple zero eigenvalue, and the other $n - 1$ eigenvalues $\lambda_1, \ldots, \lambda_{n-1}$ are real and negative; matrix $\mathcal{A}$ has n linearly independent eigenvectors. There exists one, and only one, equilibrium (steady) state, and it is reached as $t \to \infty$ for emergence of the system from any other state,

TABLE 6.7

Exchange coefficients (in reciprocal years) and steady states of 4r model (in units of Q_1)

	Q_1	Q_2	Q_3	Q_4
$a_{12} = \frac{1}{33}$, $a_{21} = \frac{1}{40}$, $a_{13} = \frac{1}{5}$, $a_{31} = \frac{1}{6}$, $a_{34} = \frac{1}{6.2}$, $a_{43} = \frac{1}{300}$	1	1.21	1.2	58.06
Preindustrial state, 10^{15} moles	51	62.2	61.7	2985.4
Limiting state, 10^{15} moles	61.2	74	73.4	3552

the rate of reaching the limiting state being determined by the eigenvalue with the lowest absolute value. The simplest model is one with four reservoirs (4r model): (1) atmosphere; (2) biosphere; (3) upper mixed layer of ocean; (4) deep ocean. Table 6.7 gives all the nonzero coefficients a_{kl} for this model (mainly according to Sawyer (1972)). For these exchange coefficients the steady state is given by the formulas

$$Q_2 = \frac{a_{12}}{a_{21}} Q_1; \qquad Q_3 = \frac{a_{13}}{a_{31}} Q_1; \qquad Q_4 = \frac{a_{13}}{a_{31}} \frac{a_{34}}{a_{43}} Q_1. \tag{6.12}$$

If the initial (preindustrial) state of the 4r model is taken to be its steady state with $Q_1 = 51 \times 10^{15}$ moles, then the corresponding values of Q_2, Q_3, and Q_4 from formula (6.12) are listed in the appropriate row of Table 6.7. Following Avenhaus *et al.* (1980), we add to the total quantity of carbon dioxide in the model $Q_0 = 3160 \times 10^{15}$ moles the amount 600×10^{15} moles equivalent to the existing supplies of fossil fuels, and we distribute the obtained sum $Q_\infty = 3760 \times 10^{15}$ moles among the reservoirs, according to formulas (6.12). The limiting state of the 4r model arrived at in this way, which is given in the last row of Table 6.7, shows that of the added 600×10^{15} moles of carbon dioxide there will remain in the atmosphere in the limiting equilibrium state only 10.2×10^{15} moles (which raises the concentration of atmospheric CO_2 just to 345 ppm), while 567×10^{15} moles escape into the deep ocean.

Of the nonlinear models, let us consider the 5r model of Revelle and Munk (1977), in which, unlike the 4r model just discussed, the biosphere is divided into two reservoirs $Q_2 = Q_2' + Q_2''$ with some biomasses participating in photosynthesis and some not participating. The evolution of these biomasses is described by the equations:

$$\frac{dQ_2'}{dt} = p_0\left(1 + \beta \ln \frac{Q_1}{Q_{10}}\right)\frac{Q_2'}{Q_{20}'} - p_0 \frac{Q_2}{Q_{20}} - \frac{Q_2'}{Q_2}\frac{d\Phi}{dt}; \tag{6.13}$$

$$\frac{dQ_2''}{dt} = - p_0 \frac{Q_2''}{Q_{20}''} + p_0 \frac{Q_2}{Q_{20}} - \frac{Q_2''}{Q_2}\frac{d\Phi}{dt}, \tag{6.14}$$

where the first term in (6.13) gives the biomass production Q_2' by photosynthesis, which is a logarithmic function of the increase in atmospheric carbon dioxide (according to Table 6.4, $p_0 \approx 53$ Gtons C/yr, and for β we can take a value of 0.05); the first term in (6.14) gives the biomass loss Q_2'' due to fires; the second terms in the equations give the exchange between biomasses Q_2' and Q_2'', and the third terms give the losses in biomass due to the destruction of forests, which releases into the atmosphere an amount of carbon dioxide described by the formula

$$\Phi = \frac{\Phi_0\Phi_\infty e^{\alpha t}}{\Phi_\infty - \Phi_0 + \Phi_0 e^{\alpha t}}, \tag{6.15}$$

where $\Phi_0 = 3.5$ Gtons C, $\Phi_\infty = 315$ Gtons C, and $\alpha = 0.03\ \text{yr}^{-1}$.

This same formula can be used to find the amount of carbon dioxide F released into the atmosphere by industry, if in it we set $F_0 = 4.5$ Gtons C and $F_\infty = 5000$ Gtons C (the corresponding maximum rate of release of CO_2 into the atmosphere, 39 Gtons/yr, will be reached in the year 2094, at which time F will be equal to $0.5F_\infty$). The variations in the total amount of carbon dioxide are equal to the variations in F:

$$(Q_1 - Q_{10}) + (Q_2 - Q_{20}) + (Q_3 - Q_{30}) + (Q_4 - Q_{40}) = F - F_0, \tag{6.16}$$

where we set $Q_{10} = 617$ Gtons C, $Q_{20}' = 200$ Gtons C, $Q_{20}'' = 2600$ Gtons C, and $Q_{30} = 864$ Gtons C. In order to determine Q_3, we must specify the buffer ratio

$$\frac{(Q_1 - Q_{10})/Q_{10}}{(Q_3 - Q_{30})/Q_{30}} = B_0 + d\,\frac{Q_1 - Q_{10}}{Q_{10}}, \tag{6.17}$$

in which it is assumed that $B_0 = 9$ and $d = 4$. Finally, for the deep ocean the following equation is used:

$$\frac{d}{dt}\frac{Q_4}{H_4} = \frac{1}{\tau}\left[\frac{Q_3 - Q_{30}}{H_3} - \frac{Q_4 - Q_{40}}{H_4}\right], \tag{6.18}$$

where the relaxation time τ for the deep ocean is assumed to be 500 years, the thickness of the upper mixed layer $H_3 = 100$ m, and the thickness of the deep layer $H_4 = 3900$ m.

Numerical integration of the above equations indicates that the presence or absence of the terms with $d\Phi/dt$ in (6.13) and (6.14) has little effect on the evolution of the atmospheric and oceanic reservoirs. Q_1 should increase to 1400 Gtons C by the end of the 21st century, but after a century it will drop to a value slightly less than Q_{10}, after which it will slowly approach Q_{10}. The sum $Q_3 + Q_4$ behaves similarly, so that ultimately all the carbon dioxide added to the atmosphere passes into the biosphere (into its two parts Q_2' and Q_2'', in proportion to their initial volumes). The latter conclusion may be changed, however, if into the model we introduce any limitations on the possible volume of the biosphere (such limitations exist in nature, since the available amounts of the nitrates and phosphates necessary for plant growth are naturally limited, as is the amount of water and, ultimately, of space as well). For example, if we assume that the plant biomass Q_2' participating in photosynthesis is invariable in time, that is, if we replace (6.13) by the condition $Q_2' = Q_{20}'$, then numerical integration of the modified equations (for $\beta = 0.3$) yields the following results: Q_1 will reach a maximum of 2889 Gtons C in the year 2180, after which it will gradually decrease asymptotically to a value of 1517 Gtons C; with a certain lag, Q_2 also increases (its maximum increment of 1132 Gtons C will be reached in 2240); $Q_3 + Q_4$ increases very slowly, the increase in the amount of carbon dioxide in the ocean prior to the 20th century being maintained at a level of 21% of the added carbon dioxide $(F + \Phi) - (F_0 + \Phi_0)$ and only in the year 2400 will it reach 48%. These results are depicted in Fig. 6.5.

Carbon dioxide has been considered here with regard to its ability, like water vapor, to absorb some of the flux of long-wave radiation ascending from the Earth's surface, and thus to contribute to the greenhouse effect in the atmosphere. A number of gaseous trace components of the atmosphere also have this ability, and in cases where their absorption bands lie inside the atmospheric transmission window (8.5–12 μm), even very small concentrations of these components can contribute appreciably to the greenhouse effect (all the more so, in view of the fact that these contributions are additive, provided that the absorption bands do not overlap). The most important of these components is ozone O_3, which has a strong absorption band (9.35–11 μm, centered at 9.569 μm) in the transmission window, as well as infrared bands at 4.75 and 14.1 μm. Also important are: the Freons (fluorochloromethanes), especially Freon-11 ($CFCl_3$), with absorption bands at 9.22 and 11.82 μm, and Freon-12 (CF_2Cl_2), with bands at 8.68, 9.13, and 10.93 μm; nitrous oxide N_2O, with bands at 4.5, 7.8, and 8.6 μm; methane CH_4, with bands at 3.3 and 7.6 μm; ammonia NH_3, with bands at 6.3 and 10.6 μm; nitric acid HNO_3, with bands at 5.6, 11.2, and 20 μm; ethylene C_2H_4, with a band at 10.5 μm; sulfur dioxide SO_2, with bands at 4, 7.35, 8.69, and 20 μm; methyl chloride CH_3Cl, with bands at 7.14, 9.85, and 13.66 μm; carbon tetrachloride CCl_4, with a band at 12.99 μm; and a number of other gases.

Wang, Yung, *et al.* (1976) estimated the orders of magnitude of the

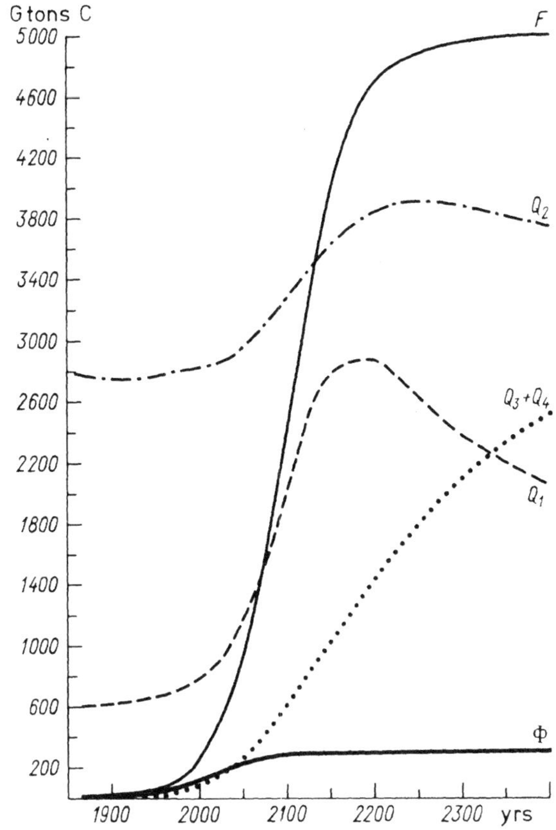

Fig. 6.5. Distribution of carbon dioxide released during combustion of fossil fuels (F) and destruction of forests (Φ), among atmosphere (Q_1), biosphere (Q_2), and ocean ($Q_3 + Q_4$), according to model of Revelle and Munk (1977), with fixed biomass Q_2' participating in photosynthesis.

greenhouse effect arising when the concentrations of the above gases are increased; for this estimate they used a one-dimensional (mean global) model of the atmosphere with a fixed temperature of the upper cloud boundary or with a fixed height of this boundary. The results obtained by these investigators are shown in Table 6.8. The concentrations in ppm in the table pertain to the surface layer of air, whereas for O_3 and HNO_3 the total content in the entire atmosphere is given, in terms of the thickness of the layer, reduced to standard temperature and pressure (at sea level). The first row of the table gives the estimated effect of doubling the water-vapor concentration in the stratosphere above 11 km, the relative humidity in the troposphere being held at 75%. For CO_2 the effect of increasing the concentration by 25% is given, rather than the effect of doubling it. For ozone an antigreenhouse effect was calculated, arising if the total O_3 content is reduced by 25%.

The table shows that a 20-fold increase in the Freon concentration (which is

TABLE 6.8

Greenhouse effect caused by increasing concentrations of gaseous constituents of atmosphere

Gas	Present concentration	Increase in concentration	Greenhouse effect, °C for fixed cloud temperature	Greenhouse effect, °C for fixed cloud height
1 H_2O	75% (relative humidity)	2-fold	1.03	0.65
2 CO_2	330 ppm	1.25-fold	0.79	0.53
3 O_3	3.43 mm	0.75-fold	−0.47	−0.34
4 $CFCl_3$	1×10^{-4} ppm	20-fold	0.54	
				0.36
5 CF_2Cl_2	1×10^{-4} ppm	″	0.54	
6 N_2O	0.28 ppm	2-fold	0.68	0.44
7 CH_4	1.6 ppm	″	0.28	0.20
8 NH_3	6×10^{-3} ppm	″	0.12	0.09
9 HNO_3	4.85×10^{-3} mm	″	0.08	0.06
10 C_2H_4	2×10^{-4} ppm	″	0.01	0.01
11 SO_2	2×10^{-3} ppm	″	0.03	0.02
12 CH_3Cl	5×10^{-4} ppm	″	0.02	
13 CCl_4	1×10^{-4} ppm	″	0.02	0.01

apparently to be expected during the next 50 years, due to the increased use of Freons in air conditioners and aerosol sprays) could enhance considerably the greenhouse effect in the atmosphere. A comparable, but somewhat lesser, effect may ensue from a doubling of the nitrous oxide concentration (which is apparently to be expected during this same period because of an increased use of nitrogen fertilizers, as well as due to the burning of fossil fuels). An even slightly lesser effect may be caused by a doubling of the concentration of methane plus ammonia, which are also of industrial origin, whereas the total greenhouse effect due to a doubling of the N_2O, CH_4, NH_3, and HNO_3 concentrations would be 1.2°C. It should be mentioned here, of course, that unlike CO_2 all the gases in rows 4–13 of Table 6.8 have comparatively short lifetimes in the atmosphere (in particular, because of the photodissociation of a number of them in the stratosphere). Consequently, an estimate of their possible concentrations in the future requires a careful calculation of the intensities of their anthropogenic and natural sources, of their dispersal in the atmosphere due to transport by air currents and diffusion, and also of the intensities of their sinks.

In addition to enhancing the greenhouse effect, some of the above-mentioned gaseous impurities in the air may be of interest for climate theory because of their ability to destroy ozone, and hence to diminish the ozone-related greenhouse effect (see row 3 of Table 6.8). This will also allow more ultraviolet solar radiation to reach the Earth's surface, with the corresponding biological consequences. With regard to this, the substances of the chlorine cycle ClO_x, the nitrogen cycle NO_x, and the hydrogen cycle HO_x, which can cause the catalytic destruction of ozone, are of great importance. Since the formation and destruction of ozone will be discussed at some length in a special section (§8), here we will confine ourselves to just a few brief comments.

The most efficient of the reactions of catalytic destruction of ozone are the main chlorine reactions

$$Cl + O_3 \rightarrow ClO + O_2 \quad \text{and} \quad ClO + O \rightarrow Cl + O_2,$$

the Cl atoms of which at heights above 20 km have been detached from molecules of carbon tetrachloride, Freon 11, or Freon 12 under the action of ultraviolet solar radiation. The main catalytic reactions of the nitrogen cycle,

$$NO + O_3 \rightarrow NO_2 + O_2 \quad \text{and} \quad NO_2 + O \rightarrow NO + O_2,$$

appear to have less effect; the NO is produced from anthropogenic nitrous oxide in the reaction

$$O\,(^1D) + N_2O \rightarrow NO + NO.$$

The main catalytic reactions of the hydrogen cycle,

$$OH + O_3 \rightarrow HO_2 + O_2 \quad \text{and} \quad HO_2 + O \rightarrow OH + O_2,$$

are also significant; the hydroxyl is produced from water vapor (in the exhausts of high-flying aircraft) via the reaction

$$O(^1D) + H_2O \rightarrow 2OH$$

and, similarly, from methane.

7. Aerosols

Aerosols can influence the climate either directly or indirectly. The direct effect is a variation in the transfer of short-wave and long-wave radiation in the atmosphere (enhancement of the absorption and scattering of radiation). The indirect effect of aerosols consists in altering the microstructure of liquid-water clouds (increasing the number of droplets per unit volume but reducing the droplet size) and their optical properties (increasing the albedo and absorptivity of the clouds).

Aerosols enter the atmosphere from the Earth's surface. They include particles of mineral dust, sea salt, volcanic ash, and combustion products of organic matter from forest fires, industry, and transport (ash and soot, condensed organic substances, and chemical derivatives of gaseous-product condensates, such as sulfates, organic nitrates, sulfuric and nitric acids), as well as derivatives of H_2S, NH_3, and terpenes of biological origin. Tentative estimates of the generation rates for aerosols of different kinds are given in Table 7.1, which was taken from Robinson (1977).

The table indicates that the anthropogenic production of aerosol now constitutes about 10% of the total production, and also that the direct production of particles (1.299 Gtons/yr) is approximately equal to the particle production via the condensation of gases (1.309 Gtons/yr); in the anthropogenic aerosol, on the other hand, condensation of gases constitutes only 2/3 of the total production.

Aerosols can leave the atmosphere in three ways: (1) condensing on the moisture itself (that is, serving as *condensation nuclei*) and turning into raindrops, which fall as precipitation; (2) being washed out, that is, being captured by falling raindrops or snowflakes; (3) settling out due to gravity, without any condensation processes.

Let us assume a mean residence time of aerosol particles in the troposphere of 8.5 days, the same as for water-vapor molecules. This figure was obtained by dividing the total moisture content of the atmosphere (24 mm of precipitation) by the mean rate of evaporation or precipitation of clouds (1036 mm/yr). Then, in the steady-state regime, the mass of aerosol in the atmosphere is found by multiplying the generation rate (2.6 Gtons/yr) by 8.5 days $\approx$ 1/43 yr; the resulting mass is 60 Mtons (where 1 Mton=10^6 tons), or about 10 μg per cm^2 of the Earth's surface. Now, by dividing this number by the weight of the air column (1 kg/cm^2), we obtain a mean specific aerosol concentration of the order of 10^{-8}. If the mass of 10 μg/cm^2 is distributed throughout the entire depth of the troposphere (about 10 km), then we get a mean volume concentration of aerosol of the order of 10 μg/m^3. Naturally, on the average, the aerosol concentration decreases rapidly with height. Therefore, near the Earth's surface it is usually several times higher than the mean values given here, whereas in the stratosphere it is tens of times less than these.

Aerosols consist of particles with a mean radius r ranging from 10^{-3} to 10^3 μm, while the main contributors to the formation of clouds are particles with radii from 10^{-2} to 10^2 μm (*condensation nuclei*); the total aerosol mass includes particles 10^{-1} to 10^2 μm in radius, while a turbid atmosphere is created by *optically active* particles 10^{-1} to 10^1 μm in radius. Figure 7.1 shows some typical particle-size distributions in a continental aerosol and a marine aerosol, together with the background aerosol of the upper troposphere. In the range of radii from 0.01 to 2.5 μm the global aerosol can be described, on the average, by a lognormal distribution

$$N(>r) = \frac{N_0}{\sqrt{2\pi}\ln\sigma}\int^{\infty}\exp\left[-\frac{(\ln r/r_0)^2}{2\ln\sigma}\right]\frac{dr}{r} \tag{7.1}$$

with parameters $N_0 = 10\ \text{cm}^{-3}$, $r_0 = 0.0725$ μm, and $\sigma = 1.86$, or by an exponential distribution

$$N(>r) = N_0\exp(r/r_0) \tag{7.2}$$

with parameters $N_0 = 8.44\ \text{cm}^{-3}$ and $r_0 = 0.075$ μm. In the narrower radius interval from 0.1 to 0.5 μm a power-law distribution

TABLE 7.1
Generation rates for aerosols of various kinds

	Type of aerosol	Generation rate, Gtons/yr	Limits
1	Natural aerosol	2.312	0.773–2.200
1a	Mineral dust	0.2	0.12–2.00
1b	Sea salt	1.0	
1c	Volcanic ash	0.004	
1d	Smoke from forest fires	0.003	0.003–0.150
1e	Sulfates from H_2S	0.204	0.130–0.200
1f	Nitrates from NO_x	0.432	0.060–0.430
1g	Ammonium from NH_3	0.269	0.080–0.270
1h	Organic aerosol from terpenes, hydrocarbons, etc.	0.2	0.075–0.200
2	Anthropogenic aerosol	0.296	0.185–0.415
2a	Carbon ash	0.036	
2b	Smoke from metallurgical factories	0.009	
2c	Products of combustion of wood	0.008	0.010–0.080
2d	" " of petroleum	0.002	
2e	" " wastes	0.004	
2f	Particles from agricultural production	0.010	
2g	Dust from cement plants	0.007	
2h	Others	0.016	
2i	Sulfates from SO_2	0.147	0.130–0.200
2j	Nitrates from NO_x	0.030	0.030–0.035
2k	Organic aerosol from terpenes, hydrocarbons, etc.	0.027	
3	Total	2.608	0.958–2.615

$$N(>r) = N_0 (r/r_0)^{-\alpha} \tag{7.3}$$

is also suitable, with parameters $N_0 = 4.34$ cm^{-3}, $r_0 = 0.1$ μm, and $\alpha = 3$.

The number concentration of aerosol particles varies widely: from 10^{-1} to 10^6 particles/cm^3. If sulfate particles 0.3 μm in radius with a density of 1.65 g/cm^3 are taken to be typical (such a particle weighs 2×10^{-7} μg), then in 60 Mtons of atmospheric aerosol there would be 3×10^{26} such particles, that is, about 5×10^7 particles per cm^2 of the Earth's surface. For a mean volume concentration of the aerosol mass equal to 10 μg/m^3, the mean number concentration of typical particles is 50 cm^{-3}. Assuming this concentration of typical particles, we obtain for the atmosphere at sea level a scattering factor per unit volume $\varrho\sigma = 0.06$ km^{-1} and a visibility in it $L = 4/\varrho\sigma$ of 67 km. This visibility is only 1/4 of that in a clean atmosphere without an aerosol, but it is 8 times the average visibility in the atmosphere over a modern American city, where the aerosol concentration may be 80 μg/m^3.

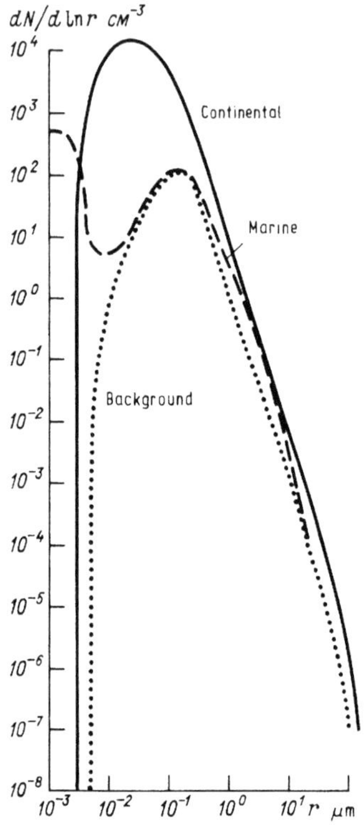

Fig. 7.1. Typical particle-size distributions taking into account continual production of very small particles in continental, marine, and background aerosols.

It is convenient to distinguish between five kinds of aerosols: mineral dust, other aerosols of continental origin, sea salt, the background aerosol of the upper troposphere, and the stratospheric aerosol.

Mineral dust enters the atmosphere from the surface of the continents, mainly during dust storms in desert regions (the latter occupy about 8% of the global surface, that is, about 1/3 of the continental area). A good example is Sahara dust, which is carried by the trade winds into the northern part of the tropical Atlantic, where the atmosphere is frequently so turbid that the setting Sun stops being visible at quite large angular distances above the horizon. Falling into the ocean, the mineral dust becomes an important element in the oceanic sedimentation process, especially in the central regions of the oceans. On land, the settling out of mineral dust created loess deposits, in particular of Pleistocene age.

A typical desert aerosol is 75% clay minerals (35% montmorillonite, 20% kaolinite, and 20% illite), 10% calcite, and 5% of each of the following: quartz, potassium nitrate, and iron compounds (limonite, hematite, and magnetite), with admixtures of certain organic substances. According to row

1a of Table 7.1, the annual production of mineral dust varies over a wide range (0.12–2.00 Gtons). The mass concentration of mineral dust in the atmosphere near the Earth's surface in the vicinity of a dust source reaches 1000 to 3000 μg/m^3, whereas far away from such sources it is only 30 to 300 μg/m^3; a typical average value of this concentration is 300 μg/m^3. Since the concentration decreases with height, mineral dust is observed predominantly in the lower half of the troposphere, up to heights of 3–5 km, and over dust-storm regions sometimes up to 5–7 km. The size distribution of mineral-dust particles usually has two maxima, one in the coarse (mostly silicate) fraction $r = 1$ to 10 μm, which appreciably affects the transfer of thermal radiation, and one in the submicron-sized fraction $r < 1$ μm, which produces the absorption of short-wave radiation. The complex index of refraction of mineral dust (its imaginary part is the absorption index k; see §5) in the visible and near infrared ranges has a typical value $m = n - ik = 1.525 - 0.005i$. The so-called *single-scattering albedo* $\omega = \sigma\,(\alpha + \sigma)$, where α and σ are the coefficients of absorption and scattering, is for a normal desert aerosol close to 0.97 in the visible range and close to 0.92 in the near infrared.

Other continental aerosols have a great number of diverse sources, the main role here being played by the condensation of gases of natural and anthropogenic origin, primarily SO_2, NH_3, NO_2, and hydrocarbons, produced by processes such as combustion, heating, and decomposition. A sizable portion of the particles of these aerosols are sulfates containing SO_4, such as $(NH_4)_2SO_4$, NH_4HSO_4, and H_2SO_4. Rows 1f–1h and 2i–2k of Table 7.1 show that the production of these kinds of aerosols is very great indeed, it being precisely here that anthropogenic sources make a considerable contribution. This aerosol spreads throughout the bottom half of the troposphere, up to a height of 5 km (higher up, it enters into the background aerosol). Its main component is submicron-sized particles, primarily in the optically active range r = 0.1 to 1 μm.

At a wavelength $\lambda = 0.53$ μm the scattering factor of the continental aerosol varies over a range $\varrho\sigma = 0.004$ to 0.6 km^{-1}, the ratio of backscattering to total scattering is $R = 12$ to 28%, the absorption coefficient $\varrho\alpha = 0.007$ to 3.2 km^{-1} (in most cases it ranges from 0.007 to 0.07 km^{-1}), the single-scattering albedo $\omega = 0.53$ to 0.87, being 0.7 on the average (for a negative correlation between ω and R), the parameter of transition from aerosol heating to cooling $(1 - \omega)/\omega R = 0.83$ to 5.2 (for a surface albedo of 0.2, its critical value is 2), the coefficient of correlation between $\varrho\sigma$ and the SO_4 content ranges from 0.24 to 0.99 (0.73 average), the index of optical properties $SO_4/\varrho\sigma = 0.05$ to 0.5 g/m^2 (0.2 g/m^2, on the average), and the aerosol optical thickness varies from 0.3 to 0.7.

Sea-salt particles enter the atmosphere as a result of evaporation of the spray produced when sea waves break. The generation of these particles is a function almost exclusively of the distribution of the wind speed at the ocean surface, and it amounts to about 40% of the total aerosol generation (row 1b of Table 7.1). Because of their large size (bimodal size distribution, as in the case of mineral dust), hygroscopicity, and solubility in water (in addition to NaCl, KCl, etc., a large part of their chemical composition is represented by $(NH_4)_2SO_4$), particles of sea salt rapidly settle out of, or are washed out of, the atmosphere, and even in the central regions of the oceans only a negligible amount are able to make it through the "cloud filter" at heights of 2 to 3 km in the troposphere. The density of marine-aerosol particles $\varrho = 2.35$ to 2.41 g/cm^3, and their mass absorption coefficient k/ϱ ranges from 0.002 to 0.036 cm^3/g, that is, it is quite high, so that these particles have considerable effect on the transfer of long-wave radiation, especially in the transmission window $\lambda = 8$ to 12 μm.

The background aerosol consists mostly of the submicron-sized fraction of mineral dust and other continental (chiefly sulfate) aerosols, which have passed through the cloud filter into the upper troposphere above 2–3 km over the oceans and 5 km over the continents, where this aerosol is distributed comparatively uniformly with a particle-number concentration of about 300 cm^{-3} at standard temperature and pressure.

The normal stratospheric aerosol forms a layer with a maximum concentration at heights of 15 to 20 km, or slightly higher in the tropics, known as the *Junge layer*. This layer consists of sulfate particles (droplets at heights of 17 to 18 km, and from 24 to 27 km solid particles of a 75% aqueous solution of sulfuric acid with inclusions of ammonium sulfate and bisulfate and certain other substances), with a mean radius of about 0.3 μm or somewhat less; their weight concentrations are of the order of 10^{-1} μg/m^3 and their number concentrations are of the order of 10^{-1} to 10 cm^{-3}. In the entire layer there are, on the average, 2×10^5 cm^{-2} particles, with a total weight of 4×10^{-2} μg/cm^2, or 2×10^5 tons in the whole stratosphere. These sulfate particles have a lifetime of around a year in the stratosphere (below 20 km the lifetime is around half a year, and above 20 km it is several years). Consequently, under steady-state conditions the matter which forms the sulfate aerosol has to enter the stratosphere from the troposphere at a rate of 2×10^5 tons/yr, or about 10^7 SO_2 molecules per cm^2 per second.

A normal sulfate aerosol can be assumed to form in the stratosphere due to the oxidation of volcanic sulfur dioxide by hydroxyl and due to the precipitation of sulfuric acid onto condensation nuclei entering from the troposphere. For this to happen, the SO_2 concentration at the tropopause must be of the order of 0.1 to 1 ppb, and the rate of oxidation of SO_2 in the stratosphere must be of the order of 10^{-8} s^{-1}. However, since it has not been demonstrated that SO_2, which is readily soluble in water, can penetrate through the cloud-rain "filter", the primary sources of sulfates were taken to be the biogenic gases H_2S, $(CH_3)_2S$, and CH_3HS, and especially carbonyl sulfide COS, an inert biogenic-volcanogenic gas which has in the troposphere specific concentrations of the order of 2×10^{-10} to 3×10^{-10}.

The normal stratospheric aerosol comprises only 1/300 of the global aerosol, and it produces just very minor optical and thermal effects (according to present estimates, it reduces the transmission of solar radiation and the heat budget of the troposphere by as little as 0.2%). However, powerful volcanic eruptions may make the amount of long-lived sulfate aerosol in the stratosphere 30 to 100 times the normal (not counting the volcanic ash, which can rain down from the stratosphere for several months). After such an eruption the amount of volcanogenic SO_2 and sulfate aerosol in the stratosphere is approximately halved each year, so that, for instance, after a 50-fold increase in the amount of stratospheric aerosol, normal conditions will be restored after 5 years.

The tremendously powerful volcanic explosions of the past, especially the eruption of Krakatoa (Indonesia) in August 1883, which threw about 20 km^2 of volcanic debris to heights in the atmosphere as great as 80 km, the eruption of Mt. Pelée (Martinique) in 1902, which destroyed the city of St. Pierre with its population of about 26,000, and the explosive eruption of Katmai (Alaska) in June 1912, led Humphreys (1913, 1929) and his followers to state their hypotheses concerning the development of climatic coolings after volcanic eruptions and even to explain the glacial periods of the Pleistocene in terms of a presumed higher frequency of volcanic explosions during that period. Attempts were made to detect such cooling-off periods by measuring the air temperature at a number of points on the Earth's surface after the indicated eruptions. However, the temperature trends proved to be different in different places, and there was not enough material for a global averaging.

Quite detailed measurements of the aerosol, radiation, and thermal parameters of the atmosphere were carried out after the eruption of Mt. Agung (Indonesia) in March 1963. This explosive eruption made the content of large aerosol particles ($r > 0.3$ μm) in the 14–20 km layer of the stratosphere tens of times higher than normal, an effect which diminished in time but was still perceptible until 1972. The direct solar radiation at the Earth's surface in the tropics was reduced, on average, by almost 4%, although the scattered radiation was found to increase slightly. During the second half of 1963 a noticeable heating of the tropical stratosphere in the 17–24 km layer took place; over Australia this heating amounted to 5–7°C (although some investigators have questioned its relation to the eruption products of Agung). However, no cooling of the lower troposphere took place.

Even more detailed measurements were carried out after the eruption of Mt. Fuego (Guatemala) in October 1974, which caused, in particular, an increase in the mean annual number of large aerosol particles ($r > 0.3$ μm) in the stratospheric layer from 15 to 20 km from 0.5×10^6 cm^{-2} in 1972 to 2.2×10^6 cm^{-2} in 1975. However, no appreciable global cooling of the lower troposphere was detected in this case either.

Supersonic air transport may prove to be an anthropogenic source of stratospheric aerosol on a par with volcanic eruptions. For a predicted intensity of stratospheric flights in 1990 amounting to about 1.5×10^6 flying hours per year, for burning of fuel with a sulfur content of 0.1%, it is estimated that the emission of sulfur dioxide into the stratosphere will be of the order of 1×10^5 tons/yr, which is comparable to the intensity of natural sources of the normal stratospheric aerosol. Consequently, supersonic air transport with the indicated flight intensity may well lead to a doubling of the amount of the normal stratospheric aerosol, but the effect on the climate of this expected source will still be many times less than the effects of massive volcanic eruptions.

In order to properly take aerosol effects into account in numerical models of the atmosphere–ocean–land system, we must include in our models the fields of the concentration and the microstructure parameters (that is, the particle-size distributions) for aerosols of various kinds (for example, the five kinds described above, as well as, perhaps, the gases from which the aerosol particles have condensed), together with the evolutionary equations describing these fields. The equations of convective diffusion can serve as such equations, provided that the microstructure is parametrized. First of all, with regard to the relative air humidity f, it must be taken into account that, for instance, in the surface air layer over the continents for $f < 0.3$, the aerosol does not contain water droplets; it will include only Aitken nuclei or large ions with $r = 0.1$ to 1 μm. If f increases from 0.3 to 0.8, however, the particles will become water-coated, and the aerosol will assume the form of a haze, the size of the largest particles being $r = 0.1$ to 0.3 μm. For $f \geqslant 0.8$ a damp haze will appear, with a continuous bimodal distribution of r having maxima in the regions 1–5 and 12–15 μm. Finally, for $f \sim 1$ there will be a fog, with particle-size maxima in the regions $r = 8$–12 and 18–25 μm. Alternatively, even kinetic equations can be used, instead of the diffusion equations, if the microstructure is calculated.

These equations, together with the boundary conditions at the Earth's surface, should describe, firstly, the sources of the aerosols (and the gases from which they originate) distributed both over the Earth's surface (and possibly depending on weather conditions, as, for instance, the generation of sea-salt particles at the ocean surface or dust particles in the desert depends on the force of the wind) and throughout the volume of the atmosphere (for instance, the condensation of particles from gases). Secondly, they should describe the dispersal of aerosols in the atmosphere, due to transport by large-scale air movements, mixing processes, and gravitational subsidence. Thirdly, these equations should take into account the modification of particles: physicochemical transformations, moisture condensation on them or evaporation from them, and coagulation. Fourthly, the equations should describe sinks of aerosol particles: their absorption at the Earth's surface and their washing out by precipitation. As yet, however, only the very first steps have been taken toward a numerical modeling of fields of aerosols. One of the few examples is the model of the stratospheric sulfate aerosol constructed by

Hamill (1976) and his colleagues at the NASA Ames Research Center (U.S.A.).

A sufficiently complete model of the atmosphere–ocean–land system has to take into account the effect of aerosols on the absorption and scattering of short-wave and long-wave radiation in the atmosphere. A brief account of these effects and the methods of calculating them was given in §5. The absorption of short-wave (solar) radiation by the tropospheric aerosol is comparable to its absorption by atmospheric gases, but, like the aerosol itself, it is quite variable; in most cases this absorption is concentrated primarily in the lower kilometer of the troposphere. The scattering of short-wave radiation by the aerosol attenuates the direct solar radiation more than the aerosol absorption does, but it takes place mostly above the layer where the absorption is concentrated.

The effect of the aerosol on the total albedo A of the earth-atmosphere system (and thus on the total thermal effect of the short-wave radiation) depends on the albedo A_s of the Earth's surface and the clouds; if the effective albedo of the aerosol layer A_e is higher than A_s, then the presence of an aerosol increases A (and hence leads to a cooling of the planet), whereas for $A_e < A_s$ the presence of an aerosol reduces A (heating of the planet). For a cloudless atmosphere over the ocean, A_s is low, so that the aerosol promotes cooling. However, since clouds (making A_s high) cover more than half of the Earth's surface, on the whole the aerosol contributes to the heating of the planet by short-wave radiation. These effects depend considerably on the zenith angle of the Sun $\vartheta_\odot$: at low $\vartheta_\odot$ (including daytime in low latitudes) an aerosol mostly reduces the value of A, whereas at high $\vartheta_\odot$ (typical for high latitudes) it increases this value.

The tropospheric aerosol can be a quite strong absorber of long-wave radiation (thereby reducing the effective radiation of the Earth's surface). Such absorption is especially important for radiation with wavelengths in the atmospheric window 8 – 12 μm (we note, in particular, the maximum of aerosol absorption in the band around $\lambda = 9.2$ μm). For a high dust content of the atmosphere, the absorption of long-wave radiation is enhanced appreciably because of multiple scattering (whereas for a more rarefied stratospheric aerosol the single-scattering approximation may be sufficient). An aerosol also emits long-wave radiation, which leads to a cooling of the aerosol layers, concurrently with their heating due to the absorption of short-wave radiation.

The detailed calculations of Luther (1974–75) have shown that the stratospheric aerosol, by affecting the transfer of short-wave and long-wave radiation, causes an overall cooling of the atmospheric layer 0–23 km for $A_s < 0.35$ and heating for $A_s > 0.35$ (on the average, the global $A_s \approx 0.28$). For any A_s the stratospheric aerosol layer itself heats up, while the troposphere cools off, on the average (in high latitudes it apparently cools off in summer and heats up in winter).

The effects of aerosols on the microstructure and optical properties of liquid-water clouds have not, however, been properly described quantitatively.

8. Ozone

The climatic importance of ozone is twofold. First of all, it contributes to the greenhouse effect, by absorbing infrared radiation in a band centered at 9.569 μm in the transmission window of the atmosphere, as well as in bands around 4.75 and 14.1 μm. Secondly, ozone promotes the heating of the lower stratosphere, by absorbing both infrared radiation and about 4% of the solar radiation (in ultraviolet bands at 0.22–0.29 and 0.31–0.36 μm); according to the approximate calculations of Manabe and Strickler (1964), ozone raises the temperature of the stratosphere by nearly 90°C. The latter effect determines the very existence of the stratosphere as well as its dynamics. Moreover, the biological importance of ozone is extremely great, in that it keeps out the ultraviolet radiation of the Sun, which would be destructive to living organisms.

DISTRIBUTION OF OZONE

The ozone distribution in the atmosphere is measured using optical methods at stations on the ground and with the aid of ozone recorders sent aloft. The ozone distribution above the level of its maximum concentration can be determined from satellite measurements of the backscattering of ultraviolet radiation (at 4 to 6 wavelengths).

The mean ozone concentration in the atmosphere is about 4×10^{-7} by volume (the equivalent thickness of the ozone layer at standard temperature and pressure is about 3 mm; at the time of the IGY in 1958–59 the ozone layer was 2.98 mm thick in the Northern Hemisphere and 3.07 mm thick in the Southern Hemisphere) or 6×10^{-7} by mass (about 3.2 Gtons of ozone in the entire atmosphere). The amount of ozone in the whole air column varies appreciably with latitude: from a minimum of 2.5–2.7 mm at the equator to maxima of 3.0–3.5 mm at the South Pole and 3.5–4.5 mm at the North Pole.

Ozone is concentrated mainly in the stratosphere, with maximum partial densities of the order of 4×10^{-10} to 6×10^{-10} g/cm^3 at heights of 15–30 km. The following distribution of ozone with height is typical: if there are 8.9×10^{18} molecules/cm^3 above 15 km, then the concentration at a height of 20 km is 4.1×10^{12} molecules/cm^3, a maximum of 5×10^{12} is reached at 25 km, and the concentration is 3.6×10^{12} at 30 km, 1×10^{12} at 40 km, 2.8×10^{11} at 45 km, 6.6×10^{10} at 50 km, and 2.4×10^{10} molecules/cm^3 at a height of 55 km. The vertical ozone distribution varies considerably with latitude: the concentration maximum in the polar regions is quite low-lying, at heights of around 16 km (and it is very narrow), whereas with the approach to the equator this height increases to 25 km (while the maximum itself becomes lower).

Observations indicate that the ozone distribution fluctuates markedly with time and that these fluctuations have very diverse periods. The variations from day to day in given regions of the stratosphere may sometimes be extremely abrupt; for instance, during the three-day period from 20 to 23 February 1970, the ozone concentration in the 13–19 km layer over Switzerland decreased to 1/8 of the initial concentration (and its total content in the air column dropped during this same time from 4.48 to 3.50 mm). Some of the day-to-day fluctuations of the total ozone content (which could be as high as 25%) resemble progressive waves of synoptic scale, with dispersion relations similar to those for Rossby waves.

Longer-period variations of the ozone content may be created by protons from large solar flares, which produce radicals of HO_x and nitric oxide NO, capable of destroying the ozone. For instance, at the time of the proton flare of 2–4 November 1969 the ozone concentration at a height of 54 km dropped to half, and the concentration at 67 km dropped to a quarter (at these heights HO_x radicals were active), after which it returned quite rapidly to the normal value. After a proton flare in August 1972 the ozone concentration at levels of 30 to 45 km (where NO is active) dropped by 16%, and this anomaly remained for 2 or 3 weeks.

Variations with even longer periods are created by *sudden warmings of the stratosphere*, which occur once every 1 to 3 years (and which make the main contribution to the year-to-year variability of the ozone content) in winter in regions between the pole and 40° latitude. These warmings manifest themselves as a lowering of a considerable part of the stratosphere by 10–40 km. They are adiabatic warmings, which over some weeks can be as high as 80°C at a level of 40 km, with a simultaneous cooling of the stratosphere in the tropics and in the other hemisphere. The ozone content then stays at a maximum for several months in the region where the warming took place (and in the hemisphere as a whole). According to the theory of Matsuno (1971), sudden warmings of the stratosphere are created by an enhancement (during a two-week period) and a penetration into the winter stratosphere of geostationary waves with wave numbers of 1 and 2. Such waves promote a poleward transfer of heat (in the Northern Hemisphere west of the high-pressure regions there is a northward flow of warm air, with a southward flow of cold air east of these regions). As a consequence, the meridional temperature gradient typical of the high latitudes changes sign. Since the poleward heat flows produce a mean air current toward the south, the Coriolis force turns the latter westward, and makes it work against the westerly wind characteristic of the winter stratosphere. When the speed u of the latter approaches zero, a "critical level" is created, at which u is equal to the wave velocity c. At this point the absorption of wave energy commences, heating the stratosphere. Obeying the formula for the thermal wind,

$$\frac{\partial u}{\partial \ln p} = - \frac{R}{af}\frac{\partial T}{\partial \theta}, \tag{8.1}$$

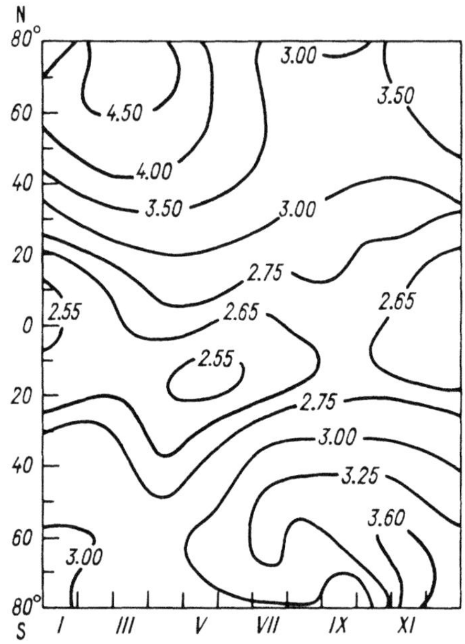

Fig. 8.1. Isolines of mean zonal amount of ozone in entire air column (in millimeters at standard temperature and pressure), as functions of latitude and time of year, according to measurements taken from 1960 to 1968.

the westerly wind changes to an easterly wind at the time of a sudden warming of the stratosphere (where R is the molar gas constant, a is the radius of the Earth, f is the Coriolis parameter, and θ is the latitude supplement).

The ozone content in the atmosphere is subject to sizable seasonal fluctuations. For the mean zonal amount of ozone in the entire air column at different latitudes, these fluctuations are shown in Fig. 8.1. It is seen that in the middle and high northern latitudes the ozone content reaches a maximum in March–April (three regions of maximum are observed on the maps at this time, the largest of which is over the northeastern U.S.S.R.) and a minimum in September–October (with an amplitude nearly half that of the maximum). In the middle and high southern latitudes the amplitude of the seasonal fluctuations is about half that of the fluctuations in the Northern Hemisphere: the maximum is reached in September (middle latitudes) and November (high latitudes), and the minimum is in February–March. In the tropics the ozone content is a minimum, with an approximately zonally uniform distribution, and the seasonal fluctuations are not very pronounced. The seasonal variations of the vertical ozone distribution have the following pattern: in spring the maximum in the polar region is much sharper than in autumn and it is further down, whereas in the tropics the spring maximum is slightly weaker.

A quasi-biennial fluctuation with a period of 26–30 months is clearly discernible in the long-period variations of the ozone distribution. This fluctuation is related, first of all, to fluctuations of movements in the tropical stratosphere, consisting in a downward motion (with the indicated period) of layers of alternating height 5–10 km thick with westerly and easterly winds. According to the theory of Holton and Lindzen (1968), this process is due to an interaction with a zonal current of upward-moving equatorial Kelvin waves, which are shifted eastward, and mixed gravity–Rossby waves, which are shifted westward; this zonal current is subject to semiannual fluctuations.

Quantitatively, the long-period trends in the variability of the ozone layer can be summarized as follows. According to the statistical data of London and Kelley (1974), from 1958 to 1961 the ozone content in the Northern Hemisphere decreased at a rate of $(4.7 \pm 1.5)\%$ per decade, while the content in the Southern Hemisphere increased at a rate of $(2.5 \pm 2.3)\%$; from 1961 to 1970 these trends had the opposite sign: in the Northern Hemisphere the ozone content increased at a rate of $(11.3 \pm 2.3)\%$ per decade, while in the Southern Hemisphere it decreased at a rate of $(2.1 \pm 1.6)\%$. According to Angell and Korshover (1976), from 1960 to 1962 the ozone content decreased by about 1%, from 1962 to 1970 it increased by 3%, from 1970 to 1972 it decreased by 1%, and from 1970 to 1974 it decreased by 2% (these investigators, and several others as well, have detected a positive correlation between the amount of ozone and the sunspot number in the 11-year cycle of solar activity). Drops in the ozone content were to be expected after the nuclear explosions of 1961–1962, which produced ozone-destroying nitrogen oxides, but the measurement data did not yield any statistically significant results with regard to this effect.

During the 1970s it was feared that nitrogen oxides in the exhausts of high-flying aircraft could reduce the size of, or even destroy, the stratospheric ozone layer. However, quantitative evaluations obtained as a result of the Climate Impacts Assessment Program have shown this effect to be minor. If the amount of fuel burned by high-altitude aircraft increases from 20–45 Mtons/yr in 1975 to 75–105 Mtons/yr in 1985, then the NO_2 emission should increase from 0.2–0.4 Mtons/yr in 1975 to 0.7–0.9 Mtons/yr in 1985 (only 5–15% of which will be produced by supersonic aircraft). This is only 0.1 to 0.4% of the main element of the natural nitrogen cycle, namely the fixation of atmospheric nitrogen by soil bacteria, which has been estimated to be 200 Mtons/yr. The main source of nitrogen oxides NO_x in the stratosphere is the oxidation of nitrous oxide N_2O produced during denitrification processes in the soil and in plants. The volume ratio of the N_2O mixture in the present atmosphere is estimated to be 2.5×10^{-7}, and the rate of N_2O oxidation in the lower stratosphere to be 0.2×10^8 to 2×10^8 molecules/(cm^2 s); simplified calculations indicate that a doubling of the N_2O production should lead to a 16% reduction in the ozone content of the stratosphere.

Even more dangerous with regard to the possible destruction of the ozone layer are, apparently, halogenated hydrocarbons. The total annual emissions of these were estimated in 1975 to be: 340 ktons of Freon-11, with an annual increment of 14%; 413 ktons of Freon-12, with an annual increment of 9.6% (equivalent to about 5.4×10^7 Cl atoms per cm^2 per second); about 100 ktons of CCl_4; and around 100–200 ktons of CH_3Cl. For comparison, the annual input of HCl to the stratosphere from volcanic eruptions is 10 to 30 ktons and from marine particles is 5 ktons. It is noteworthy that the launching of a space rocket releases chlorine compounds in a quantity equivalent to approximately 100 tons HCl, so that a single launching a week would yield 5 ktons HCl per year. Simplified calculations indicate that the emission of halogenated hydrocarbons into the atmosphere at the 1972 rate would reduce the ozone content in the stratosphere by 6.5%, with a particularly sharp drop in the O_3 concentration (by 14–17%) at heights of 35 to 40 km.

DYNAMICS OF THE STRATOSPHERE

The ozone distribution in the stratosphere provides a spatial picture of its heating, and thus of the dynamics of the stratosphere, beginning with the thermal wind (8.1). On the other hand, currents in the stratosphere tend to redistribute throughout it both ozone and the gases which affect its evolution. Consequently, the ozone and the stratospheric currents influence each other.

If the dynamic state of the stratosphere is averaged over time (but with seasonal fluctuations), the main component is seen to be the zonal circulation. The mean zonal temperature (Fig. 8.2a) is a minimum at the tropopause, the level of which rises from 8 km at the winter pole to 16 km at the equator. At the equator the absolute minimum of the stratospheric temperature is reached, below −80°C, because of dynamic factors, such as the removal of heat by currents, since radiation effects will cause a slight heating there. Higher up the temperature increases (high static stability) until it reaches a maximum of about 280 K at the stratopause (at approximately 50 km); still further up the temperature drops (and has a minimum, also of dynamic origin, at the winter pole in the lower mesosphere).

Concerning the zonal winds in the stratosphere (Fig. 8.2b), we should mention, firstly, a westerly subtropical jet stream at levels around the tropopause, which abates appreciably from winter to summer (from 30–35 down to 10–15 m/s). Secondly, there is a strong subtropical jet stream centered at a height of about 65 km in the lower mesosphere but extending throughout the entire middle and upper stratosphere; in winter it is westerly, with a speed at the center of more than 80 m/s (jet stream at the edge of the polar night); in summer it is easterly, with a speed at the stream center of about 60 m/s, with a change in the sign of the current in the equatorial zone (the above-mentioned downward motion, with a period of 26–30 months, of layers at alternating heights in which there are westerly and easterly currents, is observed in the equatorial zone). We note that these seasonal fluctuations of the currents also have a semiannual harmonic with amplitudes up to 30 m/s, a westerly

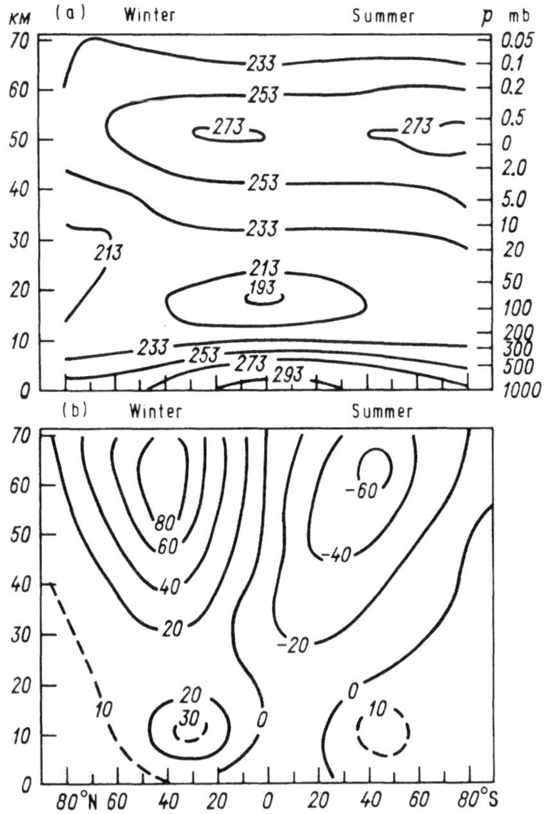

Fig. 8.2. Mean zonal temperatures in kelvins (a) and winds in m/s (b) in winter and summer in Northern Hemisphere, according to Newell (1969). Positive values = westerly wind, negative values = easterly wind.

component in spring and autumn, and an easterly component in summer and winter. The meridional circulation is much weaker. In winter it has a mostly two-celled structure, with ascent of cold air at the equator and at the pole and subsidence of warm air at latitudes of 30 to 50° (kinetic energy being converted into potential energy in the process). In summer the downward movement of air in the middle latitudes is replaced by an ascent, and the other branches of the circulation are divided up into smaller cells.

Whereas in summer the isobars and isotherms in the stratosphere are almost perfectly zonal, in winter in the Northern Hemisphere at heights from 25 to 35 km a strong geostationary nonzonal disturbance is superimposed onto them. This disturbance, with longitude wave numbers 1 and 2, produces the warm Aleutian anticyclone in the stratosphere and causes a poleward transfer of heat. Traveling waves are also observed in the stratosphere: (1) mixed gravity–Rossby waves, with longitude wave numbers of about 4, periods of 4–5 days, and phase velocities of about 25 m/s toward the west;

(2) Kelvin waves, with wave numbers of 1–2, periods of 15 days, and phase velocities of 25 m/s toward the east (waves (1) and (2) are unimportant for ozone transport); (3) diurnal high tides, with phase velocities of up to 450 m/s; (4) slow waves, with wave numbers of 1–3, periods of 1–4 weeks, and phase velocities of about ±10 m/s, which may play an important part in the transport of ozone (these are, apparently, the gradually growing, due to the baroclinic instability, long-wave "Green modes", which have maximum amplitudes in the stratosphere and which must be distinguished from the rapidly growing, shorter-wave "Charney modes", which have sizable amplitudes only in the troposphere).

The conversions of the kinetic energy and available potential energy of the zonal circulation ($\bar{K}$, $\bar{P}$) and the disturbances (K', P') in the stratosphere, according to the empirical estimates of Oort (1964) for the 100–30 mb layer of the Northern Hemisphere, have the form $K' \to (\bar{K}, P')$ and $P' \to \bar{P}$. The zonal available potential energy $\bar{P}$ produced by this is then dissipated by radiation processes (in the layer from 30 to 10 mb, containing a mesospheric jet stream with high $\bar{K}$, the processes $K' \to \bar{K} \to \bar{P} \to P'$ also take place). This means that the stratosphere is mechanically accelerated by the tropospheric kinetic energy $\bar{K}$ and K', whereas in the troposphere the differential solar heating generates $\bar{P}$ and the baroclinic instability of the waves produces the processes $\bar{P} \to P' \to K'$ with a resultant *density* of K' 15 to 20 times greater than in the stratosphere. It should be noted that the values of K' and P' in the stratosphere are a maximum in winter (and are very small the rest of the year); in winter there is thus the greatest flow of heat and ozone toward the pole, ultimately causing a spring warming of the stratosphere and a maximum ozone content in it.

Consequently, the absorption of the energy of waves entering the stratosphere from below plays a very important part in its dynamics. The amplitudes $A(z)$ of internal waves satisfy the following equation locally:

$$\frac{d^2A}{dz^2} - \frac{1}{H^2}\left[\frac{1}{4} + L^2\left(k^2 - \frac{\beta}{u - c}\right)\right] A = 0, \tag{8.2}$$

where k and c are the horizontal wave number and the phase velocity of a wave, u is the velocity of the zonal current, β is the derivative of the Coriolis parameter f with respect to the meridian, H is the thickness of the homogeneous atmosphere, and $L = (N/f)\,H$ is the Rossby radius of deformation (N being the Väisälä-Brunt frequency). Solutions which fluctuate in height (that is, waves transporting energy along the vertical) exist only on condition that

$$0 < u - c < U = \beta\left(k^2 + \frac{1}{4L^2}\right)^{-1}, \tag{8.3}$$

where U is the so-called critical Rossby velocity; thus, such solutions only exist in the layer where $c < u(z) < c + U$. According to Fig. 8.2b, in winter in the stratosphere $0 < u < 80$ m/s and long planetary waves can propagate vertically (in summer 60 m/s $< u < 0$, and no internal waves can propagate along the vertical). At the critical level $z = z_c$, where $u(z) = c$, a wave turns out to be a standing wave relative to the zonal current and, according to the linear theory, it can only become damped under the influence of dissipative factors, thereby imparting its momentum and energy to the zonal current (nonlinear effects may lead to reflection of the wave from the critical level).

PHOTOCHEMICAL THEORY

Going back to ozone itself, let us now consider the main reactions of its creation and destruction. According to the theory of Chapman (1930), these are, first of all, reactions of the oxygen cycle. The first of these is the photodissociation of molecular oxygen under the influence of ultraviolet radiation:

$$O_2 + h\nu \rightarrow \begin{cases} O(^1D) + O \ (\lambda \leqslant 0.175\ \mu m,\ z \geqslant 80\ km); \\ O + O \qquad (\lambda \leqslant 0.242\ \mu m,\ z \geqslant 25\ km), \end{cases} \tag{R1}$$

where $O(^1D)$ denotes an oxygen atom in the excited state ^1D. If the number of particles M per unit volume is designated as $[M]$, then the rate d[O]/dt of reaction (R1) can be written as $2[O_2][h\nu]$, where $[h\nu]$ is the absorbed number of quanta $h\nu$ per O_2 molecule per second, determined by the absorption cross section and by the quantum flux at the given level z. Below 30 km the latter decreases rapidly (by five orders of magnitude down to the 10-km level), so that the formation of atomic oxygen practically ceases. The second reaction is the formation of ozone via an $O + O_2$ collision with the necessary participation of some third particle M (any air molecule), which must remove the energy of 1.09 eV released during the formation of the O_3 molecule:

$$O + O_2 + M \rightarrow O_3 + (M + 1.09\ eV). \tag{R2}$$

The rate of ozone formation d$[O_3]$/dt via this reaction is proportional to $[O][O_2][M]$; the proportionality factor $k_2 = 6.6 \times 10^{-35} \exp(510/T)$ cm^6/s (if M is the nitrogen molecule). The third reaction is the photodissociation of ozone under the influence of ultraviolet and visible radiation:

$$O_3 + h\nu_1 \rightarrow \begin{cases} O_2 + O(^1D)(\lambda \leqslant 0.31\ \mu m,\ z \geqslant 20\ km), \\ O_2 + O \qquad (\lambda \leqslant 0.71\ \mu m,\ z \geqslant 0\ km). \end{cases} \tag{R3}$$

Since the rate of reaction (R2) decreases with height, while that of reaction (R3) increases, the oxygen $O + O_3$ below 60 km consists mainly of O_3, whereas above 60 km it is mainly O (a typical distribution of [O] with height at a latitude of 45° on a summer day is: 8×10^5 atoms/cm^3 at a height of 20 km, 1.2×10^8 atoms/cm^3 at 35 km, 1.4×10^{12} atoms/cm^3 at 100 km, and decreasing [O] further up).

Finally, the fourth reaction is the destruction of ozone by collisions with atomic oxygen:

$$O + O_3 \rightarrow O_2 + O_2. \tag{R4}$$

The rate constant of this reaction $k_4 = -\{[O][O_3]\}^{-1} d[O_3]/dt$ is equal to $1.9 \times 10^{-11} \exp(-2300/T)$ cm^3/s. Therefore, for the oxygen cycle, we have

$$\frac{d[O]}{dt} = 2[h\nu][O_2] + [h\nu_1][O_3] - k_2[O][O_2][M] - k_4[O][O_3];$$

$$\frac{d[O_3]}{dt} = -[h\nu_1][O_3] + k_2[O][O_2][M] - k_4[O][O_3]. \tag{8.4}$$

Thus, for photochemical equilibrium, when the left-hand parts of these equations go to zero, we get

$$[h\nu][O_2] - k_4[O][O_3] = 0;$$

$$[h\nu][O_2] + [h\nu_1][O_3] = k_2[O][O_2][M], \tag{8.5}$$

where, at heights between 15 and 40 km on the left-hand side of the second of these equations, the first term can be neglected, so that

$$[O] = \frac{[h\nu_1][O_3]}{k_2[O_2][M]} = \left\{\frac{[h\nu][h\nu_1]}{k_2k_4[M]}\right\}^{1/2};$$

$$[O_3] = \left\{\frac{k_2[h\nu][M]}{k_4[h\nu_1]}\right\}^{1/2}[O_2]. \tag{8.6}$$

Computations using these formulas showed a qualitative agreement with the observed vertical ozone profile (with a maximum at a height between 22 and 26 km). However, the total amount of ozone obtained was too high, and its latitudinal and seasonal variations turned out to be quantitatively unsatisfactory. These shortcomings are due to disturbances of the photochemical equilibrium, on the one hand, and to the fact that in (8.4) – (8.6) the reactions of ozone destruction in the chlorine, nitrogen, and hydrogen cycles are not taken into account.

The rate of attainment of photochemical equilibrium by atomic oxygen or ozone can be estimated in terms of the time $t_{1/2}$ required for its concentration to become equal to half of the equilibrium value (8.6) for zero initial concentration:

$$t_{1/2}(O) = \frac{\ln 2}{k_2[O_2][M]}; \quad t_{1/2}(O_3) = \left\{\frac{k_2[M]}{k_4[h\nu][h\nu_1]}\right\}^{1/2}. \tag{8.7}$$

The first of these formulas was obtained from the first equation of (8.4) by retaining in its right-hand part only its main (third) term. Since the time $t_{1/2}(O)$ below a height of 40 km is of the order of some seconds, the atomic oxygen is in the daytime almost always in photochemical equilibrium with the ozone (whereas at night, when $[h\nu] = [h\nu_1] = 0$, it rapidly disappears). Consequently, when calculating $t_{1/2}(O_3)$ using the second equation of (8.4), on its right-hand side concentration [O] has to be expressed in terms of $[O_3]$ using the first of the photochemical-equilibrium formulas (8.6), after which the second formula of (8.7) is obtained from this equation. Time $t_{1/2}(O_3)$ is large, but it decreases with height. Below 25 km this time is of the order of many years, so that the ozone practically never reaches its equilibrium concentration; thus, as it is transported by air currents for days and weeks, it behaves almost like a passive substance, following the air currents and their variability. At a height of 25 km $t_{1/2}(O_3) \approx 4$ months, while at a height of 40 km it is only about 20 hours; therefore, above 30–35 km the ozone regains its photochemical equilibrium in about a day and is not very dependent on the air currents (being determined mainly by the solar radiation).

According to Crutzen (1971, 1974), the destruction of ozone in the nitrogen cycle takes place primarily via the following catalytic reactions, mentioned in §6:

$$NO + O_3 \rightarrow NO_2 + O_2; \quad k_5 = -\{[NO][O_3]\}^{-1} \frac{d[O_3]}{dt}$$
$$= 9 \times 10^{-13} \exp\left(-\frac{1200}{T}\right) cm^3/s; \qquad (R5)$$

$$NO_2 + O \rightarrow NO + O_2; \quad k_6 = -\{[NO_2][O]\}^{-1} \frac{d[O]}{dt}$$
$$= 9.1 \times 10^{-12}\ cm^3/s, \qquad (R6)$$

together with the additional reaction

$$NO_2 + O_3 \rightarrow NO_3 + O_2; \ k_7 = -\{[NO_2][O_3]\}^{-1} \frac{d[O_3]}{dt}$$
$$= 9.8 \times 10^{-12} \exp\left(-\frac{3500}{T}\right) cm^3/s; \qquad (R7)$$

The importance of these reactions leads us to consider the processes of formation and destruction of nitric oxide NO and nitrogen dioxide NO_2. The main reaction of NO formation is the oxidation of nitrous oxide N_2O, which is released into the atmosphere as a result of the decomposition of nitrates and nitrites in the soil (including nitrogen fertilizers added by man) by denitrifying bacteria:

$$N_2O + O(^1D) \rightarrow \begin{cases} NO + NO; \quad k_8 = \frac{1}{2}\{[N_2O][O]\}^{-1} \frac{d[NO]}{dt} \\ = 1.0 \times 10^{-12}; \\ N_2 + O_2. \end{cases} \qquad (R8)$$

Two additional reactions of NO formation are the daytime photodissociation of NO_2 into NO and O, under the influence of radiation with $\lambda < 0.3975$ μm (the logarithmic reaction rate $J = -[NO_2]^{-1} \, d[NO_2]/dt \approx 7 \times 10^{-3}$ s^{-1}), and the oxidation of atomic nitrogen N (formed from N_2 by the action of cosmic rays) when it collides with O_2, O_3, HO, and HO_2 + M. NO can be destroyed by, in addition to reaction (R5), reactions of the type of $NO + RO_2 + M \rightarrow NO_2 + RO + M$, where $R = C_nH_{2n+1}$; in particular, for $n = 0$ we have a synthesis of nitric acid $NO + HO_2 + M \rightarrow HNO_3 + M$, taking place with a constant of the order of 10^{-29}.

Moreover, in the mesosphere NO photodissociates under the influence of radiation with $\lambda < 0.191$ μm and is destroyed by atomic nitrogen in the reaction $NO + N \rightarrow N_2 + O$. Nitrogen dioxide is formed not only by the oxidation of NO according to reaction (R5), but also via the photodissociation of nitric acid $HNO_3 + h\nu \rightarrow NO_2 + HO$ caused by radiation with $\lambda < 0.55$ μm. At the same time, it is destroyed by, in addition to reactions (R6) and (R7) and photodissociation, the reaction $NO_2 + HO + M \rightarrow HNO_3 + M$, taking place with a constant of the order of 2×10^{-30}. In view of the above-mentioned processes, it is important to consider the evolution of nitrous acid N_2O, which can oxidize as per reaction (R8) and also photodissociate into N_2 and O (^{1}D), under the influence of radiation with $\lambda < 0.315$ μm, which limits the lifetime of N_2O in the atmosphere to 118 years. The evolution of nitric acid also has to be considered, since below 25 km it is the main form of odd nitrogen, whereas above 25 km the main forms are NO and NO_2. It should be noted that at the height of maximum ozone (about 25 km) the volume mixing ratio of N_2O is of the order of 10^{-7}, that of HNO_3 is 6×10^{-9}, that of NO_2 is 4×10^{-9}, and that of NO is 7×10^{-10}.

When nitrogen oxides are taken into account, equations (8.4) for concentrations [O] and $[O_3]$, formulas (8.6) for these concentrations at photochemical equilibrium, and formulas (8.7) for the time needed to reach equilibrium all become more complicated. A nonzero equilibrium concentration of ozone turns out to be possible only provided that concentrations $[NO_2]$ are not too large, as per the condition

$$[NO_2] \leqslant [NO_2]_{cr} \approx \left\{ \frac{k_2[h\nu]\,[M]}{k_6 J} \right\}^{1/2} [O_2], \tag{8.8}$$

since for $[NO_2] > [NO_2]_{cr}$ the nitrogen oxides completely destroy the ozone. Time $t_{1/2}(O)$ turns out to be shorter than given by (8.7), and $t_{1/2}(O_3)$ is found to depend appreciably on $[NO_2]$. For $[NO_2] = 10^7$ to 10^8 molecules/cm^3, time $t_{1/2}(O_3)$ is the same as given by (8.7), whereas for $[NO_2] \sim 10^9$ molecules/cm^3 in the layer from 20 to 50 km it is only $\frac{1}{2}$ to $\frac{1}{3}$ as great.

Ozone is destroyed in the hydrogen cycle via catalytic reactions of hydrogen H, hydroxyl HO, and perhydroxyl HO_2:

$$HO + O_3 \rightarrow \begin{cases} HO_2 + O_2; & k_9 = -\{[HO][O_3]\}^{-1}\dfrac{d[O_3]}{dt} = \\ & = 1.6 \times 10^{-12} \exp\left(-\dfrac{1000}{T}\right); \\ H + 2O_2; & k_9' < 10^{-17}; \end{cases} \quad (R9)$$

$$HO_2 + O \rightarrow HO + O_2; \quad k_{10} = 1.5 \times 10^{-12} T^{1/2}; \quad (R10)$$

$$H + O_3 \rightarrow HO + O_2; \quad k_{11} = k_{10}; \quad (R11)$$

$$HO + O \rightarrow H + O_2; \quad k_{12} = 3 \times 10^{-12} T^{1/2}; \quad (R12)$$

$$HO_2 + O_3 \rightarrow HO + 2O_2. \quad (R13)$$

These reactions of conversion of O and O_3 to O_2 take place more slowly than nitrogen reactions (R5)–(R7) in the lower stratosphere, while in the upper stratosphere they are more rapid. However, even in the lower stratosphere the odd hydrogen (H, HO, and HO_2) can affect the ozone considerably (for instance, by supplementing NO and NO_2 to form HNO_3 and thereby lessening the nitrogen destruction of ozone). Thus it is necessary to consider the formation and destruction of odd hydrogen. Hydroxyl HO is produced in the stratosphere mainly by the oxidation of water, methane, and molecular hydrogen, as well as by the photodissociation of water and hydrogen peroxide. Odd hydrogen is destroyed as a result of the fast reaction $HO + HO_2 \rightarrow H_2O + O_2$ with the addition above 60 km of the reaction $H + HO_2 \rightarrow H_2 + O_2$ and below 40 km of the above-mentioned reaction $HO + NO_2 + M \rightarrow HNO_3 + M$. Consequently, the main forms of odd hydrogen below 40 km are H_2O_2 and HNO_3, while above 40 km they are HO and, ultimately, H (note that at the heights of maximum ozone the concentrations of hydroxyl and perhydroxyl are of the order of 10^6–10^7 molecules/cm^3).

The destruction of ozone in the chlorine cycle takes place, first of all, via the following fast catalytic reactions, which were mentioned in §6:

$$Cl + O_3 \rightarrow ClO + O_2; \quad (R14)$$

$$ClO + O \rightarrow Cl + O_2, \quad (R15)$$

the first of which in conjunction with the reaction

$$ClO + NO \rightarrow Cl + NO_2 \quad (R16)$$

catalytically converts $NO + O_3$ into $NO_2 + O_2$, while the NO_2 photodissociates into NO and O. Reacting with CH_4, HO_2, and H_2, atomic chlorine is converted into hydrochloric acid. The latter does not react with ozone, but atomic chlorine may be retrieved from the HCl as it photodissociates (under the influence of radiation with $\lambda \leqslant 0.29$ μm) and as it reacts with O and HO. The reaction $ClO + NO_2 + M \rightarrow ClONO_2 + M$, which proceeds in the reverse direction under the influence of radiation with $\lambda \leqslant 0.46$ μm, should also be mentioned here. Of all the forms of inorganic chlorine (Cl, ClO, HCl, $ClONO_2$, OClO, HOCl, and Cl_2) it is HCl which is predominant at all heights; only at a height of about 30 km the amount of $ClONO_2$ may be comparable (note that the volume ratio of HCl at the heights of maximum ozone are of the order of 10^{-9}; the number of ClO particles there will be 5×10^8 molecules/cm^3). Like chlorine, bromine can also bring about a catalytic destruction of ozone; bromine is used, in the form of CH_3Br, as a fumigant in agriculture and, in the form of $C_2H_4Br_2$, as an additive to gasoline. Finally, we should mention the oxidation reactions of methane and carbon monoxide, in which H_2O, HO, and HO_2 are formed and consumed (and thereby affect the hydrogen cycle of ozone destruction).

MODELS BASED ON CHEMICAL DYNAMICS

The specific concentration c of ozone, or of any of the other above-mentioned atmospheric gases, must satisfy the equation of convective diffusion with sources present:

$$\frac{\partial \varrho c}{\partial t} + \frac{\partial \varrho c v_z}{\partial z} + \frac{1}{a \sin\theta}\left(\frac{\partial \varrho c v_\theta \sin\theta}{\partial\theta} + \frac{\varrho c v_\lambda}{\partial\lambda}\right)$$

$$= -\left[\frac{\partial I_z}{\partial z} + \frac{1}{a\sin\theta}\left(\frac{\partial I_\theta \sin\theta}{\partial\theta} + \frac{\partial I_\lambda}{\partial\lambda}\right)\right] + \varepsilon^+ - \varepsilon^-, \qquad (8.9)$$

where v_z, v_θ, v_λ are the components of the velocity of the large-scale movements; I_z, I_θ, I_λ are the components of the flux of the substance in question, produced by fluctuating motions and molecular diffusion; and ε^+ and ε^- are the densities of the sources and sinks of the substance.

The usual method of parametrization of fluxes I_j consists of introducing coefficients of turbulent diffusion K_{jl}. The latter have to be so chosen as to give the best fit between the values calculated using equation (8.9) and empirical data on the concentration field:

$$I_j = -\varrho\left(K_{jz}\frac{\partial c}{\partial z} + \frac{K_{j\theta}}{\alpha}\frac{\partial c}{\partial\theta} + \frac{K_{j\lambda}}{a\sin\theta}\frac{\partial c}{\partial\lambda}\right). \qquad (8.10)$$

Sometimes it is assumed, for example, that

$$K_{zz} = K + K_z + \alpha^2 K_h; \quad K_{\theta\theta} = K_{\lambda\lambda} = K + K_h;$$

$$K_{z\theta} = K_{\theta z} = \alpha K_h, \quad K_{z\lambda} = K_{\lambda z} = K_{\theta\lambda} = K_{\lambda\theta} = 0,$$

where K is the coefficient of molecular diffusion, and α is the inclination to the horizontal of the main direction of the large-scale turbulent transfer in the lower stratosphere. If we confine ourselves to the oxygen cycle of ozone production and destruction, then we have to retain only equation (8.9), in which for $[O_3]$ densities ε^+ and ε^- are found from the right-hand side of the second of equations (8.4), and for [O] they are given by the first of the photochemical-equilibrium formulas (8.6). However, such an approach is unsatisfactory, and in fact the minimum number of gases which are to be described by equations of form (8.9) must apparently include O_3, NO, NO_2, N_2O, and HNO_3. The other gases entering into these five equations (first of all, HO and HO_2) may be described by the formulas of photochemical equilibrium.

In diagnostic calculations the velocities of the large-scale movements v_z, v_θ, v_λ in equations (8.9) are specified according to factual or climatic data, and the equations are solved for the concentrations c of the gases being considered. To simplify the calculations, a two-dimensional model is sometimes employed (a zonal model, in which $v_\lambda = 0$ and all the fields are independent of the longitude λ), or even a one-dimensional model of vertical diffusion, in which large-scale motions are ignored, the vertical exchange is parametrized by a single coefficient K_{zz}, and all the fields depend only on z; the calculations are often confined just to steady-state models.

Such methods have been used by Brasseur and Nicolet (1973), Crutzen (1971, 1974), Hesstvedt (1973), and others, to compute the components of the oxygen, nitrogen, hydrogen, and chlorine cycles and to estimate the probable effect of high-altitude air transport on the ozone. Let us mention, in particular, the two-dimensional diagnostic model worked out by Prinn *et al.* (1975) at the Massachusetts Institute of Technology (MIT), which includes 27 reactions with the participation of odd oxygen, nitrogen, and hydrogen. In this model equations (8.9) for all the long-lived components of the reactions cited were integrated on a grid with a latitude interval of 5° and a height interval of 2 km, taking the seasonal variation into account; estimates were made of the effect of a continuous emission of 1.8 Mtons/yr of NO_2 by high-flying aircraft (twice the amount forecast for 1985), corresponding to a 16% reduction of the amount of ozone. Mention should also be made of the three-dimensional diagnostic model of Hunt and Manabe (1968).

A more complete theory will have to include a concurrent determination of the concentrations of ozone and of the gases affecting it, from equations (8.9), and of the velocities of large-scale movements and the temperature field (which influences the ozone chemistry in that an increase in T reduces $[O_3]$, which leads in turn to an increase in the rates of damping of temperature waves), from the equations of hydrodynamics of the atmosphere, taking feedback into account (radiative heating due to the absorption of radiation by ozone).

An example of this is the three-dimensional quasi-geostrophic MIT model (1975), with 26 levels up to a height of 72 km. This model includes the main reactions with odd oxygen, nitrogen, and hydrogen (but with the specified fields of NO_2, HO, and HO_2). The prognostic equations are the equation for the vorticity, the linearized equation of thermodynamics, and the global mean and linearized equations for $[O_3]$; the diagnostic equations are the balance equation, the equation of hydrostatics, and the continuity equation. The heating due to the absorption of ultraviolet radiation is calculated accurately, while that due to the absorption of infrared radiation is found in the Newtonian approximation; the small-scale vertical diffusion in the lower stratosphere is assumed to be slight. The predicted fields, the tropospheric nonadiabatic heating, and the topography of the land are described spectrally in terms of six harmonics with respect to latitude and longitude, while the chemical reactions and heating in the stratosphere are calculated on a space grid. The model represents satisfactorily the factual data of Figs. 8.1 and 8.2. The mean meridional circulation in the stratosphere was obtained with ascending movements in the equatorial and polar zones and descending movements at latitudes from 30 to 40° (in the winter hemisphere). The large-scale turbulent meridional flux of ozone was found to have the opposite sign and to be much greater in size, there being an appreciable correlation between the ozone fluctuations in the lower stratosphere and in the entire air column, between the fluctuations of ozone and temperature, and between the downward and poleward ozone fluxes (which is apparently typical of linear baroclinic waves in the model with a stably stratified stratosphere over a less stable troposhere).

Schlesinger and Mintz (1979) constructed a similar model with a specified NO_2 field, without taking the hydrogen cycle into account and with a considerably more complete description of the atmospheric circulation; the absorption and emission of infrared radiation in the troposphere were calculated accurately, rather than in the Newtonian approximation, and the water-vapor field, cloud cover, and hydrologic cycle were described. Detailed calculations were carried out for January conditions according to this model. The model represents successfully the observed high correlation of ozone with tropospheric waves in the middle latitudes (correspondence of ozone maxima and minima, respectively, to wave troughs and crests), as well as, of course, the fact that the divergence of the ozone flux maintains its subequilibrium concentrations in the tropics and subtropics, while the convergence maintains the superequilibrium concentrations in the middle and high latitudes.

CHAPTER 4

THE GENERAL CIRCULATION OF THE ATMOSPHERE

9. The Equations of Dynamics of the Atmosphere

Large-scale processes in the atmosphere (synoptic and global processes) require an individual description. Since the large-scale movements are practically quasi-static and nondiverging, they can be described in an appropriate form by simplified equations of hydrodynamics (in which for the quasi-static approximation it is necessary to use the so-called "traditional approximation" for the Coriolis acceleration; in addition, since the heights $z = r - a$ are small in comparison with the distance r from the Earth's center, it is advisable everywhere to use the Earth's radius a instead of r). The equation of hydrostatics is

$$\partial p/\partial z = - g\varrho, \tag{9.1}$$

where g is the free-fall acceleration, and z is the height above the surface of the geoid (often referred to as "sea level"). Using the complement of the latitude $\theta = \pi/2 - \varphi$ and the longitude λ, and designating the velocity components as v_z, v_θ, and v_λ, we can write the continuity equation (the condition of nondivergence of the velocity field) as follows:

$$\frac{\partial v_z}{\partial z} + \frac{1}{a \sin \theta}\left(\frac{\partial v_\theta \sin \theta}{\partial \theta} + \frac{\partial v_\lambda}{\partial \lambda}\right) = 0. \tag{9.2}$$

The equations of horizontal motion will then have the form

$$\frac{\mathrm{d}v_\theta}{\mathrm{d}t} - \frac{\operatorname{ctg} \theta}{a} v_\lambda^2 - 2\Omega \cos \theta v_\lambda = - \frac{1}{a\varrho}\frac{\partial p}{\partial \theta} + f_\theta; \tag{9.3}$$

$$\frac{\mathrm{d}v_\lambda}{\mathrm{d}t} + \frac{\operatorname{ctg} \theta}{a} v_\theta v_\lambda + 2\Omega \cos \theta\, v_\theta = - \frac{1}{a\varrho \sin \theta}\frac{\partial p}{\partial \lambda} + f_\lambda, \tag{9.4}$$

where

$$\frac{\mathrm{d}}{\mathrm{d}t} = \frac{\partial}{\partial t} + v_z\frac{\partial}{\partial z} + \frac{v_\theta}{a}\frac{\partial}{\partial \theta} + \frac{v_\lambda}{a \sin \theta}\frac{\partial}{\partial \lambda}$$

is the operator of the individual derivative, Ω is the angular velocity of the Earth's rotation, and f_θ and f_λ are the components of the friction-force acceleration, expressed in terms of the tensor of the frictional stresses τ_{ij}:

$$\varrho f_\theta = \frac{\partial \tau_{\theta z}}{\partial z} + \frac{1}{a \sin \theta}\left(\frac{\partial \tau_{\theta\theta} \sin \theta}{\partial \theta} + \frac{\partial \tau_{\theta\lambda}}{\partial \lambda}\right) - \frac{\operatorname{ctg} \theta}{a} \tau_{\lambda\lambda}; \tag{9.3'}$$

$$\varrho f_\lambda = \frac{\partial \tau_{\lambda z}}{\partial z} + \frac{1}{a \sin \theta}\left(\frac{\partial \tau_{\lambda\theta} \sin \theta}{\partial \theta} + \frac{\partial \tau_{\lambda\lambda}}{\partial \lambda}\right) + \frac{\operatorname{ctg} \theta}{a} \tau_{\lambda\theta} \tag{9.4'}$$

(in the expressions for f_θ and f_λ the accelerations of the tide-producing forces can also be included).

For the velocity field described by equations (9.2)–(9.4) it is necessary to formulate the boundary conditions at the Earth's surface. Let us assume that the relief $z = h(\theta, \lambda)$ of the land surface is specified (its variations with time during epochs of growth or melting of the continental ice sheets have to be calculated separately). Then we have to require that all three velocity components go to zero: $v_z = v_\theta = v_\lambda = 0$ (conditions at the surface of drifting sea ice should be specified somewhat more accurately). At the free surface of the ocean it is necessary, as usual, to require continuity of the fluxes of mass, momentum, and energy. When considering large-scale atmospheric processes, we can assume that this surface is horizontal ($z = 0$), that the vertical velocity there goes to zero ($v_z = 0$), and that the horizontal components of the velocity, pressure, and horizontal stresses $\tau_{\theta z}$ and $\tau_{\lambda z}$ are continuous. However, equations of motion (9.3) and (9.4) are usually integrated directly only in the region above the atmospheric boundary layer, where the vertical exchange of momentum can often be neglected (that is, terms $(1/\varrho)(\partial\tau_{\theta z}/\partial z)$ and $(1/\varrho)(\partial\tau_{\lambda z}/\partial z)$ in the expressions for f_θ and f_λ are dropped); the atmospheric boundary layer itself is "parametrized" (see below).

The four equations (9.1)–(9.4) contain five unknown functions v_z, v_θ, v_λ, p, and ϱ. The equation of state for moist air,

$$p = [R_\mathrm{d} + (R_\mathrm{v} - R_\mathrm{d})\, q]\varrho T, \tag{9.5}$$

can serve as the fifth equation, where $R_\mathrm{d} = 0.287$ J/(g K) and $R_\mathrm{v} = 0.461$ J/(g K) are the gas constants of dry air and water vapor (the specific humidity q does not usually exceed 3 or 4%, while the content of liquid water and ice in clouds is usually much lower than the vapor content and can be neglected in (9.5)). This equation contains two new unknown functions T and q. To determine these, we have to invoke the equations describing the evolution of the entropy η and the specific humidity q:

$$\frac{\mathrm{d}\eta}{\mathrm{d}t} = \frac{\varepsilon + \mathscr{L}m}{T} - \left[\frac{\partial Q_z}{\partial z} + \frac{1}{a \sin \theta}\left(\frac{\partial Q_\theta \sin \theta}{\partial \theta} + \frac{\partial Q_\lambda}{\partial \lambda}\right)\right]; \tag{9.6}$$

$$\frac{\mathrm{d}q}{\mathrm{d}t} = -m - \left[\frac{\partial I_z}{\partial z} + \frac{1}{a \sin \theta}\left(\frac{\partial I_\theta \sin \theta}{\partial \theta} + \frac{\partial I_\lambda}{\partial \lambda}\right)\right]. \tag{9.7}$$

where ε is the rate of *radiative* influx of heat to a unit mass of air (the calculation of which must be considered separately, see §5) and also of dissipation of kinetic energy; m is the specific rate of condensation (or sublimation) of water vapor in clouds (if $m < 0$, then this is the rate of evaporation or sublimation); $\mathscr{L}$ is the latent heat of condensation or sublimation (for condensation $\mathscr{L} = \mathscr{L}_0 - (c_w - c_{pv})(T - T_0)$, where $\mathscr{L}_0 = 2502.5$ J/g, $c_w = 4.19$ J/(g K) is the specific heat of water, $c_{pv} = 1.81$ J/(g K) is the specific heat of water vapor at constant pressure, and $T_0 = 0$°C; for sublimation $\mathscr{L} = 2836.3$ J/g); and Q_i and I_i are the components of the small-scale fluxes of entropy and specific humidity.

The formula for the entropy should be introduced along with equation (9.6). Outside the clouds (for $m = 0$):

$$\begin{aligned} \eta &= (1 - q)\,\eta_d + q\eta_v; \\ \eta_d &= c_p \ln T - R_d \ln p_d + \text{const}; \\ \eta_v &= c_{pv} \ln T - R_v \ln p_v + \text{const}, \end{aligned} \qquad (9.8)$$

where η_d and η_v are the entropies of dry air and water vapor, and p_d and p_v are their partial pressures ($c_p = 1.003$ J/g is the specific heat of dry air at constant pressure). Pressure p_v is related to q as follows: $q = \alpha_v p_v\,[p - (1 - \alpha_v)\,p_v]^{-1}$, where $\alpha_v \approx 0.622$ is the ratio of the molecular masses of water vapor and dry air. Inside the clouds, on the other hand, the air is saturated with moisture, so that $p_v = e_m(T)$, this being the pressure of saturated water vapor, given by the Clausius–Clapeyron equation

$$T^2 \frac{d \ln e_m}{dT} = \frac{\mathscr{L}}{R_v},$$

whereas $q = q_m(T, p)$ and function m can be determined from equations (9.6) and (9.7). A more complete description of clouds can be obtained by introducing, for example, the equation for the specific water content q_w (that is, the concentration of water droplets and ice crystals):

$$\frac{dq_w}{dt} = m - n - \left[\frac{\partial I_{wz}}{\partial z} + \frac{1}{a \sin\theta}\left(\frac{\partial I_{w\theta} \sin\theta}{\partial\theta} + \frac{\partial I_{w\lambda}}{\partial\lambda}\right)\right], \qquad (9.9)$$

where I_{wi} are the components of the moisture flows, and $n = (1/\varrho)(\partial Q_w/\partial z)$ is the rate of fall of precipitation (here $Q_w = -\varrho\, q_w \bar{w}$ is the density of the vertical mass flow of water droplets and ice crystals, while $\bar{w}$ is the mean rate of fall, weighted by applying the factors $r^3 f(r)$, where $f(r)$, the probability density for cloud-particle radii r, is a function primarily of q_w).

For a number of problems it is important to know the form of the precipitation which reaches the Earth's surface, that is, whether it is solid or liquid. To do this, we must either break up the water content into the liquid-water concentration q_{w1} and the concentration of snow and ice q_{w2} and consider the two forms of precipitation n_1 and n_2, or else parametrize the precipitation types (first of all, in terms of the air temperature).

The equation of the heat budget can serve as the boundary condition at the Earth's surface for equation (9.6). We can write the heat-budget equation as

$$H_s = (1-A)\, F_s^{\downarrow} + F_L^{\downarrow} - B\,(T_s) - H - \mathscr{L}E + H_p + H_{wi}, \qquad (9.10)$$

where H_s is the heat flux into the ocean or into the active layer of the land; $F_s^{\downarrow}$ is the flux of short-wave solar radiation (direct and scattered) incident upon the surface; A is the effective short-wave albedo of the Earth's surface; $F_L^{\downarrow}$ is the flux of long-wave radiation of the atmosphere incident upon the surface; $B(T_s) = f\sigma\, T_s^4$ is the long-wave self-radiation of the Earth's surface at temperature T_s (here $\sigma = 5.669 \times 10^{-8}$ W/(m^2 K^4) is the Stefan-Boltzmann constant, and f is a multiplying factor indicating the difference between the Earth's surface and a perfect black body; the difference $B - F_L^{\downarrow}$ is called the *effective radiation* of the Earth's surface); H is the turbulent heat flux in the atmosphere at the Earth's surface (positive upward); $E = \varrho I_z$ is the evaporation rate; $H_p = c_w P(T_p - T_s)$ is the heat flux brought in by the precipitation (P is the rate of the precipitation and T_p is its temperature); and H_{wi} is the latent heat of the freezing of water or the melting of ice or snow (positive for freezing).

The formula for the saturation moisture content $q = q_m(T_s, p_s)$ at temperature T_s can serve as the boundary condition at the free surface of the ocean for equation (9.7). At the land surface, the following heat-budget equation, analogous to (9.10), can be taken as the limiting condition:

$$E_s = P - E - E_{wi} - I_s, \qquad (9.11)$$

where E_s is the source of the surface runoff; P is the rate of precipitation; E is the evaporation (evapotranspiration) rate; E_{wi} is the water flow produced by the freezing of water or, if $E_{wi} < 0$, by the melting of ice or snow ($E_{wi} = H_{wi}/\mathscr{L}_{wi}$, where $\mathscr{L}_{wi} = 334$ J/g is the latent heat of fusion of ice); and I_s is the flow of moisture into the soil.

Like the equations of motion, thermodynamic equations (9.6) and (9.7) are usually integrated only in the region above the atmospheric boundary layer. The "parametrization" of the atmospheric boundary layer is here based on the assumption that the vertical structure of this layer is determined wholly by the following: the force G and direction of the wind in the free atmosphere; the vertical differences of the potential temperature $\Theta_h - T_s$ and of the specific humidity $q_h - q_s$ across the atmospheric boundary layer; the Coriolis parameter $f = 2\Omega \cos\theta$; the buoyancy parameter $\beta = g/T_0$ (where T_0 is the mean temperature of the atmospheric boundary layer); and the roughness parameter of the underlying surface z_0. For the free surface of the ocean in the presence of a strong wind, the latter parameter is given by Charnock's formula $z_0 = 0.04\, u_*^2/g$, where u_* is the "friction velocity" in the layer of air adjacent to the water.

We thus arrive at the following formulas for the frictional stress τ, the turbulent heat flux H, and the evaporation rate E at the Earth's surface:

$$\tau = c_\tau \varrho G^2; \quad H = -c_H c_P \varrho G(\Theta_h - T_s); \qquad E = -c_E \varrho G(q_h - q_s), \tag{9.12}$$

where the coefficients of resistance c_τ, heat exchange c_H, and evaporation c_E, as well as the angle α between the direction of the wind at the upper boundary of the atmospheric boundary layer and the direction of the frictional stress at the Earth's surface, are functions of the Rossby number Ro $= G/fz_0$ (over the land) or the Charnock number Ch $= g/fG$ (over the sea), and also of the Richardson number Ri, defined as

$$\mathrm{Ri} = \left[\beta(\Theta_h - T_s) + \frac{R_v - R_d}{R_d} g(q_h - q_s)\right](fG)^{-1} \tag{9.13}$$

(strictly speaking, over the sea T_s and q_s should be replaced here by $T_s + \delta\Theta_0$ and $q_s + \delta q_0$, where $\delta\Theta_0$ and δq_0 are the jumps in the temperature and humidity values at the surface, proportional to $T_* = -H/c_p\varrho \; 1/\varkappa u_*$ and $q_* = -E/\varrho \; 1/\varkappa u_*$, respectively, where $\varkappa \approx 0.4$ is the Kármán constant). Functions $c_\tau = \Phi_1$, $\alpha = \Phi_2$, $c_H = c_E = \Phi_3$ over the land and $c_\tau = \Phi_4$, $\alpha = \Phi_5$, $c_H = c_E = \Phi_6$ over the sea can be approximated by the formulas:

$$\Phi_k = a_{kl} P_k^{a_{k2}} + a_{k3} P_k^{a_{k4}} \mathrm{Ri}^{a_{k5} P_k^{a_{k6}}} \tag{9.14}$$

(where $k = 2.5$ and $\mathrm{Ri} > 0$; $k = 1, 3, 4, 6$, and $\mathrm{Ri} \leqslant 0$);

$$\Phi_k = a_{kl} P_k^{a_{k2}} \exp(a_{k7} P_k^{a_{k8}} \mid \mathrm{Ri} \mid^{1/2})$$

(for $k = 2.5$ and $\mathrm{Ri} > 0$; $k = 1, 3, 4, 6$, and $\mathrm{Ri} \leqslant 0$); and

$$\Phi_k = a_{kl} P_l^{a_{k2}} \exp(a_{k7} P_l^{a_{k8}} \mid \mathrm{Ri} \mid^{1/2})$$

(for $k = 1, 3, 4, 6$ and $\mathrm{Ri} > 0$; $k = 2.5$ and $\mathrm{Ri} < 0$), where P_k = Ch for $k = 4, 5, 6$ and P_k = Ro for $k = 1, 2, 3$, while coefficients a_{kl} are given in Table 9.1.

10. The Dynamics of the General Circulation

The general circulation of the atmosphere is one of the component parts of the climate. Whereas a climate is defined as a statistical ensemble of states of the atmosphere-ocean-land system (see §2), *the general circulation of the atmosphere is a statistical ensemble of large-scale components of states of the atmosphere.* Large-scale components are here understood to mean those spatial inhomogeneities of the global fields describing states of the atmosphere which have scales much larger than the effective thickness of the atmosphere $H = P_s/g\varrho_s \sim 10$ km (Monin and Kolesnikova, 1965). Examples of these include global circulations, such as the zonal and monsoon circulations, and synoptic processes, such as Rossby waves, barotropic waves with typical horizontal scales of the order of $L_0 = \sqrt{gH}/f$ (Obukhov, 1946) and

TABLE 9.1
Coefficients a_{kl}

k	l = 1	2	3	4
1	0.0118	−0.14	0.795×10^{-2}	−0.34
2	1.09	−0.0605	0.135	−0.043
3	0.0176	−0.10	0.282×10^{-2}	−0.218
4	0.158×10^{-2}	−0.18	0.252×10^{-5}	−0.088
5	0.155	−0.066	0.0263	0.052
6	0.354×10^{-2}	−0.136	0.282×10^{-4}	−0.12

k	l = 5	6	7	8
1	0.975	-0.91×10^{-2}	−0.087	−0.05
2	0.306	0.01	−0.0148	0.055
3	1.05	−0.01	−0.135	−0.021
4	1.83	−0.0493	−0.066	−0.055
5	0.448	−0.0154	−0.087	−0.05
6	1.83	−0.0592	−0.098	−0.018

baroclinic waves with scales of the order of the radius of the Rossby deformation $L = NH/f$.

The atmospheric circulation exists because the heating of the atmosphere by solar radiation (direct or returned from the underlying surface) is nonuniform in space. If a warm column of air is compared with a cold air column, it is seen that, since warm air expands and its masses tend to rise, at a given height the pressure will be higher in the warm air column than in the cold column; this pressure difference causes air to move from a warm region to a cool region. Accordingly, the difference in temperature between the equator and the poles should cause air at high levels to move poleward from the equator, while at the same time there must be a compensating influx of air at low levels from the middle latitudes toward the equator, the trade winds. Similarly, the temperature differences between continents and oceans, which change sign from summer to winter, produce air movements at high levels from warm regions to cooler regions; in summer this movement is from the continents to the oceans and in winter from the oceans to the continents, the compensating opposite air flows at lower levels giving rise to the monsoons. At high levels the Coriolis force causes the poleward air flow from the equator to turn eastward, producing a westerly transport of air in the upper troposphere of the middle latitudes.

This explanation of the general circulation of the atmosphere was formulated as early as 1735 by Hadley (and later independently by the great philosopher Kant (1756) and by Dalton (1793)). Subsequently it was developed and made more complete (in particular, the additional "vortices" of the meridional circulation were introduced) by such scientists as Dove (1837), Maury (1855), Ferrel (1856, 1858), Thomson (1857), and Helmholtz (1888). However, this theory of the general circulation ultimately turned out to be incorrect. Statistical data on the atmospheric circulation were to show that the "vortices" of the meridional circulation are weak, and that their contribution to the observed meridional transfer of heat and momentum is insignificant in comparison with the total contribution of the synoptic eddies (cyclones and anticyclones).

Defant (1921) was the first to treat the synoptic eddies as large-scale turbulence, bringing about a horizontal transfer of heat from the tropics to the poles (with an effective exchange coefficient, according to his estimates, of the order of $\varrho v \sim 10^8$g/cm s). Jeffreys (1926) hypothesized that the synoptic eddies create the main meridional flow of momentum (proportional to the average product of the large-scale turbulent fluctuations of the velocity $\overline{v'_\theta \, v'_\lambda}$). V. Bjerknes (1937) established that the zonal circulation is unstable *vis-à-vis* small nonzonal perturbations, which can grow and turn into synoptic eddies. Rossby (1941, 1947) proposed that the overall effect of the synoptic eddies is to transfer momentum to latitude zones with the strongest westerlies, the jet streams, that is, to act as a negative viscosity and thus to supply kinetic energy to the mean zonal circulation. This idea, together with the explanation of the synoptic eddies as being due to a baroclinic instability of the mean zonal circulation (supplying it with potential energy), forms the basis of the modern theory of the general circulation of the atmosphere.

Due to the negative viscosity, which is created by the statistical ensemble of synoptic eddies, the so-called subtropical jet streams are formed in the cyclonic circumpolar currents or westerly transport of air in the upper troposphere of the middle latitudes. These jet streams, which are relatively narrow (their width between points where the velocity drops to half the maximum value is about 300–400 km and they are 1–2 km deep), are located at mean latitudes of about ± 35° and at heights of about 12 km (where the pressure is around 200 mb), and they have maximum velocities of 60 m/s or more.

The jet streams are baroclinically unstable. Slight disturbances arising in them can grow into Rossby waves with large latitudinal amplitudes (up to 400 km), having zonal wave numbers $k = 4$ to 8, and especially 5 to 6 (corresponding to wavelengths of about 4000 km). These waves travel eastward more slowly than does the air of the main flow (relative to which they have a westward motion with phase velocities of the order of 10 m/s).

Between alternate groups of four cyclonic and anticyclonic eddies, forming at troughs and crests of the Rossby waves, saddle regions or high-level deformation fields appear, with high-level frontal zones along their axes of contraction. Beneath the latter (in the lower troposphere and at the Earth's surface) collisions of warm and cold air masses which have been transformed (heated or cooled) by the underlying surface produce atmospheric fronts, which are thus consequences of the presence of the high-level eddies. Atmospheric fronts in the lower troposphere are, in turn, often baroclinically unstable (especially under the deltas of the high-level frontal zones). Wave disturbances arising at these fronts, as they grow, generate smaller (higher baroclinic modes) eddies at the surface (since the fronts are especially abrupt at the Earth's surface); these are cyclones and anticyclones. Sometimes the eddy formation is also facilitated by effects of the air flow around irregularities of the land relief.

Surface cyclonic eddies, once they have gone through the stage of a young cyclone with a warm sector, sometimes become occluded and penetrate into the upper troposphere, that is, they become "barotropized." The stationary (zero-vorticity) anticyclones which are final for the cyclonic series are also barotropized. The ensemble of these barotropized synoptic eddies apparently also acts as a negative viscosity on the jet streams (which are barotropically stable). Angular momentum is transferred to a jet stream from the tropics by wave disturbances, since the axes of their troughs and crests are inclined from the tropics toward the stream (here the angular momentum in front of the troughs, where it is high, is transferred to the stream along the meridian, while the angular momentum behind the troughs, where it is lower, is carried away from the stream.

One of the first to attempt a systematic application of the equations of hydrodynamics to describe the general circulation of the atmosphere was Kochin (1936). In contrast to Helmholtz, who neglected the role of viscosity in the motions of the general circulation, Kochin assumed that "in the case of the flow of the atmosphere, the viscosity forces may turn out to be great even inside the atmosphere, since the nonuniform temperature distribution can bring about large changes in the flow velocity upon transition from some point in the atmosphere to adjacent points" (here it is, of course, the eddy viscosity which is referred to, and in his examples Kochin assumed a value $\varrho\nu = 2 \times 10^4$ g/(cm s) for it). When the viscosity is taken into account, the equations describing the general circulation can be simplified with the aid of boundary-layer theory. However, it must be specified that "whereas in the usual theory of the boundary layer the volume forces of viscosity have the same order of magnitude as the inertia forces, in problems of the general circulation the inertia forces turn out to be negligible in comparison with the Coriolis force, so that in this case the volume viscosity forces must be put on an equal basis with the Coriolis force."

Kochin showed that an appropriate form of the simplified equations of hydrodynamics can be used to determine the three-dimensional fields of the wind-velocity vector and the air density, as well as the height field of the pressure, without using the equation of thermodynamics; here the surface-pressure field and the three-dimensional fields of temperature and viscosity are specified with the aid of observational data. These ideas were elaborated further by Blinova (1947), who added to Kochin's equations the equation of heat influx, in the form worked out by Kibel' (1943) and used by him to calculate thermal equilibrium in the atmosphere; see §5. Using this equation, Blinova computed the mean annual three-dimensional temperature field in the atmosphere and the field of the surface pressure, taking into account the presence of continents and oceans. This calculation included, in particular, the first theoretical derivation of the so-called centers of action of the atmosphere. Later Blinova also calculated the annual variation of the general circulation of the atmosphere.

The hydrodynamic theory of Kochin and Blinova made it possible to calculate the global atmospheric circulations (the zonal and monsoon circulations). However, the characteristics of the synoptic processes were not calculated in this theory; rather, the effects of these processes were "parametrized" by specifying the coefficients of turbulent heat conductivity (Blinova) and viscosity (Kochin). The possibility of a negative large-scale eddy viscosity was not yet allowed for. In the more complete theory the synoptic processes should not be parametrized, but rather described explicitly. This description can be statistical, whereby the methods of turbulence theory are used and the equations for the mean circulation are considered, together with the equations for the statistical moments of the synoptic fluctuations of the hydrodynamic fields (analogs of the Reynolds and Friedman–Keller equations). Alternatively, this description can be individual, use being made of the quite complete equations of hydrodynamics (9.1)–(9.7), after which the time-averaged statistical characteristics of the solutions obtained are found. The latter is a variant of the Monte Carlo method; in the theory of climate it is known as the method of numerical experiments with physicomathematical models of the general circulation. During the last two decades the second of these approaches has been applied particularly widely.

An important role in the dynamics of the general circulation of the atmosphere is played by the so-called adiabatic Lagrange invariants: hydrodynamic fields F satisfying equations of the form $dF/dt = \varepsilon_F$, the right-hand sides of which ε_F go to zero for adiabatic processes (so that then the amount of the substance having a density F does not vary with time in each fluid particle; hence the term *invariant*). The entropy itself $F = \eta$, which in general satisfies equation (9.6), is, of course, a trivial adiabatic invariant. Invariance of the entropy, $d\eta/dt = 0$, is the condition which, along with the condition of invariance of the specific humidity, $dq/dt = 0$, determines the adiabatic process. The invariant which is next in complexity is the potential vorticity $F = \Omega_*$ (Ertel, 1942); this is the scalar product of the absolute vorticity times the entropy gradient, divided by the density. There are more complicated invariants as well (in fact, there are an infinite number of them).

Invariants η and Ω_* have the following significance for the dynamics of atmospheric processes: the latter are divided into (1) slow (long-period) processes, namely global circulations and synoptic processes (Rossby waves), which are described by fields η and Ω_* and their evolution in time (the other hydrodynamic fields are "adapted" or "adjusted" to fields η and Ω_*, that is, they are defined in terms of them with the aid of *synchronous relations* which do not contain time derivatives), and (2) fast (short-period) processes, namely acoustic, surface, and internal gravitational waves and inertial-gyroscopic waves, in which adiabatic fluctuations of fields η and Ω_* are absent (at any rate, in the linear approximation).

Any function of the invariants will itself be an invariant. In particular, instead of the entropy η it is convenient to use an exponential function of it, $\Theta \sim \exp(\eta/c_p)$, known as the *potential temperature*. According to the second formula of (9.8), for dry air

$$d\eta = c_p \, d \ln \Theta = c_p \, d \ln \frac{p^{1/\varkappa}}{\varrho}; \quad \Theta = T \left(\frac{p_0}{p} \right)^{\frac{\varkappa - 1}{\varkappa}}, \tag{10.1}$$

where $p_0 = 1000$ mb is the standard atmospheric pressure at sea level, and $\varkappa = c_p/c_v \approx 0.2857$ is the ratio of the specific heats of air, at constant pressure and constant volume. The moisture content of air is always low (the specific humidity q does not usually exceed 0.03–0.04). Consequently, the entropy and potential temperature of unsaturated moist air can, in many calculations, be found using the dry-air formula (10.1). Calculations of the entropy fluctuations η' and the turbulent entropy flux Q_k are exceptions. In view of the smallness of q, from (9.8) the following approximations can be obtained:

$$\eta' \approx c_p T'/T + (\eta_v - \eta_d) \, q';$$

$$Q_k \approx \overline{\varrho \eta' \, v'}_k \approx \frac{1}{T} \left(c_p \varrho \overline{T' v'_k} + \frac{\eta_v - \eta_d}{c_p} \frac{c_p T}{\mathscr{L}} \mathscr{L} \varrho \overline{q' v'_k} \right), \tag{10.2}$$

where the second term in parentheses in the second equation (describing the contribution of the moisture flux to the entropy flux) is sometimes not small in comparison with the first term (describing the contribution of the heat flux). This must be kept in mind when the formation of the global entropy field in the atmosphere is being described quantitatively.

In the case of saturated air, which contains all the condensed moisture (moist-adiabatic processes), Θ must be replaced by the *equivalent potential temperature* Θ_e. For the approximate calculation it differs from (10.1) only in that T is replaced by the equivalent temperature $T_e = T + (\mathscr{L}/c_p)q$. If, on the other hand, the condensed moisture rapidly falls as precipitation (pseudoadiabatic processes), then the quantity $(\mathscr{L}/c_p) \, dq_m$ has to be added to the differential of the entropy, where q_m is the specific humidity at saturation. Then Θ is replaced by the *pseudopotential temperature*, which differs from (10.1) in that T is replaced by the pseudotemperature

$$T_p = T + \int_z^\infty \frac{\mathscr{L}}{c_p} \frac{\partial q_m}{\partial z} \, dz$$

(the dependence of c_p on z being taken into account).

Oort and Rasmusson (1971) used actual data for the years 1958–1963 to construct mean monthly, seasonal, and annual zonal fields of Θ in the troposphere. According to these data, Θ increases with height almost everywhere (that is, the atmosphere is hydrostatically stable; convective regions which sometimes appear in the atmosphere, especially in the tropics, constitute comparatively rare exceptions to this). In the lower troposphere Θ decreases from the equator to the pole (as an average over the year, at the 1000 mb level it drops from 299.5 K at the equator to 261.2 K at 75°N), whereas in the upper troposphere it increases (at the 50 mb level from 489.2 K at the equator to 522.7 K at 75°N); at a level just below 200 mb, $\Theta \approx 345$ K and it varies little with latitude. The mean pressure at isentropic surfaces $p_*(\Theta)$ is a useful quantity for a number of calculations as a function of η or of Θ. Table 10.1 shows some typical values of this pressure, according to Lorenz (1967). In particular, the coefficient α of the effectiveness of generation $\{\alpha T \varepsilon_\eta\}$(braces signify the integral over the mass over the entire atmosphere) of available potential energy (see §11) due to the heat influx $T\varepsilon_\eta$ is expressed in terms of this function. This coefficient is defined as

$$\alpha = 1 - \left[\frac{p_*(\Theta)}{p}\right]^{\frac{\varkappa - 1}{\varkappa}}$$

In the middle troposphere α varies from $+0.1$ at the equator to -0.25 at the pole, whereas in the upper troposphere it varies from -0.05 at the equator to $+0.05$ at the pole.

TABLE 10.1

Mean pressure at isentropic surfaces

ΘK	260	280	300	320	340	360	380	400
$p_*(\Theta)$mb	920	930	750	430	250	160	120	100

The adiabatic invariance of the potential vorticity Ω_*, as well as the equation describing its evolution for nonadiabatic processes, are derived with the aid of the hydrodynamic equation

$$\varrho \frac{\mathrm{d}}{\mathrm{d}t} \frac{\mathbf{A} \cdot \nabla\eta}{\varrho} = \nabla\eta \cdot \text{helm}\,\mathbf{A} + \mathbf{A} \cdot \nabla\varepsilon_\eta; \tag{10.3}$$

$$\text{helm}\,\mathbf{A} = \frac{\mathrm{d}\mathbf{A}}{\mathrm{d}t} - (\mathbf{A} \cdot \nabla)\,\mathbf{u} + \mathbf{A}\,\text{div}\,\mathbf{u},$$

where helm denotes the tensor (transforming vectors into vectors) linear differential operator known as the *Helmholtzian*. The condition helm $\mathbf{A} = 0$ indicates hydrodynamic conservation of the vector lines of field $\mathbf{A}$, that is, a "freezing" of the field into the moving fluid. It should be noted that the derivative with respect to time of only the component $\mathbf{e} \cdot \mathbf{A}$ of vector field $\mathbf{A}$ which is *longitudinal* relative to the *thermodynamic vertical* appears in (10.3). This is the direction $\mathbf{e} = \nabla\eta/|\nabla\eta|$ of the entropy gradient. Let us assume that field $\mathbf{A}$ in (10.3) is the absolute vorticity $\Omega_a = \text{curl } \mathbf{v} + 2\Omega$, which satisfies the equation helm $\Omega_a = \varrho^{-2}\nabla\varrho \times \nabla\varrho + \text{curl } \mathbf{f}$, where $\mathbf{f}$ is the acceleration of the viscosity forces, and let us take into account that for dry air $\eta = \eta(p, \varrho)$, so that $\nabla\eta(\nabla\varrho \times \nabla p) = 0$. Then, from (10.3) we get the equation describing the evolution of the potential vorticity $\Omega_* = \varrho^{-1}\Omega_a \cdot \nabla\eta$:

$$\varrho \frac{d\Omega_*}{dt} = \nabla\eta \cdot \text{curl } \mathbf{f} + \Omega_a \cdot \Delta\varepsilon_\eta = \text{div } (\varepsilon_\eta\Omega_a + \eta \text{ curl } \mathbf{f}) \quad (10.4)$$

(Obukhov, 1962). This equation shows that Ω_* is an invariant not only of adiabatic processes (for which $\mathbf{f} = \varepsilon_\eta = 0$) but also of those nonadiabatic processes for which the vector field $\varepsilon_\eta\Omega_a + \eta$ curl $\mathbf{f}$ is solenoidal (nondivergent).

If the quasi-static approximation (9.1)–(9.4) is used, equation (10.4) remains valid, provided in it curl $\mathbf{v} + 2\Omega$ and $\mathbf{f}$ are replaced by curl $v_h + 2\Omega_z$ and $\mathbf{f}_h + \mathbf{g}$, where subscripts h and z denote, respectively, the horizontal and vertical *vector* components of the vectors.

For unsaturated moist air $\eta = \eta\ (p, \varrho, q)$, and since $\Omega_* \approx \varrho^{-1}\Omega_{az}$ $d\eta/d\Theta\ \partial\Theta/\partial z$ the potential anchor, strictly speaking, stops being an adiabatic invariant. This difficulty can be ameliorated if η is replaced by the *pseudoentropy* $\eta_0 = \eta\ (p, \varrho, q_0)$, where q_0 is some standard specific humidity. The pseudopotential vorticity $\Omega_{*0} = \varrho^{-1}\Omega_a \cdot \nabla\eta_0$ turns out to be an adiabatic invariant of the solenoidal motions (Monin, 1974).

For adiabatic processes any two independent functions of η and Ω_* can serve as Lagrange coordinates of the air particles. The corresponding coordinate surfaces will form tubes, along which air flows without passing through the tube walls. For adiabatic processes only deformations of these tubes will occur, without disturbing their integrity; for nonadiabatic processes, on the other hand, the tube walls begin to "leak", and air can flow through them into adjacent tubes.

It is convenient to take the potential temperature Θ as one of the Lagrange coordinates. For large horizontal scales it can serve as the vertical coordinate, since the Θ = const surfaces are approximately horizontal. Thus, the vector $\nabla\Theta$ is approximately vertical, so that $\Omega_* \approx \varrho^{-1}\Omega_{az}\ d\eta/d\Theta\ \partial\Theta/\partial z$. Then, setting $\varrho = p/RT$ and $\partial\Theta/\partial z = (\Theta/T)\ (\gamma_a - \gamma)$, where γ_a and γ are the adiabatic and real vertical temperature gradients (lapse rates), we arrive at the second Lagrange coordinate

$$\tilde{\Omega} = \frac{\Omega_*}{c_p R}\ \frac{p_*(\Theta)}{\gamma_a - \gamma(\Theta)} \approx \Omega_{az}\ \frac{p_*(\Theta)}{p}\ \frac{\gamma_a - \gamma}{\gamma_a - \gamma_*(\Theta)}, \quad (10.5)$$

where $\gamma_*(\Theta)$ is the mean vertical temperature gradient at the isentropic surfaces. For large horizontal scales outside the tropical zone Ω_{az}

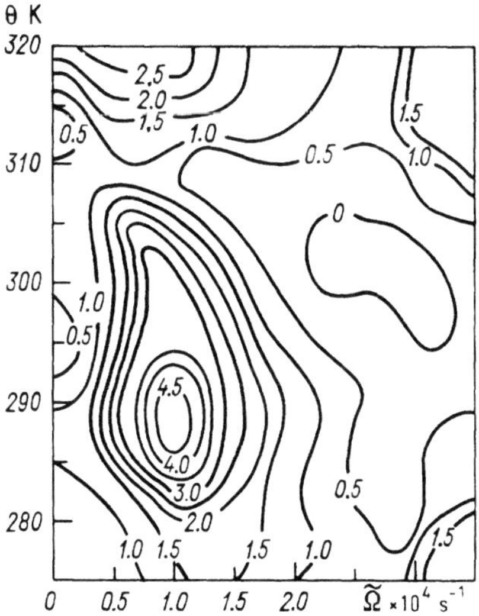

Fig. 10.1. Probability density μ (Θ, $\tilde{\Omega}$) for the period 1 to 10 April 1962, according to Karunin.

$\approx 2\Omega_z = 2\Omega \cos \Theta$, so that $\tilde{\Omega}$ can take the place of the latitude, and the most graphic portrayal of the (Θ, $\tilde{\Omega}$) tubes will ascribe to them meridional sections (for a zonal circulation these are independent of the longitude).

If $\mu (\Theta, \tilde{\Omega})\, d\Theta\, d\tilde{\Omega}$ is the fraction of the mass of the atmosphere contained in an infinitely thin (Θ, $\tilde{\Omega}$) tube, then $\mu (\Theta, \tilde{\Omega})$ will take on the meaning of the probability density for the Lagrange coordinates Θ, $\tilde{\Omega}$ of a randomly selected air particle. According to existing data, this probability is concentrated mainly in the region 280 K $< \Theta <$ 305 K; 0.5×10^{-4} s^{-1} $< \tilde{\Omega} <$ 1.5×10^{-4} s^{-1}, with a maximum located near the point Θ = 290 K, $\tilde{\Omega} = 1.0 \times 10^{-4}$ s^{-1} and a minimum in the region 297 K $< \Theta <$ 305 K; 2.2×10^{-4} s^{-1} $< \tilde{\Omega} <$ 3.2×10^{-4} s^{-1} (see example in Fig. 10.1). Function $\mu (\Theta, \tilde{\Omega})$ is one of the important characteristics of the state of the general circulation of the atmosphere. A zonal circulation having the same function $\mu (\Theta, \tilde{\Omega})$ as the actual circulation of the atmosphere can be called an equivalent zonal model of the atmosphere, EZMA for short (Obukhov).

Next we set $d/dt = (d_h/dt) + v_z (\partial/\partial z)$ and we use this relation to eliminate v_z from the pair of equations $d\eta/dt = \varepsilon_\eta$ and $d\Omega_*/dt = \varepsilon_{\Omega_*}$. Then, taking into account that the large-scale processes are both quasi-horizontal and quasi-solenoidal, so that $\Omega_* \approx \varrho^{-1}\Omega_{az}\frac{\partial\eta}{\partial z}$ and div $\mathbf{v} \approx 0$, we get

$$\frac{d_h \ln \Omega_{az}}{dt} + \left(\frac{\partial\eta}{\partial z}\right)^{-1} \frac{d_h}{dt}\frac{\partial\eta'}{\partial z} - \left(\frac{\partial\eta}{\partial z}\right)^{-2}\left(\frac{\partial^2\eta}{\partial z^2} + \frac{\partial\eta}{\partial z}\frac{\partial \ln \Omega_{az}}{\partial z}\right)\frac{d_h\eta'}{dt} = \varepsilon_1, \tag{10.6}$$

where ε_1 is a linear combination of the quantities ε_η and $\varepsilon_{\Omega *}$, and a prime denotes the part $\eta' = \eta - \eta_0$ depending on the horizontal coordinates and time. It is small compared with the steady-state values η_0, so that in the coefficients of $\mathrm{d_h}\,(\partial\eta'/\partial z)/\mathrm{d}t$ and $\mathrm{d_h}\eta'/\mathrm{d}t$ in (10.6) η can be replaced by η_0; moreover, the term $(\partial\eta_0/\partial z)\,\partial \ln \Omega_{az}/\partial z$ can be neglected in comparison with $\partial^2\eta_0/\partial z^2$. Now (10.6) becomes

$$\frac{\mathrm{d_h}}{\mathrm{d}t}\left(\ln \Omega_{az} + \frac{\partial}{\partial z}\frac{\eta'}{\partial\eta_0/\partial z}\right) = \varepsilon_1 \tag{10.6'}$$

(Gavrilin and Monin, 1969). The large-scale processes are also quasi-hydrostatic, so that

$$\frac{\partial\eta_0}{\partial z} = \left(\frac{\partial\eta}{\partial p}\right)_0\left(\frac{\partial p_0}{\partial z} - c_0^2\frac{\partial\varrho_0}{\partial z}\right) = \left(\frac{\partial\eta}{\partial p}\right)_0\frac{\varrho_0 c_0^2 N^2}{g},$$

where p_0 and ϱ_0 are the steady-state parts of the pressure and density, while c_0 and N are the velocity of sound and the Väisälä–Brunt frequency. Thus we obtain the similar expression

$$\eta' = \left(\frac{\partial\eta}{\partial p}\right)_0(p' - c_0^2\varrho') \approx -\left(\frac{\partial\eta}{\partial p}\right)_0 c_0^2\varrho'$$

$$= \left(\frac{\partial\eta}{\partial p}\right)_0(c_0^2/g)\frac{\partial p'}{\partial z}.$$

so that

$$(\partial\eta_0/\partial z)^{-1}\eta' \approx (\varrho_0 N^2)^{-1}\partial p'/\partial z,$$

and (10.6') becomes

$$\frac{\mathrm{d_h}}{\mathrm{d}t}\left(\ln \Omega_{az} + \frac{\partial}{\partial z}\frac{1}{\varrho_0 N^2}\frac{\partial p'}{\partial z}\right) = \varepsilon_1. \tag{10.6''}$$

If we introduce the stream function of the horizontal flow ψ, so that $\Omega_{az} = \Delta_h\Psi + f$, and if we take into account, finally, that the large-scale processes in the middle and high latitudes are quasi-geostrophic, so that $|\Delta_h\psi| \ll |f|$ and $p' \approx \varrho_0 f\psi$, then (10.6'') can be reduced to the approximate form

$$\frac{\mathrm{d_h}\omega_*}{\mathrm{d}t} = f\varepsilon_1; \qquad \omega_* = \mathscr{L}\psi + f; \tag{10.7}$$

$$\mathscr{L} = \Delta_h + \frac{\partial}{\partial z}\frac{f^2}{N^2}\frac{\partial}{\partial z}. \tag{10.7'}$$

The quantity ω_* can be called the potential vorticity in the quasi-geostrophic approximation. It is clear from (10.7) that ω_* satisfies the same equation as the vorticity $\omega = \text{curl}_z v_h = \Delta_h \psi$ in *two-dimensional* flows of an incompressible liquid. In such flows, in contrast to three-dimensional flows, there will be no intensification of the vorticity field due to stretching of the vortex lines. Actually, to be more definite, on the right-hand side of equation (10.7) let us take into account only the effect of viscosity, that is, let us set $f\ \varepsilon_1 = \nu \Delta_h \omega_*$. Then, for the mean kinetic energy $E = \frac{1}{2}\langle |\mathbf{v}|^2 \rangle$ and the *enstrophy* (which in this case can be called the *potential enstrophy*), that is, the mean-square vorticity $E_1 = \frac{1}{2}\langle \omega_*^2 \rangle$, we have

$$\partial E/\partial t = -2\nu E_1 \equiv -\varepsilon, \tag{10.8}$$

$$\partial E_1/\partial t = -\nu \langle |\nabla_h \omega_*|^2 \rangle \equiv -\varepsilon_1, \tag{10.8'}$$

from which it is evident that for any finite initial values of E and E_1 the enstrophy E_1 can only decrease with time.

Moreover, as $\nu \to 0$ we have $\varepsilon \to 0$, so that for high Reynolds numbers the kinetic energy E will be approximately constant in time and a spectral transfer of any sizable part of the energy from low wave numbers to high wave numbers will be impossible. At the same time, the spectral transfer of enstrophy (squared vorticity) from low to high wave numbers is possible, together with a corresponding nonzero limit of the quantity ε_1 as $\nu \to 0$. Let us define the mean values of the wave numbers over the energy and enstrophy, $\bar{k}$ (t) and $\bar{k}_1$ (t), respectively, as follows:

$$\bar{k}(t) = E^{-1} \int_0^\infty kE(k,t)\,\mathrm{d}k; \qquad \bar{k}_1(t) = E_1^{-1} \int_0^\infty k^3 E(k,t)\,\mathrm{d}k, \tag{10.9}$$

where $E(k,t)$ is the spectral energy density. Then, by applying the Cauchy–Bunyakovskii–Schwartz inequality to the integrals

$$E\bar{k} = \int_0^\infty kE(k)\,\mathrm{d}k = \int_0^\infty [kE^{1/2}(k)]\,[E^{1/2}(k)]\,\mathrm{d}k,$$

$$E_1 = \int_0^\infty k^2E(k)\,\mathrm{d}k = \int_0^\infty [k^{1/2}E^{1/2}(k)]\,[k^{3/2}E^{1/2}(k)]\,\mathrm{d}k,$$

we arrive at the following inequalities:

$$\bar{k}(t)\,\bar{k}_1(t) \geqslant k_*^2 \equiv E_1/E; \qquad \bar{k}(t) \leqslant k_*, \quad \bar{k}_1(t) \geqslant k_* \tag{10.10}$$

(Novikov, 1975). Inspection of these expressions shows that a decrease in $\bar{k}$ with time, that is, a spectral transfer of energy toward low k (a negative eddy viscosity), will be accompanied by an increase in $\bar{k}_1$, that is, by a spectral transfer of enstrophy toward high k.

At high Reynolds numbers in the inertial interval of the wave numbers k, the spectral energy density $E(k)$ must be defined, generally speaking, in terms of two parameters ε and ε_1. In this case there will be a constant length scale L_1

$= (\varepsilon/\varepsilon_1)^{1/2}$, and from dimensional arguments, the energy spectrum will have the form

$$E(k) = \varepsilon^{2/3} k^{-5/3}\, C(L_1 k), \tag{10.11}$$

where $C(\xi)$ is some function of the dimensionless wave number $\xi = L_1 k$. Let us now assume that in the vicinity of some wave number $\tilde{k}$ there is an influx of energy from outside at a rate ε and an influx of enstrophy at a rate ε_1. Then it can be shown, firstly, that the bulk of the incoming enstrophy will be transferred toward high k (and for $k \gg \tilde{k}$ it will be dissipated as a result of the viscosity), and that there will be no spectral transfer of energy in this case, so that parameter ε will be negligible and drops out of relation (10.11). Thus we have $C(\xi) \sim \xi^{-1/3}$, and (10.11) turns into an "inverse cube law":

$$E(k) = C_1 \varepsilon_1^{2/3} k^{-3}. \tag{10.12}$$

Secondly, it can be demonstrated that the bulk of the incoming energy in this case will be transferred toward low k (and will supply the regular global currents of the general circulation), whereas there will be no spectral transfer of enstrophy. Consequently, here parameter ε_1 will be negligible and drops out of relation (10.11). Thus we have $C(\xi) \sim$ const, and (10.11) becomes an "inverse 5/3 law":

$$E(k) = C\varepsilon^{2/3} k^{-5/3}, \tag{10.13}$$

already encountered in the theory of three-dimensional turbulence (where, however, the spectral energy flux ε has the opposite sign, from low k to high k). Inverse cube law (10.12) was first discovered by Ogura (1952) on the basis of spectral data on large-scale meteorological fields. It was first attributed to a spectral transfer of enstrophy (squared vorticity) by Kraichnan (1967) and Batchelor (1969). Manabe *et al.* (1970) discovered an inverse cube law in the interval of zonal wavenumbers $k = 2\pi a \cos\theta/l$ from 8 to 20 (where l is the wavelength), in kinetic-energy spectra of hydrodynamic fields, obtained by numerical integration of the equations of a physicomathematical model of the general circulation. This result was explained with the aid of an equation for the potential vorticity of type (10.7) by Charney (1971) and Gavrilin *et al.* (1972). Figure 10.2 shows a one-dimensional spectrum of the wind velocity $E_1(k)$ which was taken from the latter source. This spectrum, based on data of five studies by various investigators, demonstrates clearly the existence in the region $k > \tilde{k} \approx 8$ of an interval with an inverse cube law.

Now let us rewrite the equation describing the evolution of the quasi-geostrophic potential vorticity (10.7) in the approximation of the so-called β plane (x,y), in which the Coriolis parameter has the form $f = f_0 + \beta y$, where f_0 and β are constants, $\beta = \mathrm{d}f/\mathrm{d}y$ being the derivative with respect to the arc y of the meridian. If we take the small-scale eddy viscosity into account explicitly by introducing a coefficient ν (a term $\nu\Delta\Delta\psi$ on the right-hand side $f\varepsilon_1$ of (10.7)), and if we take the surface friction into account by introducing a term

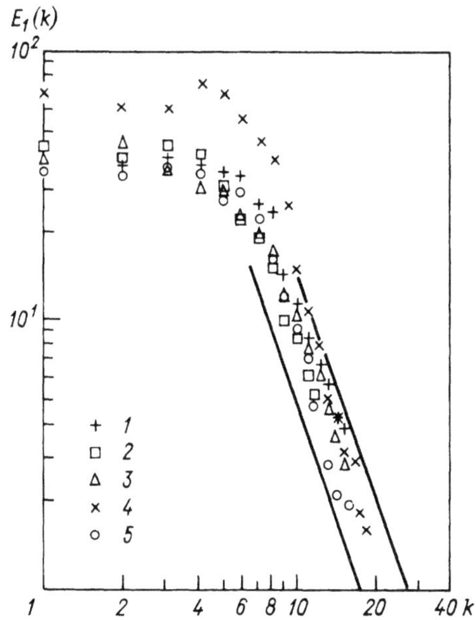

Fig. 10.2. One-dimensional spectra of wind velocity, according to data of various investigators. (1) Saltzman and Fleischer (1962), (2) Horn and Bryson (1963), (3) Kao and Wendell (1970), (4) Julian *et al.* (1970), (5) Wiin-Nielsen (1967).

$-\tau^{-1}\Delta\psi$, then we have

$$\frac{\partial \mathcal{L}\psi}{\partial t} + \frac{\partial\,(\psi, \mathcal{L}\psi)}{\partial\,(x, y)} + \beta\frac{\partial\psi}{\partial x} = \nu\Delta\Delta\psi - \tau^{-1}\Delta\psi + \varepsilon_2. \quad (10.14)$$

The second term on the left-hand side of this equation describes nonlinearly interacting eddies, the totality of which can be considered a kind of quasi-two-dimensional turbulence. The third term describes the Rossby waves. Let us assume that at the initial moment we have a random field ψ in the form of a system of tightly packed eddies with a narrow-band wave-number spectrum having a maximum at a wave number k_0. Then the relative roles of the turbulence and the Rossby waves can be gauged in terms of the ratio δ of typical values of the second term $2k_0^2U^2$ and the third term βU on the left-hand side of equation (10.14), where $U = (2E)^{1/2}$ is the mean-square velocity of motion of the fluid. This ratio turns out to be $\delta = 2k_0^2U/\beta = U/c_0$, where $c_0 = \beta/2k_0^2$ is the phase velocity of the Rossby waves with a wave number k_0. For $\delta > 1$ in field ψ turbulence will predominate, and for $\delta < 1$ Rossby waves will prevail. At the boundary $\delta = 1$ the wave number is

$$k_\beta = (\beta/2U)^{1/2}. \quad (10.15)$$

Under typical atmospheric conditions $k_\beta^{-1} \approx 1000$ km, which at a latitude of 30° corresponds to an eddy diameter $2\pi k_\beta^{-1}\cos\theta = 3100$ km. The corresponding zonal wave number for $\theta = 45°$ and $U = 15$ m/s is 3 and for $\theta = 60°$ and

U = 10 m/s it is 5. Next let us introduce a typical value $\hat{U}(k)$ of the Fourier components of a velocity field with a wave number k, with the aid of the relation $\frac{1}{2}\hat{U}^2(k) = kE(k)$. Then, for $\delta = 1$, that is, for $\hat{U}(k) = c_0(k)$, we get $[2kE(k)]^{1/2} = \beta/2k^2$, so that the boundary spectrum has the form

$$E_\beta(k) = \tfrac{1}{8}\beta^2 k^{-5}. \tag{10.16}$$

The turbulence spectra must satisfy the condition $E(k) > E_\beta(k)$, and the spectra of the Rossby-wave field must satisfy the condition $E(k) < E_\beta(k)$. The large slope of boundary spectrum (10.16) can be attributed to the rapid increase in the frequencies of the Rossby waves $\omega = -\beta k_x k^{-2}$ as their wave numbers become higher. Note, by the way, that if there is a continual influx of energy in the vicinity of a wave number $\hat{k}$, then in the region $k \ll \hat{k}$ a saturation spectrum may be formed, which, in general, depends not only on β and k, but possibly also on the position k_0 of the initial spectral maximum (since a velocity shift affecting eddies with wave numbers k is determined mostly by eddies having wave numbers around k_0).

We assume that $k_0 \gg k_\beta$, so that at the initial moment field of motion ψ represents turbulence. The initial narrow spectral maximum will obviously broaden as time goes by, that is,

$$\frac{\partial}{\partial t}\int (k - \bar{k})^2 E(k)\mathrm{d}k > 0, \tag{10.17}$$

where $\bar{k}$ is defined by formula (10.9). Since, according to (10.8′), the total enstrophy $E_1 = \int k^2 E(k)\, \mathrm{d}k$ does not increase, while the total energy does not change with time, from (10.17) we get $\partial\bar{k}/\partial t < 0$. This means that *the cascade transfer of energy over the spectrum leads to an increase in the typical sizes of turbulent eddies*, and thus to a concentration of energy in the largest eddies. The energy spectrum $E(k, t)$ in this case evolves in time in a self-similar manner:

$$E(k, t) = E^{3/2}\, t\varphi\left(E^{1/2} kt\right) \tag{10.18}$$

(Batchelor, 1969), where $\varphi(\xi)$ is a function which is normalized according to the condition $\int_0^\infty \varphi(\xi)\,\mathrm{d}\xi = 1$. The evolution in time of the energy $E(t)$, the enstrophy $E_1(t)$, and the mean wave number $\bar{k}(t)$ is described by the equations:

$$E(t) = E(\infty) - A\nu t^{-1}; \tag{10.19}$$

$$E_1(t) = At^{-2}; \qquad A = \int^{\infty} \xi^2\varphi(\xi)\,\mathrm{d}\xi; \tag{10.19′}$$

$$\frac{\partial}{\partial t}\frac{1}{\bar{k}(t)} = BU; \qquad B = \int_0^\infty \xi\varphi(\xi)\,\mathrm{d}\xi. \tag{10.19″}$$

Consequently, the predominant wave number of the eddies $\bar{k}$ and their typical frequency $\bar{\omega} \sim v\bar{k}$ are seen to decrease with time, and in a time of the order of $(BUk_\beta)^{-1} \sim [B(\beta U)^{1/2}]^{-1}$ which is independent of k_0 the wave number $\bar{k}$ reaches a value k_β and the eddies are transformed into Rossby waves (Rhines, 1975). After this, $\bar{k}$ continues to decrease, because of resonance interactions between these waves, but much more slowly than before (B decreases from 3×10^{-2} to between 3×10^{-3} and 6×10^{-3}), since such interactions require both a superposition of three waves in space and satisfaction of the resonance relations between their wave numbers and frequencies (the frequencies decrease as a result of such interactions because in the resonance triad the wave with the highest frequency imparts energy to the other two waves).

After the transformation of eddies into waves, in view of the dispersion relation for the waves $\omega = -\beta k_x k^{-2}$ there will be a tendency for inhomogeneities of field ψ to move westward (negative sign of ω). At the same time, the decrease of the frequencies causes these inhomogeneities to become anisotropic, in that their meriodional wavenumbers k_y enjoy a preferential growth; in other words, zonal currents are created. At the limit, these will be steady-state zonal currents in narrow latitude zones, with westward or eastward directions that vary from zone to zone; to be sure, scattering of the wave energy in space, together with decay of the waves, will keep this limit from actually being reached in the Earth's atmosphere. The tendency of the wave interactions to impart energy predominantly to other directions in space rather than to other wavelengths also leads to a sharp reduction of B in (10.19″) at the transition from turbulence to waves. As a result of this sharp reduction of B for $\bar{k} \sim k_\beta$, the eddies do not grow to reach global dimensions (as would have been the case had the Rossby-wave radiations not appeared and had the turbulence been left to itself).

The decrease in B due to the β effect, that is, the slowing down of the spectral transfer of energy to large scales, results in a slowing of the spectral transfer of enstrophy to small scales. Therefore, because of the dissipation of enstrophy, the spectrum must lose its small-scale structure, become narrower, and thus become steeper as well. In addition to having a sizable effect at low wave numbers $k < k_\beta$, the β effect is also a source of anisotropy at high wave numbers $k > k_\beta$, with the meridional anisotropy at low wave numbers creating a zonal anisotropy at high wave numbers.

Now let us take into account the three-dimensional nature of operator $\mathscr{L}$ in equation (10.14), defined by formula (10.7). Its structure produces a difference in the behaviour of waves with lengths $\bar{l}$ less than the Rossby deformation radius $L = NH/f$, on the one hand, for which the vertical interaction between different fluid layers plays a minor role and the layers evolve approximately independently of one another, and waves with lengths $l > L$, on the other hand, in which different fluid layers interact strongly along the vertical, that is, they behave like a single layer, making these waves effectively barotropic. Thus, as the mean wave number $\bar{k}(t)$ decreases, reaching the value $2\pi/L$, there will be a tendency for the eddies to become "barotropized", that is, for the differences between currents in them at different depths to become less.

Consequently, the evolution of the eddies consists in their growth, together with the development of tendencies toward westward movement, anisotropization (elongation along latitude circles), and barotropization. The opposite tendencies can be created by, firstly, the generation of small eddies via internal processes of baroclinic instability of the large-scale currents; in particular, very long ($l \gg L$) baroclinic Rossby waves turn out to be unstable, and this appears to be especially the case for meridional motions in these. Secondly, flow around irregularities of the land relief may cause the generation of enstrophy in the form of intermediate-scale topographic eddies which are uniformly distributed in space (as well as baroclinization of the motion). A relief spectrum of form $E_h\ (k) \sim k^{-3/2}$ or k^{-2} (the latter is the "white noise" in the slope spectrum) creates a tendency toward equalizing the kinetic-energy spectrum of the currents. At the same time, eddies with scales larger than those of the initial eddies of the relief roughness enhance the β effect: f/h isolines assume the role of latitude circles, wave number $k_h = [h\ \nabla\ (f/h)|/2U]^{1/2}$ replaces k_β, and anisotropization manifests itself in a tendency toward flow along f/h isolines. The interaction of the generation and dissipation processes, and of the above-mentioned opposite tendencies, determines the statistical regime of the synoptic eddies and Rossby waves which is characteristic of the general circulation of the atmosphere.

11. The Energetics of the General Circulation

Solar heat is the main energy source for the general circulation of the atmosphere. As shown in §4, it reaches the Earth at an average rate of about 1356 W/m^2 = 1.946 cal/ (cm^2 min), which is equivalent to 1.74×10^{14} kW over the entire cross section of the Earth. A fraction A of this radiation (the planetary albedo) is reflected back into space. Satellite measurements indicate A to be 0.28, on the average (previously it was thought to be 0.35 to 0.40, which made the energy budget of the climate 10 to 17% too high; now all such climatic estimates have to be revised). The assimilated radiation is 1.25×10^{14} kW, or a mean daily amount per unit area of the Earth's surface $4\pi a^2$ equal to 244 W/m^2 = 0.35 cal/(cm^2 min). This is the same as the mean power of the Earth's thermal radiation into space; it corresponds to a radiation temperature of about -17°C, which is actually reached in the atmosphere at a height of about 5 km.

The internal energy of the entire atmosphere $I = \int c_v T\ dm$ (where $dm = \varrho\ dV$ is a mass element and dV is a volume element) has been estimated to be 8.6×10^{23} J, or 1.7×10^9 J/m^2. The potential energy $\mathcal{P} = \int gz\ dm$ (which in the quasi-static approximation is $\int RT\ dm$, since then

$$\int gz\varrho \, dz = -\int z \, dp = \int p \, dz = \int RT\varrho \, dz)$$

is 3.6×10^{23} J, or 7.2×10^{8} J/m^2. Their sum $\Pi = I + \mathcal{P} = \int (c_v + R)T \, dm = \int c_p T \, dm = 1.22 \times 10^{24}$ J, or 2.42×10^{9} J/m^2, is called the *labile energy*.

The portion P of the labile energy which can be converted into kinetic energy $K = \int (v^2/2) \, dm$ via adiabatic processes (where the total energy $\mathcal{E} = \Pi + K$ does not vary with time) is known as the *available labile energy*. In order to find the unavailable labile energy $\Pi_1 = \min \Pi$ as a minimum of a functional of the atmospheric distributions of the independent thermodynamic parameters, assuming adiabatic variations of these, it is necessary to solve some variational problem. Lorenz (1955) proposed that, in a state with minimum labile energy, all the thermodynamic parameters are constant along flow lines of the adiabatic motion, so that for dry air they may be functions only of the potential temperature Θ. This applies, in particular, to the pressure p. Then Π_1 is the value of the labile energy Π which remains after the atmosphere has arrived adiabatically at a state with constant pressure along any isentropic surface $p = p_*(\Theta)$ (and with a stable stratification). Since

$$\Pi = \int c_p T \, dm = \int c_p \Theta \left(\frac{p}{p_0} \right)^{(\varkappa - 1)/\varkappa} dm,$$

we have

$$\Pi_1 = \int c_p \Theta \left[\frac{p_*(\Theta)}{p_0} \right]^{(\varkappa - 1)/\varkappa} dm. \tag{11.1}$$

Now, since $\mathcal{E} = \Pi_1 + P + K$ and Π_1 are invariable for adiabatic processes, it is clear that the sum $P + K$ is also an adiabatic invariant. Moreover,

$$P = \Pi - \Pi_1 = \int \frac{c_p \Theta}{p_0^{(\varkappa - 1)/\varkappa}} \left\{ p^{(\varkappa - 1)/\varkappa} - [p_*(\Theta)]^{(\varkappa - 1)/\varkappa} \right\} dm. \tag{11.2}$$

This integral can be expressed approximately in terms of the weighted-mean variance of the pressure at isentropic surfaces, and thus also in terms of the weighted value of the variance of the temperature at isobaric surfaces σ_T^2:

$$P \approx \frac{1}{2} c_p \left\{ \frac{\gamma_a \sigma_T^2}{(\gamma_a - \bar{\gamma}) \bar{T}} \right\}, \tag{11.2'}$$

where the bar indicates averaging over the isobaric surface (Lorenz, 1955). According to the estimate of Oort (1964), the available labile energy is 5.5×10^{6} J/m^2, that is, only about 0.2% of the entire labile energy of the atmosphere.

Oort found an even lower value for the kinetic energy of the atmosphere: $K = 1.5 \times 10^{6}$ J/m^2, which accounts for only about a quarter of the available labile energy. This gives a kinetic energy per unit mass of atmosphere of 140 m^2/s^2, which corresponds to a mean velocity of atmospheric motions of about 17 m/s (the following estimates of Borisenkov are also of interest: in the Northern Hemisphere $K = 1.9 \times 10^{20}$ J in summer and 4×10^{20} J in winter; in the Southern Hemisphere $K = 3.9 \times 10^{20}$ J in summer and 7.1×10^{20} J in winter).

If the atmosphere is brought to a state of neutral stratification by means of adiabatic processes (a state in which isentropic surfaces are vertical, that is, θ is independent of z), then the labile energy in this state Π_z can be found by replacing the factor $[p_* (\Theta)]^{(\varkappa - 1)/\varkappa}$ in the integrand of (11.1) by the mean value of $p^{(\varkappa - 1)/\varkappa}$ over the mass of the vertical air column. This mean value is equal to

$$\frac{\varkappa}{2\varkappa - 1}\, p_s^{(\varkappa - 1)/\varkappa},$$

where p_s is the surface pressure. Thus we have

$$\Pi_2 = \int \frac{\varkappa}{2\varkappa - 1} c_p \Theta \left(\frac{p_s}{p_0}\right)^{(\varkappa - 1)/\varkappa} dm. \tag{11.3}$$

The difference $\Pi_2 - \Pi_1$, known as the *macrostability parameter*, determines the amount of kinetic energy liberated or consumed during an adiabatic transition from a given atmospheric stratification to a neutral stratification. It is approximately equal to the weighted-mean value of the vertical gradient of the potential temperature $(-\partial\Theta/\partial p)$ over the whole thickness of the atmosphere (Lorenz, 1960).

When studying the hydrodynamic stability of atmospheric movements, it is useful to consider the adiabatic invariant $\mathcal{E} + \mathcal{J}$, where $\mathcal{J}$ is a measure of the inhomogeneity of the field of the potential vorticity Ω_*, expressed in terms of the integral $\int \varphi\, dm$ of the function $\varphi(\Omega_*, \Theta)$, which is quadratic in Ω_*. For small adiabatic perturbations of the motion, the principal part of functional $\mathcal{E} + \mathcal{J}$ is invariant in time; if this principal part is positive-definite, then the motion will be stable with respect to small perturbations.

Two thirds of the total energy of solar radiation (1.25×10^{14} kW) is used for direct heating of the atmosphere and the upper layers of the oceans and continents, while the remaining one third goes to evaporate moisture from the Earth's surface (assuming a global evaporation of 5.26×10^{20} g/yr and a latent heat of vaporization of 2.4×10^3 J/g, a total of 4.2×10^{13} kW is expended to power the evaporation process).

Only a very small portion of the assimilated radiation is used to generate the kinetic energy of atmospheric motions: according to present estimates, the rate of generation of kinetic energy in the atmosphere as a whole $\partial K/\partial t$ is about 2×10^{12} kW, or 4 W/m^2, that is, only 1.6% of the total rate of the assimilated radiation. In view of this low "efficiency" of the atmospheric "heat engines", the kinetic energy amounts to only a tiny fraction (around 0.06%) of the total energy of the atmosphere. As an average per unit mass of atmosphere, the rate of generation of kinetic energy is 4 cm^2/s^3. The mean specific rate of dissipation of kinetic energy to generate heat because of friction must be of about the same order of magnitude, and in fact different investigators have estimated it to be from 2.3 to 10.4 cm^2/s^3. The typical time required to generate the kinetic energy of the atmosphere (and also for it to decay because of viscosity) is of the order of $[(\mathbf{1}/\boldsymbol{K})\,(\boldsymbol{\partial K}/\boldsymbol{\partial t})]^{-1} \sim 10^{21}$ J/2 $\times 10^{12}$kW $\sim 5 \times 10^5$ s, or about a week.

Consequently, according to Oort's estimates, the principal energy cycle of the atmosphere consists in a generation of the available labile energy $P = 5.5 \times 10^6$ J/m^2 as a result of heat influxes at a rate $\partial P/\partial t = 2.3$ W/m^2, followed by a conversion to kinetic energy $K = 1.5 \times 10^6$ J/m^2 at this same rate, and finally a dissipation of the latter to form heat, again at this same rate $\partial K/\partial t = 2.3$ W/m^2.

Let us now write local equations for the transfer in the atmosphere of kinetic energy with a specific density $v^2/2$, of potential energy in the gravitational field with a specific density gz, and of internal energy with a specific density $c_v T$. These equations, expressed in a tensor notation suitable for transformation to either spherical or Cartesian coordinates, have the form:

$$\frac{\partial}{\partial t}\varrho\frac{v^2}{2} + \nabla_\alpha\left(\varrho\frac{v^2}{2}v^\alpha + pv^\alpha - \tau^{\alpha\beta}v_\beta\right) = \varrho\mathbf{v}\cdot\mathbf{g} + p\,\mathrm{div}\,\mathbf{v} - \frac{1}{2}\tau^{\alpha\beta}\varphi_{\alpha\beta}; \qquad (11.4)$$

$$\frac{\partial}{\partial t}g\varrho z + \nabla_\alpha(g\varrho z v^\alpha) = -\varrho\mathbf{v}\cdot\mathbf{g}; \qquad (11.5)$$

$$\frac{\partial}{\partial t}c_v\varrho T + \nabla_\alpha\left(c_v\varrho T v^\alpha + H^\alpha + \frac{\partial\chi}{\partial q}I^\alpha\right) = -p\,\mathrm{div}\,\mathbf{v} + \frac{1}{2}\tau^{\alpha\beta}\varphi_{\alpha\beta}, \qquad (11.6)$$

where $\tau^{\alpha\beta}$ is the tensor of the viscous stresses, $\varphi_{\alpha\beta}$ is the tensor of the rates of deformation, H^α and I^α are the fluxes of heat and moisture, and $\chi = c_v T + p/\varrho$ is the specific enthalpy. The term $\varrho\mathbf{v}\cdot\mathbf{g}$ in (11.4) and (11.5) describes the interchange of kinetic and potential energy due to the work of Archimedes bouyancy forces. The term p div $\mathbf{v}$ in (11.4) and (11.6) describes the interchange of kinetic and internal energy due to the adiabatic heating or cooling of air as it contracts or expands. Finally, the term $\frac{1}{2}\tau^{\alpha\beta}\varphi_{\alpha\beta}$ in (11.4) and (11.6) describes the effect of the viscous dissipation of kinetic energy into heat. The sum of equations (11.5) and (11.6) gives the local equation of transfer of labile energy. The sum of all three of the equations (11.4)–(11.6) gives the law of conservation of total energy

$$\frac{\partial}{\partial t}\varrho\left(\frac{v^2}{2}+gz+c_vT\right)+\nabla_\alpha\left[\varrho\left(\frac{v^2}{2}+gz+c_vT\right)v^\alpha+\right.$$

$$\left.+pv^\alpha+H^\alpha+\frac{\partial\chi}{\partial q}I^\alpha-\tau^{\alpha\beta}v_\beta\right]=0. \tag{11.7}$$

It is of interest to calculate separately the values of $\bar{K}$, $\bar{P}$, and K', P' and the rates of conversion of kinetic energy K and available labile energy P for the zonally averaged circulation (to be denoted here by a bar over the appropriate letters) and for deviations from it, that is, for nonzonal processes (to be denoted by primes). For the two forms of kinetic energy $\bar{K}$ and K', from the equations describing the dynamics of the atmosphere we get the following local equations:

$$(\bar{K}):\quad \frac{\partial}{\partial t}\left(\varrho\frac{\bar{v}^2}{2}+\overline{\varrho'\mathbf{v}'}\cdot\bar{\mathbf{v}}\right)+\nabla_\alpha\left[(\overline{\varrho v^\alpha}+\overline{\varrho' v'^\alpha})\frac{\bar{v}^2}{2}+\right.$$

$$\left.+\bar{\mathbf{v}}\cdot(\overline{\varrho' v'}\cdot\bar{v}_\alpha+\overline{\varrho\mathbf{v}'v'^\alpha})+\overline{p v'^\alpha}-\bar{\tau}^{\alpha\beta}\bar{v}_\beta\right]$$

$$=\left[\overline{\varrho\mathbf{v}'v'_\alpha}\,\nabla\bar{v}^\alpha-\overline{\varrho' v'_\alpha}\left(\overline{\mathbf{v}'\cdot\nabla v'^\alpha}+\frac{1}{\bar{\varrho}}\overline{\xi\nabla_\alpha p}-\frac{1}{\bar{\varrho}}\overline{\xi\nabla_\beta\tau^{\alpha\beta}}\right)\right]+$$

$$+(\overline{\varrho\mathbf{v}}+\overline{\varrho'\mathbf{v}'})\cdot\mathbf{g}+p\operatorname{div}\bar{\mathbf{v}}-\frac{1}{2}\bar{\tau}^{\alpha\beta}\bar{\varphi}_{\alpha\beta};$$

$$(K'):\quad \frac{\partial}{\partial t}\varrho\frac{\overline{v'^2}}{2}+\nabla_\alpha\left(\frac{1}{2}\overline{\varrho v'^2 v^\alpha}+\frac{1}{2}\overline{\varrho v'^2 v'^\alpha}+\overline{p' v'^\alpha}-\overline{\tau'^{\alpha\beta}v'_\beta}\right)$$

$$=-\left[\overline{\varrho\mathbf{v}'v'_\alpha}\cdot\nabla\overline{v^\alpha}-\overline{\varrho' v'_\alpha}\left(\overline{\mathbf{v}'\cdot\nabla v'^\alpha}+\frac{1}{\bar{\varrho}}\overline{\xi\nabla_\alpha p}-\right.\right.$$

$$\left.\left.-\frac{1}{\bar{\varrho}}\overline{\xi\nabla_\beta\tau^{\alpha\beta}}\right)\right]+\overline{p'\operatorname{div}\mathbf{v}'}-\frac{1}{2}\overline{\tau'^{\alpha\beta}\varphi'_{\alpha\beta}},$$

where the notation $\xi=\bar{\varrho}/\varrho$ has been introduced. The right-hand sides of these equations describe the energy conversions. The terms in square brackets on the right-hand sides which have opposite signs for $(\bar{K})$ and (K') describe the conversions back and forth of the kinetic energy of the zonally averaged circulation and the nonzonal processes; the other terms on the right-hand sides describe the mutual conversions of kinetic and available potential energy (concerning these, see immediately below). The first, and apparently principal, term in square brackets, $\overline{\varrho\mathbf{v}'v'_\alpha}\cdot\nabla\bar{v}^\alpha$, describes the so-called barotropic instability of the zonally averaged circulation. Contrary to accepted ideas, it is apparently positive, that is, it corresponds to a transfer of kinetic energy on the average from nonzonal processes to the zonally averaged circulation (negative macroturbulent viscosity), as was explained in the previous subsection. The local equations for the two forms of labile energy $\bar{\Pi}$ and Π' are then

$$(\bar{\Pi}): \quad \frac{\partial}{\partial t}\bar{\varrho}\,(g\bar{z} + c_v\bar{T}) + \nabla_\alpha\left[(g\bar{z} + \bar{c}_v\bar{T})\,(\bar{\varrho}\bar{v}^\alpha + \overline{\varrho' v'^\alpha})\right]$$

$$= \left[\overline{\varrho'\mathbf{v}'}\cdot\nabla\,(g\bar{z} + c_v\bar{T}) - \overline{\varrho\mathbf{v}'\cdot\nabla\,(gz' + c_vT')} - \right.$$

$$\left. - \overline{\xi\,\nabla_\alpha\left(H^\alpha + \frac{\partial\chi}{\partial q}I^\alpha\right.}\right] - \overline{\varrho\mathbf{v}}\cdot\mathbf{g} - \overline{\xi p\,\mathrm{div}\,\mathbf{v}} + \frac{1}{2}\,\overline{\xi\tau^{\alpha\beta}\varphi_{\alpha\beta}}\;;$$

$$(\bar{\Pi}'): \quad \frac{\partial}{\partial t}\,(g\overline{\varrho' z'} + c_v\overline{\varrho' T'}) + \nabla_\alpha\left[(g\overline{\varrho' z'} + c_v\overline{\varrho' T'})\,\bar{v}^\alpha + \right.$$

$$\left. + \overline{(gz' + c_vT')\,\varrho v'^\alpha} + \bar{H}^\alpha + \overline{\frac{\partial\chi}{\partial q}I^\alpha}\right]$$

$$= -\left[\overline{\varrho'\mathbf{v}'}\cdot\nabla(g\bar{z} + c_v\bar{T}) - \right.$$

$$\left. -\varrho\,\overline{\mathbf{v}'\cdot\nabla\,(gz' + c_vT')} - \overline{\xi\,\nabla_\alpha\left(H^\alpha + \frac{\partial\chi}{\partial q}I^\alpha\right)}\right] -$$

$$- \overline{\varrho'\mathbf{v}'}\cdot\mathbf{g} - \overline{(1-\xi)\,p\,\mathrm{div}\,\mathbf{v}} + \frac{1}{2}\,\overline{(1-\xi)\,\tau^{\alpha\beta}\varphi_{\alpha\beta}}\,.$$

The right-hand sides of these equations, like the equations for $(\bar{K})$ and (K'), also describe energy conversions. The terms in square brackets on the right-hand sides, as well as the other terms containing ξ, have opposite signs in the equations for $(\bar{\Pi})$ and (Π'). Consequently, they describe the mutual conversions of labile energy of the zonally averaged circulation and the nonzonal processes; the other terms on the right-hand sides describe the mutual conversions of available potential energy and kinetic energy. The first, and apparently principal, term in square brackets, $\overline{\varrho'\mathbf{v}'}\cdot\nabla\,(g\bar{z} + c_v\bar{T})$, describes the so-called baroclinic instability of the zonally averaged flow. According to empirical data, it is negative, on the average, that is, it corresponds to a transfer of labile energy from the zonally averaged state to nonzonal processes, which represents for the latter the main energy source. The kinetic energy of the zonally averaged circulation $\bar{K}$ is exchanged with the available labile energy of the zonally averaged state $\bar{P}$ at a rate $\bar{\varrho}\bar{v}\cdot\mathbf{g}$ and with the available labile energy of the nonzonal processes P' at a rate $\overline{\varrho'\mathbf{v}'}\cdot\mathbf{g} + \bar{p}\,\mathrm{div}\,\bar{\mathbf{v}} - \frac{1}{2}\,\bar{\tau}^{\alpha\beta}\varphi_{\alpha\beta}$. The kinetic energy of the nonzonal processes

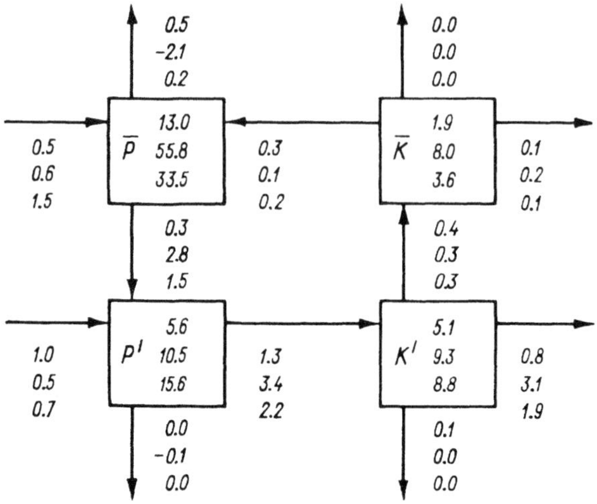

Figure 11.1 Energy conversions in atmosphere of Northern Hemisphere from 1958 to 1963, according to Oort and Peixoto (1974).

K' is exchanged with P' at a rate $\overline{p' \operatorname{div} \mathbf{v}'} - \frac{1}{2}\overline{\tau'^{\alpha\beta}\varphi_{\alpha\beta}}$ (there is no direct exchange with $\bar{P}$).

Figure 11.1 shows some empirical estimates of $\bar{K}$, $\bar{P}$ and K', P' and the rates of their conversion in the atmosphere of the Northern Hemisphere, according to Oort and Peixoto (1974). The energy values given here are in 10^5 J/m^2 and the rates of energy conversion are in W/m^2; upward and downward arrows pointing outward indicate energy exchange with the Southern Hemisphere, and outward arrows to the right indicate transformations of $\bar{K}$ and K' to P' via viscous dissipation of kinetic energy. Numbers at the top correspond to July, numbers in the middle to January, and numbers at the bottom to the entire year. The main branch of the energy cycle has the form $\bar{p} \rightarrow P' \rightarrow K' \rightarrow \bar{K}' \rightarrow (\bar{P}, P')$. It should be noted that in winter the energy cycle is much more intense than in summer.

CHAPTER 5

THE WORLD OCEAN

12. The Equations of Dynamics of the Ocean

In the ocean, as in the atmosphere, synoptic and global processes require individual description. They can also be characterized with the aid of the simplified equations of hydrodynamics (9.1)–(9.4), (9.3′), (9.4′), if $\varrho = \varrho_w$ is taken to be the density of seawater. The boundary conditions for the velocity field at the ocean floor, the relief of which $z = -h_w(\theta, \lambda)$ is assumed to be specified, remain the same as those for the velocity of air currents at the land surface (here, by analogy with the atmospheric boundary layer, we can introduce a "parametrized" near-bottom boundary layer). However, the boundary conditions at the ocean surface $z = \zeta(\theta, \lambda, t)$ now have to be specified more precisely than was the case for the atmosphere (since the quantity $\zeta(\theta, \lambda, t)$ plays a significant role in the ocean dynamics, and it is an unknown function which must be determined from these boundary conditions): here it can no longer be assumed that $\zeta = 0$ everywhere, but for large-scale processes ζ will be small (of the order of decimeters, as a rule). The kinematic boundary condition at the ocean surface can then be written as

$$v_z = \frac{\partial \zeta}{\partial t} + \frac{v_\theta}{a}\frac{\partial \zeta}{\partial \theta} + \frac{v_\lambda}{a \sin \theta}\frac{\partial \zeta}{\partial \lambda} - \frac{1}{\varrho_w}(P - E - E_{wi}). \qquad (12.1)$$

The dynamic boundary conditions can with good accuracy be reduced to the requirement of continuity of the tangential stresses and the pressure. In the latter of these requirements the effects of surface tension, in particular, are neglected, since they are considerable only for processes of very small scale (capillary waves), whereas for the large-scale processes being considered here they are of practically no importance. At the same time, gravity waves are not excluded by these boundary conditions; these can be filtered out of the fluctuations of ζ produced by the quantity $P - E - E_{wi}$, by fluctuations in atmospheric pressure and by gusts of wind, with the aid of the so-called quasi-solenoidal approximation, which is, by the way, more accurate than the quasi-geostrophic approximation. In the quasi-solenoidal case the evolution equation for the divergence of the horizontal velocity has to be replaced by a synchronous relation, the so-called "balance equation". In view of the smallness of ζ, the boundary conditions at the ocean surface can be "shifted" from level $z = \zeta$ to level $z = 0$; the condition of pressure continuity will then take the form

$$p = p_a + g\varrho_w\zeta \qquad \text{for } z = 0. \qquad (12.2)$$

At the ocean shore the horizontal flow velocity must go to zero everywhere, except at the mouths of rivers, where the velocity component normal to the shore equals the mean (over the section of the river) flow velocity in the river. The effects of the runoff of small rivers and streams must, of course, be parametrized; the effect of the formation of icebergs also has to be considered separately. Sometimes, by the way, the horizontal exchange of momentum is taken into account in equations (9.1)–(9.4) just in the near-shore boundary layers, and then at their inner boundaries only the velocity component normal to the shore has to go to zero (once again, except at the mouths of rivers).

Equations (9.1)–(9.4) must be supplemented by the equation of state for seawater:

$$\varrho_w = \varrho_w(T, p, S) \approx \varrho_{w0}\left[1 - \frac{p}{A + Bp + Cp^2}\right]^{-1}, \tag{12.3}$$

where ϱ_{w0}, A, B, and C are certain functions of the temperature T and the salinity S (representable as polynomials of low powers of T°C and $S^{1/2}$; the empirical function $\varrho_w(T, p, S)$ is tabulated). For small vertical movements of the fluid particles, equation (12.3) can be linearized and written in the form

$$\varrho_w - \varrho_0 \approx -\alpha_0\varrho_0(T - T_0) + \gamma_0\varrho_0(S - S_0). \tag{12.3'}$$

Equation of state (12.3) contains two new unknown functions of T and S. These can be determined with the aid of the equations of evolution of entropy η_ω and salinity S:

$$\frac{d\eta_w}{dt} = \frac{\varepsilon}{T} - \left[\frac{\partial Q_z}{\partial z} + \frac{1}{a\sin\theta}\left(\frac{\partial Q_\theta \sin\theta}{\partial\theta} + \frac{\partial Q_\lambda}{\partial\lambda}\right)\right], \tag{12.4}$$

$$\frac{dS}{dt} = -\left[\frac{\partial J_z}{\partial z} + \frac{1}{a\sin\theta}\left(\frac{\partial J_\theta \sin\theta}{\partial\theta} + \frac{\partial J_\lambda}{\partial\lambda}\right)\right], \tag{12.5}$$

where ε is the rate of heat influx to a unit mass of seawater, both because of the absorption of radiative energy (in the upper layer of the ocean) and as a result of the dissipation of kinetic energy (in particular, of tidal energy), and Q_i and J_i are the components of the entropy and salinity (salt) fluxes. Equation (12.4) must be supplemented by the formula for the entropy $\eta_w = \eta_w(T, p, S)$, which in differential form can be written as

$$d\eta_w = \frac{c_p}{T}\,dT - \frac{\alpha}{\varrho}\,dp + \frac{\partial\mu}{\partial T}\,dS, \tag{12.6}$$

where

$$\alpha = \frac{1}{\varrho_w}\left(\frac{\partial\varrho_w}{\partial T}\right)_{p,\,S}$$

is the thermal expansion coefficient, and μ is the chemical potential of seawater (this is at present known only to the accuracy of an arbitrary linear function of the temperature with constant coefficients). Sometimes equation (12.4) is approximated by the heat conduction equation. The equation of the heat budget (9.10) can be taken as the boundary condition at the ocean surface for equation (12.4), it being necessary to parametrize the dependence of the albedo A on the force of the sea waves and the height of the Sun. For equation (12.5) the boundary condition is the equation of the salinity budget

$$J_s = - S (P - E - E_{wi}), \qquad (12.7)$$

where J_s is the salt flux into the depths of the ocean.

The requirements that the normal components of the heat and salt fluxes be zero, except in hydrothermal regions at the bottom, can serve as the boundary conditions for thermodynamic equations (12.4) and (12.5) at the ocean floor and at the shore. The hydrothermal regions lie mainly in the axial zones of mid-oceanic ridges, where it may be necessary to take into account the geothermal heat flux, and here and there the deposition of salts. Other exceptions to the requirement of zero normal components of the heat and salt fluxes are the mouths of rivers and places where icebergs break away, since flows of fresh water and heat exist there (and are significant in a number of regions, in the Arctic, for instance).

The upper layer of the ocean, including the upper mixed layer with the picnocline and the seasonal thermocline beneath it, may be parametrized, for example, on the basis of the approximate similarity of the vertical profiles of temperature and salinity in it. This is described by the formula of Kitaigorodskii and Miropol'skii:

$$T = \begin{cases} T_s & \text{for } \zeta \geqslant z \geqslant -h_0, \\ T_s - (T_s - T_1) F_T\left(\dfrac{h_0 + z}{h_0 - h_1}\right) & \text{for } -h_0 \geqslant z \geqslant -h_1, \end{cases} \qquad (12.8)$$

where h_0 is the thickness of the upper mixed layer, h_1 is the depth of the lower boundary of the thermocline, T_1 is the temperature at depth h_1, and F_T is some quasi-universal function for which the above investigators obtained the empirical formula

$$F_T(\xi) = \frac{8}{3}\xi - 2\xi^2 + \frac{1}{3}\xi^4. \qquad (12.9)$$

Similar formulas are used for the salinity S, but in this case parameters T_S, T_1, S_S, S_1, h_0, and h_1 have to be determined using equations constructed with the aid of thermodynamic equations (12.4) and (12.5), the boundary conditions, and additional semiempirical relations (including the equation for the turbulent energy and the formula for the integral rate of dissipation of turbulent energy).

Finally, to round out our equations of the dynamics of the ocean, we must add the equations of evolution of the ice cover, describing the freezing and melting of ice (and taking into account, in particular, the fall of precipitation onto the ice surface), its drift, and the formation of leads and hummocks.

13. The Stratification of the Ocean

The state of the ocean is determined primarily by atmospheric factors: the frictional stress of the wind on the ocean surface τ_0, the heating or cooling (the vertical heat flux H_S of formula (9.10)), and the moisture exchange (the difference between the precipitation rate P and the evaporation rate E). For some processes (chiefly biochemical processes) the exchange of oxygen and carbon dioxide is also important. The tide-producing effects of the Moon and Sun are also significant, as are the discharge of rivers and the melting of icebergs, the settling out of aerosols, geothermal heat and volcanic exhalations at the ocean floor, and tsunamigenic effects of undersea earthquakes. But it is atmospheric effects which shape the stratification of the ocean and which create the wind-driven currents and thermohaline circulation. These processes will be discussed in the present chapter.

STRATIFICATION

This is the formation of layers of different density in the gravitational field, due to the compressibility of seawater. The dependence of the water density ϱ_w on the temperature T, the salinity S, and the pressure p is described by empirical formula (12.3). This dependence is such that with an increase in p or S or a drop in T (down to some temperature of maximum density $T_1 = T_1 (p, S) \approx 4°C$) the water density increases. It should be noted that, with an increase in p or S, both T_1 and the melting temperature $T_2 = T_2 (\varrho, S)$ decrease, T_1 more rapidly than T_2, and for $p < p_1 \approx 270$ atm (or 27×10^6 Pa) and $S < S_1 \approx 24‰$ there is a temperature interval $T_1 > T > T_2$ within which the dependence of ϱ on T has the opposite sign. For the climate of the Earth, it is an exceptionally important fact that water is an anomalous liquid: it expands as it freezes, making ice lighter than water and able to float on the water surface. If water were a normal liquid, it would contract as it freezes, making the ice sink to the bottom and ultimately fill vast regions of the ocean.

Heating from above (by contact and due to strong absorption by the water of the light penetrating it) and freshening of the seawater (due to precipitation, discharge of rivers, and melting of ice) can affect only the very thin uppermost layer of the ocean, some tens of meters in thickness. This is because of the hydrostatic stability of the heated or freshened layer, which keeps it from mixing independently with the underlying water. Moreover, the forced mixing produced by breaking surface waves does not penetrate deeply (while the mixing at turbulent spots forming in places where the surface waves are hydrodynamically unstable is generally very weak and apparently acts extremely slowly).

The mixing of the upper layer of the ocean creates at its lower boundary a layer of density discontinuity with $\partial \varrho_w/\partial z \approx 10^{-6}$ g/cm^4. This is one and a half orders of magnitude greater than the density gradients in the sizable, but somewhat rare, surface temperature inversions, where $\partial T/\partial z \approx 1$°C/m and $\partial \varrho_a/\partial z \approx 4 \times 10^{-8}$ g/cm^4. In the formulation with the aid of the Väisälä–Brunt frequency N, which is more significant dynamically, the layer of density discontinuity is seen to be stratified much more stably ($N \approx 3 \times 10^{-2}$ s^{-1}) than the troposphere as a whole, in which $\partial T/\partial z \approx 6.5$°C/km and $N \approx 10^{-2}$ s^{-1}; at the same time it is less stably stratified than strong atmospheric inversions ($N \approx 1.7 \times 10^{-1}$ s^{-1}). The ubiquitous spreading of the layer of density discontinuity throughout the ocean, together with the rarity of strong inversions in the atmosphere, explain why internal waves propagate much more widely in the ocean than in the atmosphere.

However, the layer of density discontinuity is thin, so that its stratification, that is, the effect of heating (or freshening) of the ocean water from above, has very little influence on the mean stratification of the ocean. On the other hand, in regions where the ocean surface is cooled (or becomes more saline), the water from the surface must descend into the depths, producing a so-called thermohaline circulation of the water, that is, a mixing effect which destabilizes the stratification. This is similar to the heating of the troposphere from below, though, of course, with the proviso that the ocean water cannot be cooled below the freezing point T_2. Consequently, dynamically speaking, the ocean as a whole can be said to be cooled from above, in contrast to the troposphere, which is heated from below. At the same time, in the ocean there is no effect analogous to the stabilizing effect of the heating of the troposphere from above (by absorption of ultraviolet solar radiation in the ozone layer of the stratosphere). In other words, in the ocean there are no external sources of cooling from below, and as a result its mean stratification, albeit stable, is an order of magnitude closer to adiabatic than the mean stratification of the troposphere: on the average, over the whole depth of the ocean, $\bar{N} \approx 10^{-3}$ s^{-1}. This circumstance is of great dynamic importance: as will be shown below, because of it synoptic vortices and Rossby waves in the ocean are much smaller, much slower, and much more long-lived than those in the atmosphere.

GENERATION OF THERMOHALINE CIRCULATION

In order to judge the effects of the increase (or decrease) in the density of the ocean water as it cools (heats) or becomes more (less) saline, let us consider the vertical mass flux at the ocean surface associated with these processes (Monin, 1970). This flux is defined by the formula

$$M_s = J_s + \frac{\alpha}{c_w} H_s, \tag{13.1}$$

where J_s and H_s are the fluxes of salt and heat into the interior of the ocean (see formulas (12.7) and (9.10)), α is the coefficient of thermal expansion, and c_w is the specific heat of water. Mass flux $M_s > 0$ if this flux is directed downward, thereby reducing the bouyancy and assisting the thermohaline subsidence of surface water (on the chart of mean annual values of M_s given by Agafonova *et al.* (1972), M_s has the opposite sign).

The heat budget of the ocean $(-H_s)$ is made up mostly of the latent heat of vaporization, the heat of effective radiation, and the turbulent heat exchange with the atmosphere; according to Albrecht (1949), these have a ratio, on the average, of 51 : 42 : 7. The annual values of $(-H_s)$ range from +100 kcal/cm^2, or $+41.9 \times 10^3$ MJ/m^2 (heating up of ocean) in the equatorial Pacific to -100 kcal/cm^2, or -41.9×10^2 MJ/m^2 (cooling off of ocean), in the Gulf Stream and Kuroshio. The mean zonal values of $(-H_s)$, according to Stepanov (1974) are shown in Fig. 13.1. They are seen to be positive in the tropical zone between 30°N and 15°S and negative outside of this region.

The water budget of the ocean is made up mostly of the difference $P - E$ between the precipitation and the evaporation. For the World Ocean as a whole, this difference is negative: present estimates give $P \approx 4.12 \times 10^{20}$ g/yr and E $\approx 4.53 \times 10^{20}$ g/yr, or a difference of 0.41×10^{20} g/yr (about 9% of E), which is compensated by the discharge of rivers. The annual values of $P - E$ range from +200 g/cm^2 in the western equatorial Pacific to -150 g/cm^2 in the subtropical Atlantic. The mean zonal values of $P - E$ are positive in the equatorial zone between 10°N and 5°S, as well as at latitudes higher than 40°, and they are negative in the tropics and subtropics.

The resulting annual values of M_s are positive from the tropics to the middle latitudes (with maxima as high as 20 g/cm^2 in the Gulf Stream and Kuroshio regions), and probably in the Arctic and Antarctic as well, and they are negative (dropping to -15 g/cm^2) in the equatorial zone, as well as along the eastern shores of the Pacific. Typical values of $M_s \sim 5$ g/cm^2 are yielded by a quite high estimate of the rate of generation of kinetic energy of the thermohaline circulation in a unit mass $g\varrho_w^{-1} M_s \approx 1.5 \times 10^{-4}$ cm^2/s^3 (for comparison, we note that for an intensity of tide-producing forces of 2.76×10^6 MW, estimated using astronomical methods, per unit mass of the ocean (total mass of 1.37×10^{24} g), the corresponding value is 2×10^{-5} cm^2/s^3; measurements of the small-scale oceanic turbulence indicate that the rate of dissipation of kinetic energy per unit mass varies from 1–10^{-1} cm^2/s^3 at the ocean surface to 10^{-3} – 10^{-4} cm^2/s^3 at the layer of density discontinuity, and probably to even lower values in the depths of the ocean). The values of H_s, J_s, and M_s can be expected to have a high seasonal variability, and the actual regions of generation of the thermohaline circulation with maximum positive values of M_s are probably to be sought on the winter charts.

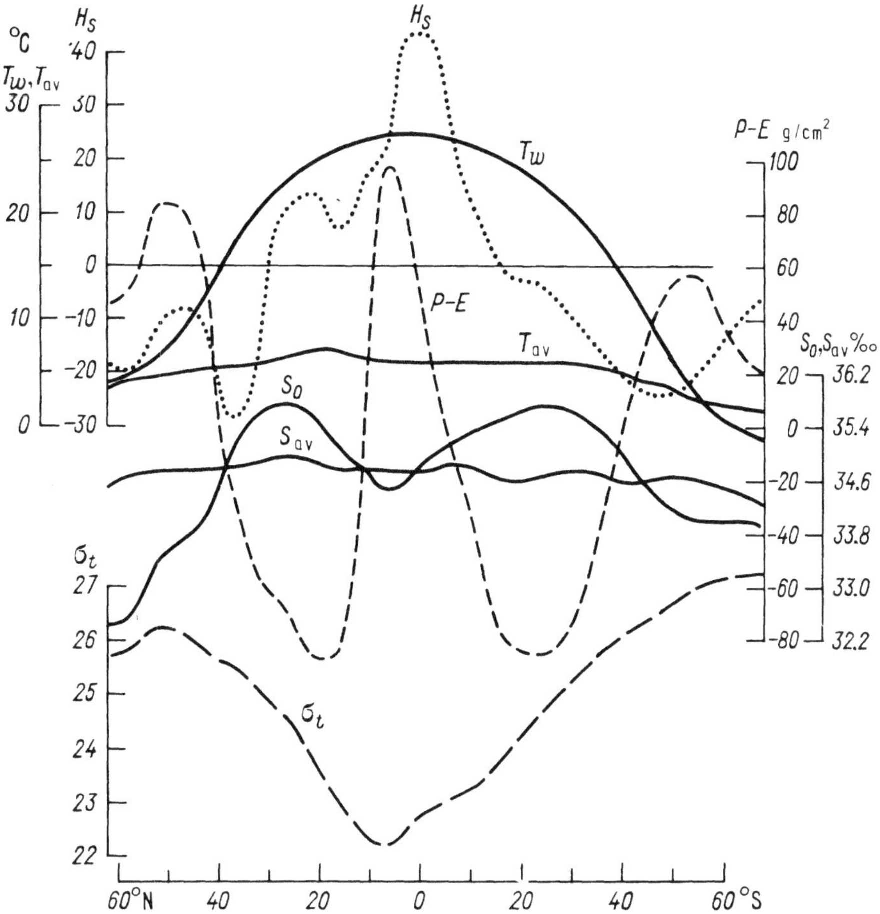

Fig. 13.1. Zonal climate of ocean, according to Stepanov (1974). H_s is annual heat budget in kcal/cm² (where 1 kcal/cm² = 41.9 MJ/m²); T_w is temperature of ocean surface; T_{av} is mean temperature over depth of ocean; $P - E$ is annual water budget; S_0 is salinity at ocean surface; S_{av} is mean salinity over depth of ocean; σ_t is relative density at ocean surface.

TEMPERATURE

The temperature field of the surface water of the ocean (T_w field) is of great importance with regard to the effect of the ocean on the atmosphere, that is, with regard to the long-term variations in weather and climate produced by the state of the ocean. This temperature field has considerable spatial variability, with a wide range of scales: *global variability*, with amplitudes δT_w of degrees or tens of degrees (including the latitudinal variability and sharp contrasts at warm and cold surface currents); *synoptic variability*, with amplitudes $\delta T_w = 10^{-1}$ to 10°C and with scales $l = 10^1$ to 10^2 km (produced by atmospheric effects, such as storms, precipitation, evaporation, and cloudiness, as well as by synoptic oceanic processes, especially upwellings and cyclonic vortices); *mesoscale variability* ($l = 10^2$ to 10^4 m, $\delta T_w \approx 10^{-1}$ °C); and *small-scale variability*, produced by hydrological fronts, discharges of rivers, surface films, internal waves, individual clouds, ice, icebergs, etc. ($l \leqslant 10^1$ m, $\delta T_w = 10^{-2}$ to 10^{-1}°C).

Field T_w is also quite variable in time, being characterized by: *a variability measured in years* (mainly the integral of the thermal effects of the ocean-atmosphere interaction); *a seasonal or long-term variability*, measured in months (also the integral of effects of the ocean-atmosphere interaction and possibly due to the variability of the principal ocean currents and upwellings); *a synoptic oceanic variability*, measured in months or weeks (produced by synoptic vortices in the ocean); *a synoptic atmospheric variability*, measured in days (produced by storms, etc.); *a mesoscale variability*, measured in hours or minutes (produced by internal waves and clouds); and *a small-scale variability*, measured in minutes, seconds, or fractions of a second (produced by surface waves and turbulence).

The variability of field T_w makes it necessary to monitor, and then predict, this field. For the operational monitoring of the temperature field on a global scale, satellite techniques are employed. Measurements of T_w using satellites are now carried out in the infrared range $\lambda = 0.7$ to 14 μm, and possibly in the microwave range $\lambda = 0.1$ to 21 μm as well. The radiation picked up by a satellite over a point $\mathbf{x}$ of the ocean is represented, in general, as $I_\lambda(\mathbf{x}) = I_\lambda^0(M\mathbf{x})\,\mathcal{P}_\lambda(M\mathbf{x}) + I_\lambda^a(M\mathbf{x})$, where I_λ^0 is the radiation of the ocean surface, $\mathcal{P}_\lambda$ is the transmission function of the atmosphere, I_λ^a is the radiation of the atmosphere, and $M\mathbf{x}$ is the area of the ocean surface, centered on $\mathbf{x}$, which is resolvable with the satellite radiometer. In the infrared range it must be taken into account that I_λ^0 originates from only a very thin surface layer of the water (for $\lambda = 8$ to 12.5 μm this layer is about 0.02 m thick), the temperature T_w of which may differ considerably (by several tenths of a degree or more) from the temperature typical of the upper mixed layer. Here we can set

$$I_\lambda^0(M\mathbf{x}) = B_\lambda[T_w(M\mathbf{x})];$$

$$I_\lambda^a(M\mathbf{x}) = \int B_\lambda[T(M\mathbf{x}, z)]\, d_z \mathcal{P}_\lambda(M\mathbf{x}, z), \qquad (13.2)$$

where $B_\lambda(T)$ is the radiation function (approximately equal to the Planck function). The transmission function of the atmosphere in the spectral range $\delta\lambda$ can be defined as the ratio $P = \langle I_\lambda \rangle / \langle I_\lambda^0 \rangle$, where the angle brackets signify averaging over λ with a weighting function characterizing the spectral sensitivity of the radiometer. According to empirical data obtained with the Soviet satellites *Cosmos 149* and *Cosmos 243*, in the range $\lambda = 8$ to 12 μm function P increases from 0.7 at $T_w \approx 0$°C to 0.8 at $T_w \approx 30$°C. Values of P calculated on the basis of aerological soundings of the atmosphere turn out to be too high (and they decrease with increasing T_w), possibly because the absorption of radiation by the aerosol and by thin high-level clouds was not taken into account. Consequently, empirical atmospheric corrections are now being worked out for several ranges of $\delta\lambda$ (on American NOAA satellites four ranges are used: 0.5–0.9 μm, 0.75–1.0 μm, 3.6–4.1 μm, and 10.5–12.5 μm), providing an rms accuracy in determining the absolute values of T_w which is of the order of $\sigma \leqslant 0.5$°C (for T_w variations along the satellite trajectory, the accuracy will naturally be somewhat higher).

In the SHF range $I_\lambda^0 = \varkappa_\lambda T_w$, where $\varkappa_\lambda = 1 - R_\lambda$ is the radiation coefficient (and R_λ is the reflection coefficient). The radiation coefficient is a function of the polarization of the radiation, the nadir angle, the state of the ocean surface (waviness, amount of foam, surface films, ice), and the salinity of the water; this complicates considerably the quantitative interpretation of SHF measurements. However, since radio waves are not absorbed much in the atmosphere, even in clouds (for $\lambda > 3$ cm the optical depth of the atmosphere τ is low, so that the transmission function $\mathscr{P} = e^{-\tau}$ is close to unity), the SHF range is an "all-weather" range. The directly measured radio brightness temperature of the ocean surface T_r is the sum of the emission of the surface, attenuated as it passes through the atmosphere, $T_1 = T_w\,(1 - R)\,e^{-\tau}$, the radiation of the atmosphere itself, $T_2 = T_a\,(1 - e^{-\tau})$, and the atmospheric radiation reflected by the ocean surface and then attenuated as it passes through the atmosphere, $T_3 = T_2\,Re^{-\tau}$. For $\lambda > 3$ cm terms T_2 and T_3 are small, so that $T_w \approx T_r\,(1 - R)^{-1}$. The first experimental SHF measurements of T_w were made in 1968 with *Cosmos 243* (for $\lambda = 3.4$ and 8.5 cm, with a sensitivity of 2°C).

An introduction to the theory of predicting the large-scale variability of the T_w field will be presented in the next section.

Since the mean annual T_w field is approximately zonal, the mean zonal values of the field plotted in Fig. 13.1 provide a fairly good representation. The mean temperature of the ocean surface is 17.82°C, which is 3.6°C higher than the mean air temperature at the Earth's surface. A comparison of the curves for H_s and T_w in Fig. 13.1 indicates that the values of T_w are not closely correlated with the local heat budget H_s; this is evidently because of warm and cold ocean currents.

The thermohaline circulations produce a temperature stratification of the ocean such that T decreases with depth down to values close to the minimum winter water temperatures in the coldest regions on the ocean surface. The temperature stratification of the ocean has the following typical features: (1) an upper mixed layer, in which T varies little with depth, having a thickness of about 100 m in the tropics and in high latitudes a summer thickness of 10–20 m and a winter thickness of hundreds of meters (sometimes extending even to the bottom); (2) a layer of discontinuity or seasonal thermocline tens of meters thick, in which T decreases with depth by several degrees; (3) a main thermocline with a tentative lower boundary at a depth of 1500 m, in which T decreases with depth smoothly, but progressively more slowly, reaching 10.3–11.2°C at a depth of 300 m, 4.0–4.8°C at a depth of 1000 m, and 2.7–3.5°C at a depth of 1500 m; (4) a deep layer, in which T diminishes very gradually with depth down to 1.0–1.5°C. The mean temperature over the entire depth of the ocean T_{av} = 5.7°C. This is 22.7°C higher than the mean temperature over the mass of the atmosphere (−17°C), so that the ocean is, considered as a whole, much warmer than the atmosphere. Figure 13.1 also shows the mean zonal values of T_{av}. The temperature stratifications of the Atlantic Ocean and the standard atmosphere are shown in Fig. 13.2.

SALINITY

The latitudinal zonality of the mean annual salinity field S_0 of the surface waters of the ocean is less pronounced than the zonality of the T_w field; the salinity field has superimposed on it, in particular, regions of freshening of the seawater where great rivers discharge into the ocean. Nevertheless, the mean zonal values of S_0 plotted in Fig. 13.1 agree fairly well with the local water budget $P - E$. The mean value of S_0 is 34.84‰. The values of S_0 and T_w are statistically correlated quite closely with each other: cold surface waters are less saline than warm surface waters, which makes the density variations smaller; there are, to be sure, some regional exceptions to this rule.

The salinity stratification of the ocean has the following typical features: (1) an upper mixed layer; (2) a seasonal halocline, in which S increases appreciably with depth; (3) a subsurface layer of increased salinity, the highest over the vertical, at depths of 100 to 250 m; (4) an intermediate layer of reduced salinity, the lowest over the vertical, at depths of 600 to 1000 m, which is compensated hydrostatically by the greater influence on the water density of the decrease in T with depth; (5) a main halocline at depths to 1500–2000 m, characterized by a gradual rise in S with depth; (6) a deep layer of approximately constant S.

The mean salinity over the entire depth of the ocean is S_{av} = 34.71‰. Figure 13.1 shows the mean zonal values of S_{av}. In the equatorial zone and at latitudes higher than ±40°, it is seen that $S_{av} > S_0$, while in the subtropics and extraequatorial tropics $S_{av} < S_0$. Consequently, appreciable regional variations in the salinity stratification can exist. Actually, the surface minimum of S is absent in tropical waters. In polar and subpolar waters, on the other hand, this minimum is quite pronounced, whereas layers (3) and (4) are absent, so that S increases with depth throughout. In the North Atlantic S is found to decrease monotonically with depth; in Indo-Malayan waters S has a single maximum (at a depth of 600 m); the salinity of the Mediterranean goes through two maxima, one at the surface and one at depths of 500 to 1000 m. This diversity is explained by the fact that a stable stratification of the density $\varrho_w(z) = \varrho_w[T(z), p(z), S(z)]$ can be created by various combinations of

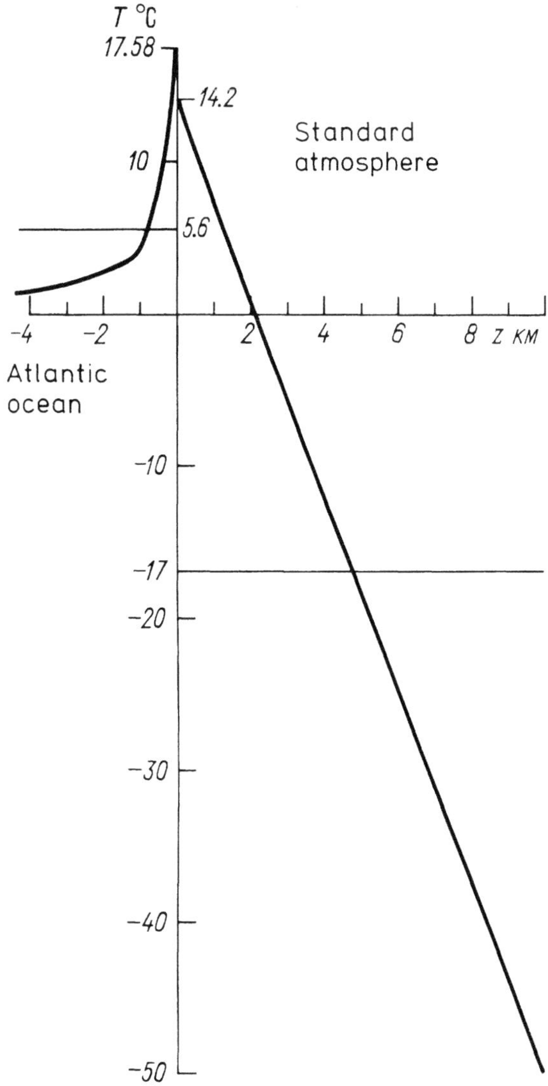

Fig. 13.2. Temperature stratifications of Atlantic Ocean and standard atmosphere.

profiles $T(z)$ and $S(z)$ satisfying the inequality

$$-\alpha \frac{\partial T}{\partial z} + \gamma \frac{\partial S}{\partial z} \leqslant \frac{c_p - c_v}{c_v} \frac{g}{c^2}, \tag{13.3}$$

where α and γ are coefficients in a formula of type (12.3′) for the water density, c_p and c_v are the specific heats of water at constant pressure and constant volume, and c is the velocity of sound; an equals sign in (13.3)

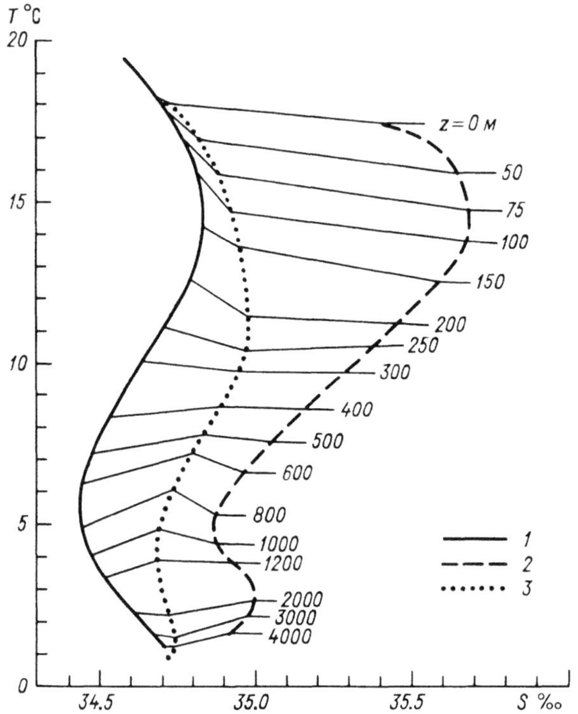

Fig. 13.3. Mean *T–S* curves for Pacific (1), Atlantic (2), and Indian (3) Oceans.

indicates neutral stratification. The combinations of $T(z)$ and $S(z)$ profiles can be portrayed conveniently by so-called *T–S* curves, plotted on graphs in coordinates *S* and *T* on which different depths z are given as points. The mean *T–S* curves for the three oceans are shown in Fig. 13.3. The *T–S* curves deviating the most from these mean curves are those of polar waters (to the left and downward), subpolar waters (to the left), and North Atlantic and Mediterranean waters (to the right).

DENSITY

The relative (reduced to standard atmospheric pressure) density of water

$$\sigma_t - 1000 \left[\frac{\varrho_{w0}(T, S)}{\varrho_{w0}(4^0, 0)} - 1 \right],$$

where ϱ_{w0} is the coefficient in (12.3). This quantity, and even more so the total density ϱ_w, almost everywhere increases with depth. Therefore, σ_t is a minimum at the surface of the ocean. Its mean value over this entire surface is 24.74, and the mean annual zonal values (see Fig. 13.1) are a minimum ($\sigma_t = 22.18$) at latitudes from 10 to 5°N, where, along with the high temperature, the freshening effect of the precipitation in the Intertropical Convergence Zone serves to reduce the water density. The stratification of σ_t has the following typical features: (1) an upper mixed layer; (2) a layer of density discontinuity or seasonal pycnocline, where $\delta\sigma_t/\delta z \approx$ 1/10 m; (3) a principal pycnocline down to a depth of about 1500 m, in which σ_t increases gradually with depth (by about 1.5 units); 4) a deep layer where σ_t increases very gradually with depth. The main contribution to this stratification is made by temperature effects (the principal exception being the Arctic, where the layer of discontinuity is produced mainly by the salinity).

BIOGENIC ELEMENTS

Let us consider briefly the stratification of the so-called biogenic elements: carbon compounds, dissolved oxygen, and compounds of silicon, nitrogen, and phosphorus. These may be of importance in problems of interactions between the oceanic biosphere and the climate. The mean concentrations of the biogenic elements in the waters of the World Ocean (in mg/l) can be expressed numerically as follows: 29.19 C, 5.46 O_2, 2.13 Si, 0.45 N, and 0.07 P. The forms of the carbon compounds found in seawater were discussed in detail in §6. 95% of the carbon is present in dissolved inorganic compounds, the so-called carbonate system, about 80% of which takes the form of the bicarbonate ion HCO_3^-. The pH of the seawater in the tropics and subtropics decreases with depth, on the average, from 8.25 at the ocean surface to a minimum of 7.80–7.85 at depths of 500 to 1000 m, after which it rises to 7.90 at a depth of 1500 m, to remain approximately constant further down. In the polar regions the pH at the surface is 8.10 to 8.05, and it decreases monotonically with depth. The Alk/Cl ratio at the ocean surface increases, on the average, from 0.121 in the tropics and subtropics to 0.124–0.126 in the polar regions; it increases monotonically with depth, reaching 0.128–0.129 at the bottom. The partial pressure of CO_2 in the water at the ocean surface increases from 2.9×10^{-4} –3.0×10^{-4} atm (29 –30 Pa) in the tropics to 3.2×10^{-4} atm (32 Pa) in the Arctic and 3.6×10^{-4} (36 Pa) in the Antarctic. On the whole, it increases with depth, reaching a maximum of 7.6×10^{-4} – 7.9×10^{-4} atm (76–79 Pa) at depths of 500 to 1000 m and in the near-bottom waters; this maximum is only 5.9×10^{-4} atm (59 Pa) in the Arctic.

The ocean water is undersaturated with oxygen, since oxygen in the sea is consumed in the oxidation of substances and the respiration of organisms (at a rate of 0.11 mg/ (liter yr), on the average, so that it is renewed in 50 years). In the surface waters the O_2 concentration increases, on the average, from 4.4–4.6 ml/l in the equatorial zone to 7.0–7.9 ml/l near the poles (especially in the Antarctic). At intermediate depths z_m, there is usually a minimum of O_2, equal to c_m (and sometimes 2, or even 3, minima)(see Table 13.1).

More than 95% of the Si in the ocean water is present in the form of metasilicic and polysilicic acids, the water being highly undersaturated with these substances, since in the upper 50–100 m layer silicic acid is extracted from the water by organisms and enters into the composition of the skeletons of diatoms (which make up 77% of the oceanic phytoplankton), radiolarians, and siliceous sponges. Below 200 m siliceous skeletons start to dissolve, so that the concentration of silicic acid increases with depth (it increases monotonically everywhere except in the Mediterranean intermediate water of the Atlantic, which has an SiO_2 minimum at depths between 1000–1200 and 1400–1700 m). In the surface water at latitudes 35°N to 50°S (and in the entire North Atlantic, except Labrador) the concentration of dissolved silicic acid does not exceed 10 μmoles Si/l, and in the Pacific it increases to 40 μmoles toward the north and 60 μmoles toward the south. In the bottom water this concentration increases to 30–50 in the North Atlantic, 120–140 in the Indian Ocean, and 130–160 μmoles Si/l in the Pacific, especially in its northern regions.

TABLE 13.1

Minimum oxygen concentration in ocean water

Region	z_m m	c_m ml/l
Atlantic Ocean		
Equatorial zone	400	1–2.5
Northern subtropics	>800	3.5–4
Southern subtropics	>1400	4.2–4.4
Arctic	>600–400	5.5–6
Antarctic	>600	4.5
Near-bottom water		4.4–5.9
Pacific Ocean		
Northern tropics	<600–400	0.1–0.5
Northern subtropics	>1400	0.5
Southern subtropics	>2400	3.4–3.5
Northern circumpolar regions	<800–600	<0.5
Antarctic	<800–600	>4
Near-bottom waters		3.5–4.6

There is a large amount of dissolved molecular nitrogen in the ocean, but a significant biogenic role is played only by bound nitrogen, both organic (more than 95% of which is present in dissolved organic matter) and inorganic (ions of nitrate NO_3^-, nitrite NO_2^-, and ammonium NH_4^+). In the photosynthesis layer organic nitrogen predominates. The oxidation of organic matter yields first ammonium, concentrated mainly in the photosynthesis layer, then nitrites (in the layer of density discontinuity), and finally nitrates. A further oxidation of the organic matter may take place via the reduction of nitrates to free nitrogen (denitrification) and, if there is a shortage of nitrates as well, via the reduction of sulfates SO_4^{2-} to free sulfur (sulfate reduction), nitrogen being released in the form of ammonia NH_3.

Consequently, nitrates predominate below the photosynthesis level. In the surface waters their concentrations are a minimum in the tropics and subtropics (less than 1 μmole N/l), except for the eastern equatorial Pacific (partial maximum as high as 15–20 μmoles, caused by equatorial upwelling), whereas toward Antarctica and in the North Pacific these concentrations increase to more than 25; in the North Atlantic, on the other hand, they remain low. At intermediate depths z_M the nitrate concentration reaches a maximum c_M. In the Atlantic $c_M = 25$ to 30 μmoles N/l; the depth $z_M \approx 800$ m in the north and in the tropics, while in the southern subtropics $z_M > 1400$ m and down toward Antarctica $z_M < 400$ m. In the Pacific the nitrate concentration is considerably higher: in the northern subtropics $z_M > 1800$ m and $c_M > 45$ μmoles N/l, in the southern subtropics $z_M > 2400$ m and c_M varies from 35 to 40 μmoles N/l, and in polar waters z_M decreases.

Phosphorus is exceptionally important for the biosphere, since it is a component of the biological "fuel" ATP and also of the phospholipid base of cell membranes. It is present in the ocean water both in organic matter (more than 95% is in dissolved matter) and in inorganic forms (mainly in phosphates, salts of orthophosphoric acid H_3PO_4, both undissociated and in the form of ions $H_2PO_4^-$, HPO_4^{2-}, and PO_4^{3-}).

In highly productive regions of the Pacific the concentrations of organic and inorganic phosphorus are as follows:

Depth, m	0–100	100–500	500–1000	1000–4000
c_M, μmoles P/l				
Organic	0.5	0.4–0.3	0.2	0.1–0.05
Inorganic	1.0	2.0–2.5	3.2	3.0–2.8

In unproductive regions there is very little organic phosphorus present, even in the photosynthesis layer. The phosphate concentrations in the surface water reach a minimum in the tropics and subtropics (less than 0.2 μmole) except for the eastern equatorial Pacific. At intermediate depths the phosphate concentration reaches a maximum, slightly less deep and more abrupt than the nitrate maximum, with concentrations of about 2 μmoles P/l in the Atlantic and more than 3 μmoles P/l in the northern half of the Pacific. At depths of around 2000 m the phosphate concentrations have a broad minimum, and then they increase slightly toward the ocean floor.

MAIN THERMOCLINE

To conclude this section, let us present an approximate theory of the main thermocline, according to Kamenkovich (1958) and Welander (1959). Local

Cartesian coordinates (x, y, z) will be used, with the x axis directed toward the east, the y axis toward the north, and the height z upward from the upper boundary of the thermocline (that is, from the lower boundary of the so-called Ekman surface boundary layer of the ocean). The corresponding velocity components are u, v, and w. The premises of the approximate theory are as follows: (1) the thermocline is assumed to be quasi-stationary; (2) horizontal currents in it are geostrophic; (3) density perturbations $\varrho_w' = \varrho_w - \varrho_0$ depend only on the temperature perturbations $T' = T - T_0$, that is, only the first term on the right-hand side of formula (12.3′) is taken into account (ϱ_0 is here taken to be constant); (4) the equation for the entropy (12.4) is replaced by the equation of convective heat conduction with vertical turbulent exchange. With these simplifications equations (9.1)–(9.4) and (12.4) become:

$$\frac{\partial p'}{\partial z} = \varrho_0 g \alpha_0 T'; \qquad fv = \frac{1}{\varrho_0}\frac{\partial p'}{\partial x}; \qquad fu = -\frac{1}{\varrho_0}\frac{\partial p'}{\partial y};$$
$$\frac{\partial u}{\partial x} + \frac{\partial v}{\partial y} + \frac{\partial w}{\partial z} = 0; \qquad u\frac{\partial T'}{\partial x} + v\frac{\partial T'}{\partial y} + w\frac{\partial T'}{\partial z} = K\frac{\partial^2 T'}{\partial z^2}, \tag{13.4}$$

where, as usual, $p' = p - p_0$ is the pressure perturbation and f is the Coriolis parameter. As limiting conditions, at the upper boundary of the thermocline $z = 0$ we will assume that the temperature $T_s'(x, y)$ and the vertical velocity $w_s(x, y) = \mathrm{curl}_z(\tau_0/f)$ are specified, while at the ocean floor $z = -H(x, y)$ the temperature T_H' is known (and with no loss of generality assumed to be zero) and the condition of nonpassage $w = -u(\partial H/\partial x) - v(\partial H/\partial y)$. The first four equations of (13.4) enable us to express all the unknown quantities in terms of a single function M, as follows:

$$u = -\frac{1}{f}\frac{\partial^2 M}{\partial y \partial z}; \qquad v = \frac{1}{f}\frac{\partial^2 M}{\partial x\,\partial z}; \qquad w = \frac{\beta}{f^2}\frac{\partial M}{\partial x};$$
$$p' = \varrho_0\frac{\partial M}{\partial z}; \qquad T' = \frac{1}{g\alpha_0}\frac{\partial^2 M}{\partial z^2}, \tag{13.5}$$

where, as usual, $\beta = df/dy$. By substituting these formulas into the last equation of (13.4) and into the boundary conditions, we obtain the following equations and boundary conditions for function M:

$$Kf\frac{\partial^4 M}{\partial z^4} + \frac{\partial(\partial^2 M/\partial z^2, \partial M/\partial z)}{\partial(x, y)} - \frac{\beta}{f}\frac{\partial M}{\partial x}\frac{\partial^3 M}{\partial z^3} = 0;$$
$$z = 0: \frac{\partial^2 M}{\partial z^2} = g\alpha_0 T_s'; \qquad M = \frac{f^2}{\beta}\int_0^x w_s\,dx;$$
$$z = -H: \frac{\partial^2 M}{\partial z^2} = 0; \qquad \frac{\beta}{f}\frac{\partial M}{\partial x} = \frac{\partial(H, \partial M/\partial z)}{\partial(x, y)}. \tag{13.6}$$

If the vertical velocities w, w_s are measured on a scale W, the horizontal distances x, y on a scale L, and the temperatures T' on a scale Θ, then from the third equation of (13.5) we get a scale fL^2W for M and from the fifth equation we get the so-called *advective* vertical scale $H_a = (fL^2W/g\alpha_0\Theta)^{1/2}$ for z. In addition, differentials of the bottom relief will be measured on a scale H_w. If we convert to the corresponding dimensionless variables in (13.6), we are left with only two parameters: $\varepsilon = H_d/H_a$ (where $H_d = K/W$ is the *diffusive* vertical scale) for the first term (with the higher derivative) in equation (13.6) and $\delta = H_a/H_w$ in the boundary conditions at the ocean bottom. For the following values typical of the principal thermocline: $W \approx 10^{-4}$ cm/s, $K \approx 10^{-1}$ cm^2/s, $\Theta \approx 10°$, and taking into account that $L \approx 6 \times 10^3$ km, $f \approx 10^{-4}$ s^{-1}, $g \approx 10^3$ cm/s^2, and $\alpha_0 \approx 2.5 \times 10^{-4}$ K^{-1}, we obtain $H_d \approx 10$ m and $H_a \approx 400$ m, so that $\varepsilon \approx 1/40$.

Consequently, except in the boundary layers at the upper limit of the principal thermocline and at the ocean bottom, the first term in equation (13.6) can be neglected. This means that throughout most of the main thermocline the vertical advection of heat is equalized by the horizontal advection. Equation (13.6) is then third order in z, so that for it we can construct solutions satisfying three of the four boundary conditions of (13.6). In order to eliminate the discrepancy in satisfying the fourth boundary condition, there must exist a corresponding boundary layer within which the first term of equation (13.6), that is, the effect of turbulent heat conduction, plays an appreciable role.

Let us consider, for instance, the upper boundary layer and let us denote its thickness as H_{bl}. From the upper boundary conditions it is clear that the vertical variations M_z of quantity M must be of order $g\alpha_0\Theta\, H_{bl}^2$, while the horizontal variations M_h are, once again, of order fL^2W. Then the ratio of the orders of magnitude of the three terms in equation (13.6) will be

$$1 : \frac{M_z}{M_h}\frac{H_{bl}}{H_d} : \frac{H_{bl}}{H_d} .$$

Now, if the first two terms were comparable in value, that is, if $M_zH_{bl} \sim M_hH_d$, then it would follow that $H_{bl} \sim (fL^2K/g\alpha_0\Theta)^{1/3} \sim 100$ m, and the third term in (13.6) would be an order of magnitude greater than the first two, which is impossible. Thus, the third term must be comparable with the first term, that is, we should set $H_{bl} = H_d$. The relative order of magnitude of the second term turns out to be

$$\frac{M_z}{M_h} = \frac{g\alpha_0\Theta H_d^2}{fWL^2} \sim 10^{-3}.$$

Thus, in the boundary layer the second term in equation (13.6) (that is, the effect of horizontal advection) can be neglected, so that in this layer the vertical advection of heat is counteracted by the turbulent heat transfer. Since in this layer $w \approx w_s$ varies little with depth, equation (13.6) in it can be rewritten approximately as

$$K \frac{\partial^4 M}{\partial z^4} - w_s \frac{\partial^3 M}{\partial z^3} = 0. \tag{13.6'}$$

The solution of this equation satisfying the upper boundary conditions has the form

$$M = \frac{f^2}{\beta} \int_0^x w_s \, \mathrm{d}x + Az + g\alpha_0 T_s \frac{z^2}{2} - \\ - B\left[1 + \left(\frac{w_s}{K}\right)^2 \frac{z^2}{2} - \exp\left(\frac{w_s z}{K}\right)\right], \tag{13.7}$$

where A and B are arbitrary functions of x and y. The temperature then has the form

$$T' = T_s' - \frac{B}{g\alpha_0}\left(\frac{w_s}{K}\right)^2 \left[1 - \exp\left(\frac{w_s z}{K}\right)\right]. \tag{13.8}$$

It must, naturally, increase with height, so that for $w_s > 0$ (in regions of upwelling) we have $B > 0$; for $w_s < 0$ (in regions of subsidence of the water), on the other hand, we have to take $B \leqslant 0$, although for usual boundary layers such a solution is abnormal.

Let us give another particular solution of the complete equation (13.6), similar to the solution of Needler (1967, 1971):

$$M = \frac{f^2}{\beta} \int_0^x w_s \mathrm{d}x + Az - \frac{f^2 g\alpha_0}{b^2} T_s' \left[1 - \exp\left(\frac{bz}{f}\right)\right], \tag{13.9}$$

where b is an arbitrary constant, and A is a function of x and y which satisfies the equation

$$\frac{\partial(A, \ln T_s')}{\partial(x, y)} = \frac{b^2}{f} K - bw_s + \frac{\beta g\alpha_0}{b} \frac{\partial T_s'}{\partial x}. \tag{13.10}$$

This solution already satisfies the upper boundary conditions of (13.6). For $b > 0$ (in the Northern Hemisphere) and large bH it also satisfies approximately the first of the bottom conditions. The second of the bottom conditions may be satisfied for certain special forms of function $w_s(x, y)$, that is, of the wind field (for $H = \text{const}$, for instance, these forms are found from the condition $\partial M/\partial x = 0$ for $z = -H$).

Of course, partial solution (13.9) has the following shortcomings: the turbulent thermal conductivity K plays only a secondary role, and diffusive boundary layers, which can naturally be expected to exist, are absent in this solution. Nevertheless, solution (13.9) was found to describe fairly well the features of the vertical distributions of u, v, and T in the ocean (in the western

Atlantic, in particular, according to Needler's calculations), including the appreciable variations of these distributions with depth in the principal thermocline, about one kilometer thick, and their approximate constancy further down (in the abyssal zone). Solution (13.9) can be used as a first approximation for a numerical solution of the complete problem (13.6) using the iteration method. A generalization taking the salinity S into account can be made with the aid of the equations

$$u \frac{\partial T'}{\partial x} + v \frac{\partial T'}{\partial y} + w \frac{\partial T'}{\partial z} = K_T \frac{\partial^2 T'}{\partial z^2};$$

$$u \frac{\partial S'}{\partial x} + v \frac{\partial S'}{\partial y} + w \frac{\partial S'}{\partial z} = K_S \frac{\partial^2 S'}{\partial z^2};$$

$$g(\alpha_0 T' - \gamma_0 S') = \frac{\partial^2 M}{\partial z^2}, \qquad (13.11)$$

where u, v, w are given by formulas (13.5), and for $z = 0$ and $z = -H$ the boundary conditions for w, T', and S' are specified. It should be noted, too, that in some works the large-scale advection (left-hand sides of equations (13.11)) was replaced by the horizontal turbulent exchange (then, however, the model of the thermocline depended considerably on the choice of the exchange coefficients).

The uppermost layer of the ocean (mixed layer + seasonal thermocline) requires a much more detailed description, and it will be the subject of the following section. Finally, it should be mentioned that no mathematical models accounting for the vertical distributions of biogenic elements in the ocean have as yet been constructed.

14. The Upper Layer of the Ocean

Atmospheric effects (frictional stress of wind τ_0 and fluxes of heat H_s and moisture $P - E$) produce capillary and gravity waves on the ocean surface and give rise to drift currents, a mixed layer, and a seasonal thermocline in the upper layer of the ocean.

The capillary waves have a very small scale. Actually, the frequencies of linear capillary-gravity waves on the surface of a deep ocean are given by the formula $\omega = (gk + \gamma k^3)^{1/2}$, where γ is the surface tension, equal to 72.5 cm^3/s^2 for a water–air interface (pure water) at 20°C. Note, however, that the surface tension may vary considerably, due to the presence of surface films, which affect the roughness of the water surface and thus the transfer of momentum from the air to the water; such films also influence the evaporation of water and may be of climatic significance.

Hence it appears that capillary effects predominate only at low wavelengths $l < 2\pi(\gamma/g)^{1/2} \approx 1.7$ cm. They contribute to the formation of the effective roughness z_0 of the wavy water surface, which can in general be represented as

$$z_0 = \frac{u_{*a}^2}{g} F\left(\frac{\gamma^{3/4}}{g^{1/4} v_w}, \frac{u_{*a}^3}{g v_a}, \frac{g\sigma\xi}{u_{*a}^2} \right), \tag{14.1}$$

where u_{*a} is the friction velocity in the air layer next to the water (approximately 0.02–0.03 of the wind speed G at the upper limit of the atmospheric boundary layer and increasing in value with a decrease in log $(g/fG) = 6.0$ to 3.5), and $\sigma_\xi \approx 0.2\ U^2/g$ is the mean-square height of the gravity waves (U being the wind speed at the height of a ship's mast, about 10 m). The first argument of function F is the Reynolds number of the capillary waves Re_γ, the second is the Reynolds number of the air layer next to the water Re_a, and the third differs just by a numerical multiplier from the quantity $(c_0/u_{*a})^2$, where $c_0 \approx (g\sigma_\xi)^{1/2}$ is the mean effective phase velocity of short gravity waves, which contribute appreciably to the roughness of the water surface (the ratio c_0/u_{*a} increases as the waviness develops). For very low Re_a we have $F \sim (10\ \mathrm{Re}_a)^{-1}$ and $z_0 \sim 0.1\ v_w/u_{*w}$. For high Re_a in the case of developing waviness $(1 \leqslant c_0/u_{*a} \leqslant 10^2)$ we have

$$F \sim \left(\frac{c_0}{u_{*a}} \right)^2 \exp\left(- \frac{\varkappa c_0}{u_{*a}} \right),$$

while in the case of developed waviness $(c_0/u_{*a} \geqslant 1)$ function F turns out to depend only on Re_γ, so that we get $z_0 = m(\mathrm{Re}_\gamma)u_{*a}^2/g$, where usually $m \sim 0.04$ (formula of Charnock, 1955).

The energy-bearing components of developed wind waves have phase velocities c of the order of the wind speed U (usually 10–30 m/s), lengths $l \approx 2\pi U^2/g \approx 60$ to 600 m, periods $t \approx 2\pi\ U/g \approx 5$ to 20 s, and heights $\sigma_\xi \approx 0.2U^2/g \approx 2$ to 20 m. Their mean energy per unit horizontal area $E = g\varrho_w\sigma_\xi^2$ and momentum $M = E/c$ are of the same order of magnitude as the air layer of thickness l next to the water. The rate of transfer of energy from wind to waves can be evaluated approximately using the formula $\delta E/\delta t \approx 10^{-1}\ \tau_0\ (g\sigma_\xi)^{1/2}$ (Kitaigorodskii, 1970). The corresponding rate of transfer of momentum to the waves, in fractions of the total momentum flux τ_0, is

$$\frac{1}{\tau_0} \frac{\partial M}{\partial t} \sim 10^{-1} \frac{(g\sigma_\xi)^{1/2}}{c} \approx 10^{-1} \frac{(0.2U^2)^{1/2}}{U} \approx 5 \times 10^{-2}.$$

Thus, the bulk of the momentum flux is imparted to drift currents. According to Longuet–Higgins (1969), the energy E_1 lost by the waves during a wave cycle t, for breaking of the wave crests, is of the order of $E_1 \sim 10^{-4}\ E$. This is equivalent to a rate of energy dissipation per unit mass $\varepsilon \sim E_1/\varrho_w lt \approx 10^{-7}\ gU \approx 1 \times 10^{-1}$ to $3 \times 10^{-1}\ \mathrm{cm^2/s^3}$, which agrees fairly well with the estimates based on turbulence measurements, cited in the previous section during the discussion of the generation of thermohaline circulations.

The vertical momentum flux at the water surface must remain continuous, that is, $\tau_0 = \varrho_a u_{*a}^2 = \varrho_w u_{*w}^2$. Moreover, the friction velocity in the water $u_{*w} = (\varrho_a/\varrho_w)^{1/2} u_{*a} \approx \frac{1}{30} u_{*a}$ for a typical value of $u_{*a} \approx 30$ cm/s (where $\tau_0 \approx 0.1$ Pa) is of the order of $u_{*w} \approx 1$ cm/s. Note that equality of the vertical turbulent heat fluxes $q \approx c_p \varrho_a u_{*a} \delta T_a = c_w \varrho_w u_{*w} \delta T_w$ gives

$$\delta T_w = \frac{c_p}{c_w} \left(\frac{\varrho_a}{\varrho_w} \right)^{1/2} \delta T_a \approx 10^{-2} \delta T_a,$$

so that almost undectectable vertical temperature differentials δT_w in the upper mixed layer (of the order of 0.01°/10 m, say) can cause considerable heat fluxes there. The vertical scale of the stratification is given by the formula of Monin and Obukhov (1953):

$$L_w = \frac{u_{*w}^3}{\varkappa (g/\varrho_w)\, M_s}, \tag{14.2}$$

where M_s is the vertical mass flux defined by formula (13.1). In the upper mixed layer, for $u_{*w} = 1$ cm/s and $M_s = 5$ g/(cm^2 yr), this scale is 150 m. Since for $q_a = q_w$ we obtain

$$\frac{L_a}{L_w} \sim \alpha T_a \frac{c_p}{c_w} \left(\frac{\varrho_w}{\varrho_a} \right)^{1/2} \sim \frac{1}{2},$$

a scale $L_a = 75$ m signifies an almost neutral stratification, and the stratification of the upper mixed layer must be recognized as such. The vertical scale of the *Ekman* boundary layer in the ocean, in which the eddy viscosity and the Coriolis force are appreciable, can be found, as for the atmosphere, using the formula

$$h_E = \varkappa_0 \frac{u_{*w}}{f}, \tag{14.3}$$

where, according to atmospheric measurements, $\varkappa_0 \approx 0.2$, so that for $u_{*w} \approx 1$ cm/s and $f \approx 10^{-4}$ s^{-1} we get $h_E \approx 20$ m. The response time of the currents in the Ekman boundary layer, with respect to wind effects, $\varrho_w h_E v_{av}/\tau_0 \approx 3$ h (where $v_{av} = u_{*w}/\varkappa_0 \approx 5$ cm/s is the mean drift-current velocity over the depth of the Ekman boundary layer), that is, it is short in comparison with the time of synoptic variations of the wind, so that the currents in the Ekman layer can be considered quasi-stationary. Therefore, the equations of horizontal motion for the Ekman boundary layer, in the same system of coordinates as (13.3), can be written as

$$-fv = -g \frac{\partial \zeta}{\partial x} + \frac{1}{\varrho_w} \frac{\partial \tau_x}{\partial z}; \qquad fu = -g \frac{\partial \zeta}{\partial y} + \frac{1}{\varrho_w} \frac{\partial \tau_y}{\partial z}. \tag{14.4}$$

First let us consider a pure drift current ($\zeta = 0$), in which case the x axis has the direction of the driving wind (vector $\boldsymbol{\tau}_0$), so that for $z = 0$ we will have $\tau_{x0} = -\varrho_w u_{*w}^2$, $\tau_{y0} = 0$. Then, if equations (14.4) are averaged over the thickness of the Ekman boundary layer, we obtain $v_{av} = -\dfrac{\tau_{x0}}{f\varrho h_E} = \dfrac{u_{*w}}{\varkappa_0}$ and $u_{av} = 0$. This result has already been used above; it should be noted that the mean drift current over the depth is perpendicular to the driving wind. The velocity of a surface drift current U_d and the angle α_d between it and the x axis can be determined with the aid of the familiar resistance law for the Ekman layer:

$$\frac{U_d}{u_{*w}} = \frac{1}{\varkappa}\left[A^2 + \left(\ln\frac{u_{*w}}{fz_0} - B\right)^2\right]^{1/2};$$

$$\sin\alpha_d = \frac{A}{\varkappa}\frac{u_{*w}}{U_d}, \tag{14.5}$$

where $A \approx 4.3$ and $B \approx 1.1$. For instance, for $z_0 = 0.1\, \nu_w/u_{*w}$, $u_{*w} = 1$ cm/s, $\nu_w = 0.012$ cm^2/s, and $f = 10^{-4}$ s^{-1}, we get $U_d \approx 40$ cm/s and $\alpha_d = 16°$.

Now let us take into account not only the drift current, but also the gradient current produced by inclinations of the ocean level $\nabla\zeta$. Using a semiempirical theory of turbulence, in (14.4) we set $\tau_x = \varrho_w\, K(\partial w/\partial z)$ and $\tau_y = \varrho_w\, K(\partial v/\partial z)$, where K is the eddy viscosity. First, in order to get a qualitative picture, we assume that K = const. Then the solution of equations (14.4) for an ocean of constant depth H has the form

$$u + iv = H\frac{\tau_{x0} + i\tau_{y0}}{\varrho_w \mu K}\frac{\sinh\mu(z'+1)}{\cosh\mu} + $$
$$+ i\frac{g}{f}\left(\frac{\partial\zeta}{\partial x} + i\frac{\partial\zeta}{\partial y}\right)\left(1 - \frac{\cosh\mu z'}{\cosh\mu}\right)$$
$$\approx H\frac{\tau_{x0} + i\tau_{y0}}{\varrho_w \mu K}e^{\mu z'} + $$
$$+ i\frac{g}{f}\left(\frac{\partial\zeta}{\partial x} + i\frac{\partial\zeta}{\partial y}\right)[1 - e^{-\mu(z'+1)}], \tag{14.6}$$

where $z' = z/h$; $\mu = (1+i)/E$; $E = h'_E/H$ is the Ekman number, which is low (making possible the approximate equality in (14.6)); and $h'_E = (2K/f)^{1/2}$ is the new vertical scale of the Ekman boundary layer (for $h'_E = 20$ m and $f = 10^{-4}$ s^{-1} we have $K = 200$ cm^2/s). Function $\zeta(x, y)$ must be found from the condition

$$w(z = -H) = \int_{-H}^{0}\left(\frac{\partial u}{\partial x} + \frac{\partial v}{\partial y}\right)dz = 0,$$

with the required boundary conditions at the ocean shore Γ. Equations (14.4) without the nonlinear terms and horizontal turbulent exchange, allow us to require at the ocean shore only nonpassage as an average over the depth, that is, equality to zero of that component S_n of the so-called *total flow* S which is normal to Γ. The total flow is defined as

$$S_x + iS_y = \int_{-H}^{0} (u + iv)\,dz = \frac{\tau_{x0} + i\tau_{y0}}{if\varrho_w}\left(1 - \frac{1}{\cosh\mu}\right) +$$

$$+ \frac{igH}{f}\left(\frac{\partial\zeta}{\partial x} + i\frac{\partial\zeta}{\partial y}\right)\left(1 - \frac{\tanh\mu}{\mu}\right) \approx \frac{\tau_{x0} + i\tau_{y0}}{if\rho_w} +$$

$$+ \frac{igH}{f}\left(\frac{\partial\zeta}{\partial x} + i\frac{\partial\zeta}{\partial y}\right)\frac{\mu - 1}{\mu}\;. \qquad (14.7)$$

It should be noted that, in view of the nondivergence of the three-dimensional velocity and the zero boundary conditions for w, the total flow is also nondivergent: $\partial S_x/\partial x + \partial S_y/\partial y = 0$. Finally, we will give an expression for the Ekman vertical velocity w, obtained from the continuity equation with the aid of the approximate formula (14.6):

$$w = \text{curl}_z\left(\frac{\tau_0}{\varrho_w f}\right) + \frac{g\beta z}{f^2}\frac{\partial\zeta}{\partial x} - \text{div}\left\{\frac{\tau_{x0} + i\tau_{y0}}{i\varrho_w f}\, e^{\mu z'}\right.$$

$$\left. - \frac{gH}{i\mu f}\left(\frac{\partial\zeta}{\partial x} + i\frac{\partial\zeta}{\partial y}\right) e^{-\mu(z'+1)}\right\}, \qquad (14.8)$$

the first two terms of which describe w outside of the boundary layers, where the current is geostrophic (the term without an exponent in approximate formula (14.6)). Therefore, even the geostrophic vertical velocity depends explicitly on τ_0, so that the upper Ekman boundary layer must be considered when determining this velocity. Vertical velocity (14.8) makes possible "Ekman pumping", namely an exchange of heat and foreign matter in the water between the Ekman boundary layer and the lower-lying layers; this exchange may be of climatic significance.

The above calculation for K = const is merely qualitative in nature. For the calculation to be quantitatively accurate, K cannot be assumed constant (or any specified function of z). Instead, it has to be determined with the aid of additional equations, the principal one usually being a semiempirical equation for the kinetic energy of turbulence per unit mass $b = \frac{1}{2}(\sigma_u^2 + \sigma_v^2 + \sigma_w^2)$. This equation can be written as follows:

$$\frac{\partial b}{\partial t} = K\left[\left(\frac{\partial u}{\partial z}\right)^2 + \left(\frac{\partial v}{\partial z}\right)^2\right] + \frac{g}{\varrho_w}M - \frac{1}{\varrho_w}\frac{\partial B}{\partial z} - \varepsilon, \quad (14.9)$$

where the first term on the right-hand side describes the generation of turbulent energy by means of the work of Reynolds stresses $(\tau/\varrho_w)(\partial u/\partial z)$, the second term represents the effect of the work of Archimedes forces, where

$$M = -\overline{\varrho'_w{}'} = \alpha\varrho_w\overline{T'w'} - \gamma\varrho_w\overline{S'w'} =$$

$$= -\alpha\varrho_w K_T \frac{\partial T}{\partial z} + \gamma\varrho_w K_s \frac{\partial S}{\partial z} \tag{14.10}$$

is the vertical mass flux, the magnitude of which at the ocean surface M_s is given by formula (13.1), the third term represents the vertical diffusion of turbulent energy (where $B = -\varrho_w K_b(\partial b/\partial z)$ is the vertical flux of turbulent energy), and the fourth term is its viscous dissipation.

The coefficients of turbulent exchange k, K_T, K_s, K_b and the quantity ε can, according to Kolmogorov (1942), be represented in terms of the turbulent energy b and its vertical scale l (the "mixing length") with the aid of the following relations of dimensional analysis:

$$K = lb^{1/2}; \quad K_T = c_T lb^{1/2}; \quad K_s = c_s lb^{1/2}; \quad K_b = c_b lb^{1/2};$$
$$\varepsilon = c_\varepsilon l^{-1} b^{3/2}, \tag{14.11}$$

According to experimental data, $c_b \approx 1$ and $c_\varepsilon \approx 0.88$, while coefficients $c_T \approx c_s$ are functions of the dimensionless depth z/L_w, where L_w is the vertical scale of stratification, or else functions of the dynamic Richardson number Rf or the kinematic Richardson number Ri. The latter are defined by the formulas

$$\mathrm{Rf} = -gM\left(\tau \frac{\partial \mathbf{u}}{\partial z}\right)^{-1} = c_T \mathrm{Ri};$$
$$\mathrm{Ri} = g\left(\alpha \frac{\partial T}{\partial z} - \gamma \frac{\partial S}{\partial z}\right)\left[\left(\frac{\partial u}{\partial z}\right)^2 + \left(\frac{\partial v}{\partial z}\right)^2\right]^{-1}, \tag{14.12}$$

and for stable stratification it appears that $c_T \ll 1$.

Now the only thing left is to find the mixing length l. It is sometimes defined using the formula of Prandtl (1942) $l = -\varkappa_z$, where $\varkappa \approx 0.4$, or else it is assumed that $l = ch_0$, where h_0 is the thickness of the upper mixed layer and $c \approx 0.05$. Alternatively, some intermediate formula can be used, for example, $l = -\varkappa\Psi\,(\partial\Psi/\partial z)^{-1}$, where $c_1 \approx 25$. Finally, in some works it is assumed that $l = -\varkappa\Psi\,(\partial\Psi/\partial z)^{-1}$, where, generalizing the familiar formula of von Kármán for the boundary layer $\Psi = \partial u/\partial z$, in the manner of formulas (14.11) it is taken that $\Psi \sim l^{-1}b^{1/2} \sim K^{-1/2}\varepsilon^{1/2}$. Now, if ε is found from equation (14.9) for $\partial b/\partial t = 0$, we have

$$\Psi = \left[\left(\frac{\partial u}{\partial z}\right)^2 + \left(\frac{\partial v}{\partial z}\right)^2 - c_T g\left(\alpha \frac{\partial T}{\partial z} - \gamma \frac{\partial S}{\partial z}\right) + \right.$$
$$\left. + \frac{1}{K} \frac{\partial}{\partial z} K \frac{\partial b}{\partial z}\right]^{1/2}. \tag{14.13}$$

Instead of constructing formulas for l, Kochergin *et al.* (1975–1976), using an idea of Kolmogorov, constructed a semiempirical equation for the rate of dissipation of turbulent energy ε. This equation, which is analogous to (14.9), has the form

$$\frac{\partial \varepsilon}{\partial t} = 1.38 \frac{\varepsilon}{b} K \left[\left(\frac{\partial u}{\partial z} \right)^2 + \left(\frac{\partial v}{\partial z} \right)^2 \right] + 1.4 \frac{\varepsilon}{b} \frac{g}{\varrho_w} M +$$

$$+ \frac{\partial}{\partial z} K \frac{\varepsilon}{\partial z} - 1.4 \frac{\varepsilon^2}{b}, \tag{14.14}$$

and, used in conjunction with (14.9) and the formula $K = 0.08\, b^2 \varepsilon^{-1}$ following from (14.11), it makes it unnecessary to introduce l. The results of numerical calculations of the structure of the upper layer of the ocean, using these equations under both steady-state and nonsteady-state conditions, proved to be very close to the field data. The turbulence can be described even more completely without introducing the exchange coefficients K, if we use the equations for all the single-point second moments of the fluctuations u', v', w', T', and S', derived from the equations of hydrodynamics, the third moments and the other "extraneous" unknowns in these equations being expressed with the aid of semiempirical formulas, as was done by Monin (1965) for the atmospheric boundary layer.

We noted above that the response time of currents in the Ekman layer to wind effects (measured in hours) is short in comparison with the time required for synoptic variations of the wind (measured in days), and thus the equations of motion (14.4) for the Ekman boundary layer were taken to be steady-state. Since the relaxation time $b\varepsilon^{-1}$ of the turbulence in the upper layer of the ocean is even shorter (for $b^{1/2} = 1$ to 10^{-1} cm/s and $\varepsilon = 10^{-1}$ to 10^{-4} cm^2/s^3, this time is of the order of 10^1 to 10^2 s), the equations of dynamics of the turbulence (14.9) and (14.14) can, all the more so, be taken to be steady-state. On the other hand, the time $h_0 \delta T\, (H_s/c_w \varrho_w)^{-1}$ of response of the upper layer of the ocean to thermal effects of the atmosphere (where $h_0 \approx 20$ m is the thickness of the upper mixed layer, $\delta T \approx 1$°C is the temperature differential in the seasonal thermocline, and $H_s \approx 70$ W/m^2 is the heat flux from the atmosphere, so that $H_s/c_w \varrho_w \approx 2 \times 10^{-3}$ cm K/s) turns out to be of the order of 10^6 s ≈ 12 days. Consequently, variations of the thermal structure of the upper layer of the ocean caused by atmospheric effects must be described by unsteady-state equations. The response time of the salinity $h_0\, \delta S[(E - P)S/\varrho_w]^{-1}$ is even longer: for $\delta S \approx 0.3$‰, $E - P \sim 200$ g/(cm^2 yr), and $S \approx 3.5 \times 10^{-2}$, it is of the order of a month. If, when describing the evolution of the upper layer of the ocean under the influence of atmospheric effects, we neglect the contributions of salinity, large-scale advection, and horizontal turbulent exchange, then the evolution of the thermal structure can be described by the equation

$$\frac{\partial T'}{\partial t} = - \frac{\partial Q}{\partial z} = \frac{\partial}{\partial z} K_T \frac{\partial T'}{\partial z}\ ; \quad Q = \overline{T'w'}. \tag{14.15}$$

To describe the deepening of the seasonal thermocline (the increase in the thickness h_0 of the upper mixed layer which takes place in autumn and winter, and also during storms), Barenblatt (1978) proposed using a self-similar solution of equation (14.15) for $K_T \approx$ const in the form of a steady traveling heat wave:

$$T'(z,t) = \begin{cases} T'_s & \text{for } 0 \geqslant z \geqslant -h_0(t) \\ & = -(h_{00} + u_0 t), \\ T'_s \exp\left\{ \dfrac{u_0}{K_T} [h_0(t) + z] \right\} & \text{for } z \leqslant -h_0(t), \end{cases} \tag{14.16}$$

where $u_0 = \partial h_0/\partial t > 0$ is the rate of deepening (submergence) of the lower boundary of the upper mixed layer, assumed here, like T'_s, to be quasi-constant. This solution is stable relative to small perturbations. It shows a good fit with the laboratory data of Linden (1975) and with the observational data obtained aboard the weather ship *Papa* in the Pacific (50°N, 145°W), plotted in Fig. 14.1. Consequently, this solution can be taken as theoretical evidence in favor of the similarity hypothesis of Kitaigorodskii and Miropol's-kii (12.8). Formula (12.9) can serve as an approximation for (14.16) when choosing the effective final thickness of the seasonal thermocline $h_1 - h_0 = (8/3)(K_T/u_0)$. Since in Linden's experiments $h_1 - h_0 \approx 3$ cm and $u_0 \approx 10^{-3}$ cm/s, the value of K_T was of the order of 10^{-3} cm²/s. This is somewhere between the values of the molecular ($\sim 10^{-5}$ cm²/s) and turbulent (~ 10 cm²/s) thermal diffusivities. This accounts for the presence in the seasonal thermocline of horizontally spreading turbulent spots, forming in regions where internal waves become unstable. At *Papa* station $h_1 - h_0 \sim 10$ m and $u_0 \sim 10^{-3}$ cm/s, so that there $K_T \sim 1$ cm²/s.

For a more detailed description of the evolution of the upper layer of the ocean, K_T and u_0 are not specified, but instead they are determined with the aid of additional equations. In the simplest models we can use equations (14.15) and (14.9), integrated over the thickness of the upper mixed layer. Then, in (14.15) we can set $Q = \overline{T'w'}\ \frac{F_s}{c_w \varrho_w} e^{\gamma z}$, where F_s is the vertical flux of solar radiation at the ocean surface, and $\gamma = 0.2$ to 0.05 m^{-1} is the coefficient of its attenuation in seawater; the left-hand side of (14.9), $\partial b/\partial t$, can be equated to zero. In one of the first such models, devised by Kraus and Turner (1967), the seasonal thermocline was idealized as a temperature jump at a depth $z = -h_0(t)$ from $T'_s(t)$ to $T'_1 = T'_s - \delta T =$ const. Then, integration of equations (14.15) and (14.9) gives

$$h_0 \frac{\partial T'_s}{\partial t} = Q_1 - Q_s - \frac{\partial h_0}{\partial t}\,\delta T; \tag{14.17}$$

$$\int_{-h_0}^{0} \left(\frac{\tau}{\varrho_w} \cdot \frac{\partial \mathbf{u}}{\partial z} + \frac{\alpha g F_s}{c_w \varrho_w} e^{\gamma z} - \varepsilon \right) dz - \frac{B_s - B_1}{\varrho_w} = -\frac{\alpha g h_0}{2} \left(Q_1 + Q_s - \frac{\partial h_0}{\partial t}\,\delta T \right), \tag{14.18}$$

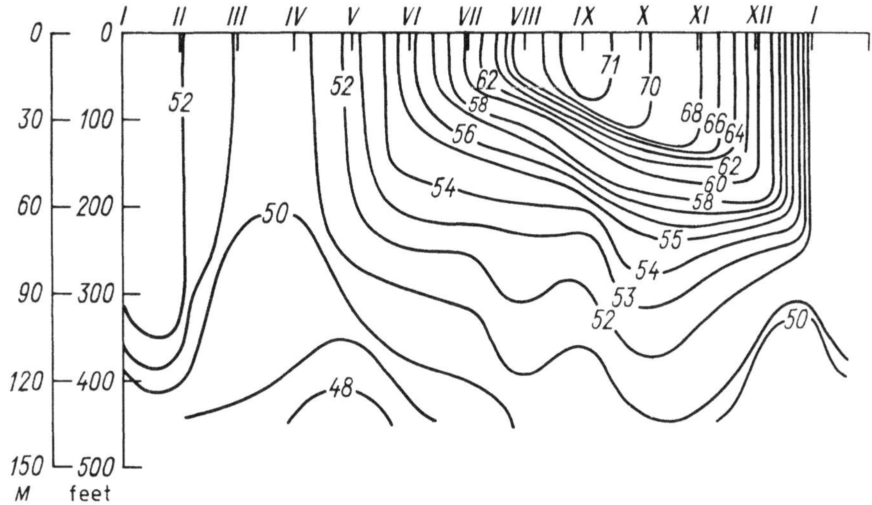

Fig. 14.1. Seasonal variation of temperature (°F) in upper layer of ocean, according to observations made aboard weather ship *Papa* (50°N, 145°W). t°C = 5/9 (t_F − 32).

apropos of which Kraus and Turner assumed that

$$Q_1 = \begin{cases} 0 & \text{for } \dfrac{\partial h_0}{\partial t} > 0, \\ \dfrac{\partial h_0}{\partial t}\ \delta T & \text{for } \dfrac{\partial h_0}{\partial t} \leqslant 0, \end{cases} \tag{14.19}$$

where $c_1 < c_2 \approx 10$. However, the proportionality assumed here to exist between the integral dissipation and generation of turbulent energy is questionable. Therefore, Resnyanskii (1975) suggested assuming factor c_1 in the third formula of (14.19) to be proportional to h_0, setting

$$c_1 = \left(c_3 + c_4 \left| \frac{L_w}{h_E} \right| \right) \frac{h_0}{h_E} .$$

Calculations made on the basis of this theory provided a satisfactory reconstruction of the observed seasonal variations of the upper mixed layer. Other variants of the semiempirical equations for $T_s'(t)$ and $h_0(t)$ have also been proposed.

If, instead of idealizing the seasonal thermocline as a temperature discontinuity, we make use of universal temperature profile (12.8), then integration of equation (14.15) over z gives

$$[h_1 - a_T(h_1 - h_0)] \frac{\partial T_s'}{\partial t} + a_T(T_s' - T_1') \frac{\partial h_0}{\partial t} = Q_1 - Q_s, \tag{14.20}$$

where $a_T = \int_0^1 F_T(\xi)\, d\xi$ (using (12.9), we get $a_T = 11/15$). A similar equation (14.20′) can be written for the salinity. Now, if (14.9) is integrated with the aid of the last two formulas of (14.19), we have

$$\frac{\alpha g h_1^2}{2}\left[1 + 2\left(\frac{h_1 - h_0}{h_1}\right)^2 b_{\mathrm{T}}\right]\frac{\partial T_s'}{\partial t} +$$

$$+ \alpha g\,(2b_{\mathrm{T}} - a_{\mathrm{T}})\,(T_s' - T_1')\,(h_1 - h_0)\,\frac{\partial h_0}{\partial t}$$

$$= (c_2 - c_1)\,\frac{\varrho_{\mathrm{a}}}{\varrho_{\mathrm{w}}}\,u_{*\mathrm{a}}^3 - \frac{\alpha g F_s}{c_{\mathrm{w}}\varrho_{\mathrm{w}}\gamma}\,(1 - \mathrm{e}^{-\gamma h_1}) - \alpha g h_1 Q_1, \qquad (14.21)$$

where $b_{\mathrm{T}} = \int_0^1 \xi F_{\mathrm{T}}(\xi)\, d\xi (= 4/9)$. To take the salinity into account, here and in the following formula we will everywhere replace $\alpha T'$ by $\alpha T' - \gamma S'$. Zilitinkevich *et al.* (1976) used the above approach in their global model of an interacting atmosphere and ocean (to which we will return in Chap. 10). They assumed c_1 to be a function of the integral Richardson number

$$\mathrm{Rf} = -g\int_{-h_0}^{0} M\,\mathrm{d}z\left[\int_{-h_1}^{0}\tau\cdot\frac{\partial\mathbf{u}}{\partial z}\,\mathrm{d}z - (B_s - B_1)\right]^{-1}$$

$$= -\frac{\alpha g\varrho_{\mathrm{w}}}{c_2\varrho_{\mathrm{a}}u_{*\mathrm{a}}^3}\left[\frac{F_s(1 - \mathrm{e}^{-\gamma h_0})}{c_w\varrho_{\mathrm{w}}\gamma} + h_0 Q_s + \frac{h_0^2}{2}\frac{\partial T_s'}{\partial t}\right], \qquad (14.22)$$

and they selected a piecewise-linear function $c_1 = c_2(m_1 - m_2\,\mathrm{Rf})$ with different values of m_2 for positive and negative Rf. For specific calculations it is convenient to solve equations (14.20), (14.20′), and (14.21) (taking into account formula (14.22) and the effects of salinity) for the time derivatives $\partial T_s'/\partial t$, $\partial S_s'/\partial t$, and $\partial h_0/\partial t$. Using a similar approach in the global model of an interacting atmosphere and ocean, it is advisable to go from T_s' and S_s' to the mean temperature over the upper layer of the ocean $\bar{T} = T_1' + (1 - (h_1 - h_0/h_1)a_{\mathrm{T}})\,(T_s' - T_1')$ and to the analogous mean salinity $\bar{S}$. Then, equations akin to (14.20) – (14.22) can be derived, using instead of (14.15) the general equation $((\mathrm{d}/\mathrm{d}t) - K_h\Delta_h) \times T' = -\partial Q/\partial z$ and an analogous expression for S', mean products of type $\overline{uT'}$ being replaced by products of the means $\overline{u}\overline{T}$. Such equations will be presented in Chap. 10.

15. The General Circulation of the Ocean

SPACE–TIME SPECTRUM

Motions in the ocean can be classified as follows, on the basis of their spatial scales l and time scales τ: (I) small-scale and intermediate-scale; (II) large-scale. In turn small-scale motions can be divided into: (1) turbulence, producing vertical mixing ($l = 10^{-4}$ to 10^1 m, $\tau = 10^{-3}$ to 10^2 s); (2) vertical layering microstructure ($l_z = 10^{-3}$ to 10^1 m, $\tau \sim 1$ min to 10 h); (3) acoustic waves ($l = 10^{-2}$ to 10^2 m, $\tau = 10^{-5}$ to 10^0 s); (4) surface capillary waves ($l = 10^{-1}$ to 10^0 cm, $\tau = 10^{-2}$ to 10^{-1} s); (5) surface gravity waves ($l = 10^{-2}$ to 10^2 m, $\tau = 1$ to 10 s). Intermediate-scale phenomena include: (6) internal gravity waves ($l = 10^{-1}$ to 10^3 m, $\tau = 10$ s to 1 day); (7) inertial fluctuations ($l_h = 1$ to 10 km, $\tau = 0.5$ to 1 day); (8) tides ($l = 10^0$ to 10^3 km, $\tau = 0.5$ to 1 day and long periods).

Finally, large-scale motions include: (9) synoptic processes, such as frontal and free oceanic eddies and Rossby waves ($l = 10^1$ to 10^2 km, $\tau \sim 10^1$ to 10^2 days); (10) disturbances produced by atmospheric synoptic processes ($l = 10^2$ to 10^3 km, $\tau \approx 1$ day); (11) seasonal fluctuations, mainly monsoon currents ($l \approx 10^3$ km, $\tau \approx 1$ yr); (12) principal (quasi-steady) currents in the upper layers of the ocean ($l = 10^3$ to 10^4 km, with fluctuation periods that are synoptic, seasonal, or even longer); (13) slow thermohaline currents throughout the entire depth of the ocean, shaping its gross stratification ($l = 10^3$ to 10^4 km, $\tau = 10^0$ to 10^2 yr). The statistical ensemble of large-scale motions forms the *general circulation of the World Ocean.*

According to Ozmidov (1965), there are three intervals of scales l in which the main generation of the kinetic energy of oceanic motions takes place. The first of these is the small-scale interval $l = 10^0$ to 10^2 m, in which surface gravity waves originate (the rate of influx of energy per unit mass of ocean in its upper 100-meter layer $\varepsilon_1 \approx 10^{-1}$ cm^2/s^3, so that an amount $F_1 \sim 10^3$ g/s^3 reaches a unit horizontal area; the mixing coefficient in this case is of the order of $K_1 \sim 10$ cm^2/s. The second is the intermediate-scale interval $l = 10^2$ to 10^3 m, in which are generated internal waves, inertial fluctuations, and shallow-water tides ($\varepsilon_2 \approx 10^{-3}$ cm^2/s^3, $F_2 \approx 5 \times 10^2$ g/s^3, $K_{h2} \approx 10^3$ cm^2/s). Finally, the third is the large-scale interval $l \approx 10^1$ to 10^4 km, in which synoptic eddies and the major ocean currents originate ($\varepsilon_3 \approx 10^{-4}$ cm^2/s^3, $F_3 \approx 10^1$ g/s^3, $K_{h3} \approx 10$ cm^2/s). Consequently, we can expect to find maxima of the space–time spectral density of the kinetic energy per unit mass in the regions of gravity waves (5)–(6), inertial and tidal fluctuations (7)–(8), synoptic processes (9), and major currents (12).

Integration of the space–time spectrum over all the scales l yields the time-dependent spectral density $E(\omega)$ describing the kinetic-energy distribution per unit mass over the fluctuation frequencies. As an example, Fig. 15.1 gives (on a log-log scale) the spectra $\omega E(\omega)$ cm^2/s^2 of currents at depths of 100, 400, 700, and 1400 m in the frequency interval $\omega = 10^{-4}$ to 7.5×10^{-1} cycles/h or periods $\tau = 4$ to 4×10^3 h. These spectra were plotted from data of Soviet buoy measurements made as part of the POLYMODE program in the Sargasso Sea, in a 300 × 300 km area centered at 29°N, 70°W; the measurements were made for 13 months, from July 1977 to August 1978, and averaged over 19 buoy stations. For all the specified depths the spectra revealed, first, a most interesting (with regard to the general circulation of the

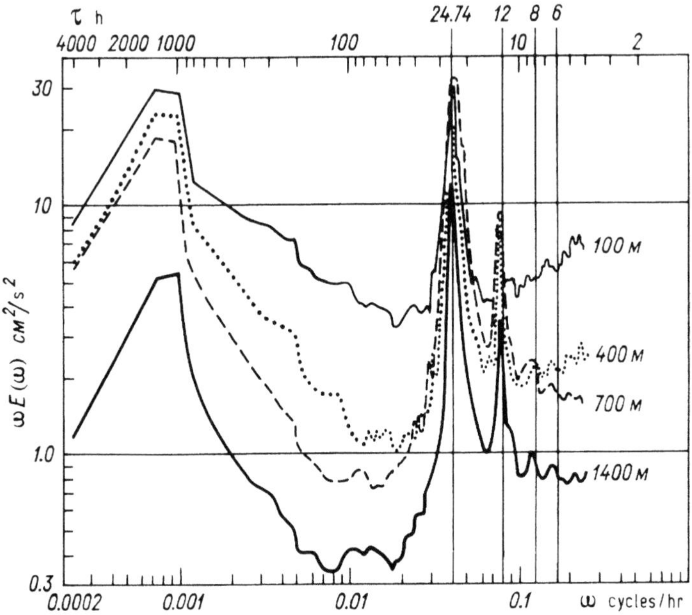

Fig. 15.1. Mean time spectra of kinetic energy of currents at depths of 100, 400, 700, and 1400 m in test area of POLYMODE program (27.5–30.5°N, 68.5–71.5°W).

ocean) definite maximum in the region of synoptic processes (9) at periods of about 57 days with values of ωE decreasing with depth from 31 – 24 cm²/s² at depths of 100–400 m to 19 cm²/s² at 700 m, and then to 6 cm²/s² at 1400 m. Second, the spectra revealed a broad minimum at periods from 2 to 8 days with ωE from 3.6 cm²/s at 100 m to 1.2–0.7 cm²/s² at 400–700 m and then to 0.4 cm²/s² at 1400 m (although at some buoys in this spectral interval comparatively slight secondary maxima were observed, apparently produced by atmospheric synoptic processes).

Third, these spectra have a very high sharp maximum at the inertial period (7), which is here, on the average $\tau \approx 24.7$ hr. The values of ωE range from 32 cm²/s² at a depth of 100 m to 12 cm²/s² at 1400 m; on individual spectra the top of this maximum may be divided into two parts, since the diurnal tidal period is involved as well as the inertial period. In addition, there is a somewhat less high, but even narrower, maximum at the semidiurnal tidal period (8) (where ωE decreases from 9.5 cm²/s² at 100 m to 3.6 cm²s² at 1400 m). These maxima are separated by a quite deep minimum around a period of 14 hr, where the ωE values range from 4 cm²/s² at 100 m to 1 cm²/s² at 1400 m. Past the tidal maximum, the values of ωE at depths of 100–400 m increase with the frequency, possibly due to bumping of the surface buoys, which thus (unlike deeply submerged buoys) can be used to measure the low-frequency variability of the currents in the dynamically very important surface layers of the ocean. Further down, from 700 to 1400 m, ωE decreases, everywhere going through not very pronounced maxima at periods of about 6 and 8 hr (possibly subharmonics of the inertial and tidal periods).

CIRCULATION OF SURFACE WATERS

We will begin our description of the general circulation of the ocean with a discussion of the major ocean currents, and in particular the currents at the ocean surface. "Bottle post" data, observations of vessel drift, and measurements with moored buoy stations all demonstrate that there exists at the ocean surface a system of large-scale currents, which are quasi-stationary in the sense that they are always present in certain water areas, although in places they are subject to sizable seasonal and synoptic fluctuations. These currents can be correlated closely with the dynamic topography of the ocean surface, that is, with the distribution of the heights of this surface above a deep level with a pressure of 15×10^6 Pa (these heights are calculated with the aid of the equation of hydrostatistics, on the basis of data of hydrological stations on the vertical distributions of the water density): isolines of the dynamic heights of the ocean surface coincide approximately with streamlines of the large-scale currents.

In particular, trough axes of the dynamic topography show a good fit with lines of divergence of surface currents, at which there occurs a separation or divergence of the surface water, and thus an attendant upwelling of the deep water. On the other hand, ridge axes of the dynamic topography correspond to lines of convergence, at which the surface waters pile up or come together, with an attendant downwelling into the depths. The lines of divergence and convergence divide the map of surface currents into quasi-latitudinal dynamic zones. From south to north on the map, we can discern the following zones: Antarctic Divergence (AD), Antarctic Convergence (AC) (or, alternatively, South Polar Front), South Subtropical Convergence (SSTC) (or Subantarctic Front), South Tropical Divergence (STD), North Tropical Convergence (NTC) (this is the equatorial line of dynamic symmetry, which is shifted northward somewhat, due to the asymmetry of the Southern and Northern Hemispheres), North Tropical Divergence (NTD), North Subtropical Convergence (NSTC), North Polar Convergence (NPC) (or North Polar Front), and Subpolar Divergence (SPD).

The South Equatorial Current and the North Equatorial Current play extremely important roles in the World Ocean. These proceed from the STD and NTD to the NTC with an appreciable westward component (conforming exactly to the Trade Winds in the atmosphere). Their total volume transport, for instance, at 150°E longitude, has been estimated to be 130×10^6 m^3/s.

South of the STD, with its axis on the SSTC, and north of the NTD, with its axis on the NSTC, in the southern and northern halves of the oceans gigantic anticyclonic gyres of water curve around the corresponding atmospheric anticyclonic "centers of action" (intensifying from winter to summer). In the Northern Hemisphere these are the Azores and Honolulu Gyres, and in the Southern Hemisphere the St. Helena, Mauritius, and South Pacific Gyres. The periods of circulation of the water in these are measured in years (for a radius of 2500 km and a current velocity of 10 cm/s at the periphery of the gyre, the period is 5 years). The western branches of these gyres form strong narrow coastal currents: the Gulf Stream in the Azores Gyre, the Brazil Current in the St. Helena Gyre, the Madagascar Current in the Mauritius Gyre, the Kuroshio in the Honolulu Gyre, and the East Australian Current in the South Pacific Gyre. In the eastern branches of the gyres, on the other hand, no intensification of the coastal currents occurs (this east-west asymmetry of the gyres is known as the "beta effect", concerning which, see below).

There is no subtropical anticyclone in the northern Indian Ocean, and an abrupt seasonal (monsoon) variability of the wind there is observed, and during it a variation of the ocean currents as well. At the time of the winter northeasterly monsoon (November–March) a relatively weak cyclonic gyre arises there, which includes both the North Equatorial Current (that turns southward near Africa) and the East Equatorial Countercurrent in the latitude zone from 3°N to 10°S (with a maximum in February). During the winter southwesterly monsoon (May-September) a strong anticyclonic gyre arises there, which includes the South Equatorial Current, that swings northward near Africa to form the enhanced Somali current (or East Africa Coast Current) and then eastward as the Monsoon Current, only to merge with the Equatorial Countercurrent, which at this time is shifted to the north (with a maximum in July).

Poleward from the subtropical convergences, cyclonic gyres are formed in the ocean beneath the corresponding cyclonic wind systems. In the Southern Ocean this takes the form of the West Wind Drift (or Antarctic Circumpolar Current), the current with the greatest volume transport in the World Ocean (sometimes apparently exceeding 210×10^6 m^3/s). In the North Atlantic a cyclonic oceanic gyre arises under the Icelandic low, and in the North Pacific one arises under the Aleutian low.

An enormous anticyclonic oceanic gyre is present in the Arctic Basin, which in the Amerasian Subbasin has a period of 4 years. The eastward Transarctic Current flows around its Asiatic periphery, becoming the East Greenland Current, which transports arctic water to the Atlantic (while an opposite transport of Atlantic water to the Arctic is effected by the Norway Current, which then divides into the North Cape Current and the Spitsbergen Current). The water budget of the Arctic Basin is estimated to be 182×10^3 km^3/yr: inflow of 112×10^3 through Fram Strait, 35×10^3 through North Cape – South Cape, 30×10^3 through Bering Strait, 3.8×10^3 river discharge, and 1×10^3 km^3/yr excess of precipitation over evaporation; outflow of 124×10^3 through Fram Strait, 57×10^3 through Canadian straits, and ice transport 1.3×10^3 km^3/yr.

The largest surface currents have typical speeds of tens of cm/s and volume transports of tens of millions of m^3/s. Stepanov estimates (on the basis of the density field, according to the model of Sarkisyan) that the mean velocities of surface currents range from 19.3 cm/s in the Indian Ocean to 12.3 cm/s in the Pacific and 11.6 cm/s in the Atlantic. According to the calculations of Galerkin and Gritsenko (1980) for the Pacific, of the total 100 cm^2/s^2 mean kinetic energy per unit mass of the surface currents, 77% corresponds to zonal motions (of which in turn 64% are westward and 36% eastward) and 23% corresponds to meridional motions (of which in turn 60.6% are northward and 39.4% southward). Of the zonal motions strong (more than 20 cm/s) westward flows transport 40.6% of the energy and occupy 11% of the area (these are primarily equatorial currents), while strong eastward flows transport 12.7% of the energy and occupy 3.7% of the area (these are the equatorial countercurrents). Of the meridional motions strong northward currents transport 4.2% of the energy and occupy 0.4% of the area, while strong southward currents transport 13.6% of the energy and occupy 1.2% of the area.

The fact that the directions of the major surface currents coincide with those of the prevailing winds, together with the fact that their seasonal variability is highest in regions of maximum seasonal variability of the winds (the monsoon regions of the Indian Ocean), combine to show that these currents are produced primarily by the winds. The corresponding convergences and divergences (as well as variations in atmospheric pressure, precipitation and evaporation, and thermohaline expansion and contraction of the water) create the above-mentioned dynamic topography of the ocean surface, with height differences of the order of decimeters. The greatest upward deviations from the equilibrium level of the geoid are observed at the western peripheries of the oceans, especially in the subtropics, while the greatest downward deviations are in the circumpolar regions. The height difference between the North Subtropical Convergence and the Subpolar Divergence is 170 cm in the Atlantic and 120 cm in the Pacific. The height differences of the ocean surface and of other isobaric surfaces in the ocean produce horizontal pressure differences in the depths that give rise to barogradient deep currents.

DEEP CIRCULATION

Calculations made on the basis of the density field, together with a few measurements, indicate that the circulation of the subsurface and intermediate water down to a depth of 1500 m and a temperature of about 3.5°C is in general the same as the surface circulation, except that it becomes weaker with depth (the tropical circulation disappears almost completely, while subtropical anticyclones are shifted poleward somewhat). The circulation is such that intermediate water of reduced salinity is transported from polar fronts to the subtropics and tropics, while water with a higher temperature moves to the subpolar regions.

The narrow (± 2.5° latitude) highly saline streams of the equatorial deep countercurrents, flowing eastward at depths of 50–300 m under the westward equatorial currents at the surface, constitute a special case of subsurface currents. These streams, the Cromwell Undercurrent in the Pacific, the Lomonosov Undercurrent in the Atlantic, and the Tareev Undercurrent in the Indian Ocean, have maximum velocities of up to 150 cm/s and volume transports of up to 40×10^6 m^3/s (note that the Tareev Undercurrent is well-defined only during the time of the winter monsoon). The undercurrents are produced by barogradient effects, by westward-equatorial surge-related elevations of the ocean level; at the equator the eastward direction of the currents is stabilized by the Coriolis force (see below), while the uppermost water layer is "blown" to the west by the tradewind.

According to approximate calculations, at depths greater than 1500 m the water circulation (except for the West Wind Drift) is already more or less separate from the surface circulation and the wind field above the oceans. In direction it is predominantly opposite to the surface circulation (it includes the deep countercurrent under the Gulf Stream, the recirculation in the South Atlantic and the Indian Ocean, the cyclonic circulation in the northern mid-latitudes of the Pacific, and the anticyclonic circulations further north). Consequently, at the transition depths of 1.5–2 km the circulation slackens, and then intensifies somewhat toward the ocean floor (apparently this is mostly as a result of thermohaline factors) and begins to follow the isobaths of the bottom relief.

The Antarctic bottom water in the Southern Ocean moves eastward with the West Wind Drift (which apparently penetrates to the bottom). In the Atlantic it gets to the north chiefly via the western basins, up to 40°N, where it confronts the North Atlantic deep water and the Arctic bottom water; at depths greater than 4 km the Antarctic bottom water and the North Atlantic deep water move in opposite directions. In the Indian Ocean the Antarctic bottom water occupies all the deep basins. In the Pacific the main flow of the Antarctic bottom water is along the Kermadec and Tonga Trenches; at 10°S latitude it has a branch which goes off to the east, reaching the southeastern part of the northern half of the Pacific. North of the equator the main flow splits up and moves along troughs, reaching approximately to the northern tropics. The velocities of these flows range from 0.1 to 1 cm/s.

On the basis of typical meridional sections of the T and S fields, Stepanov used the equations of convective diffusion of T and S to estimate the mean (over the entire ocean) absolute values of the meridional velocity $\langle |v| \rangle \approx 2.4$ cm/s, the vertical velocity in the upper layer of the ocean $\langle |w| \rangle \approx 5 \times 10^{-5}$ to

10×10^{-5} cm/s, and the velocity in the deep layers, which is an order of magnitude higher. More precise calculations carried out by Galerkin and Gritsenko for the Pacific gave rms zonal velocities σ_u of 13.1, 7.5, 4.1, and 3.0 cm/s at depths of 0, 100, 1500, and 4000 m; these are about $1\frac{1}{2}$ times greater than the corresponding meridional velocities (the σ_v values are, respectively, 7.0, 4.7, 3.0, and 2.3 cm/s). Averaged over the entire depth of the ocean, the velocity of the quasi-steady currents was found to be 4.5 cm/s, which corresponds to a density of kinetic energy (without synoptic vortices) of about 1 J/m^3. This is only 1/200 of the value in the atmosphere, the reason being that the ocean receives the kinetic energy of quasi-steady currents mainly from the atmosphere, the "coupling" being quite weak. The density of internal energy $\varrho_w c_w T$ in the ocean in 1.2×10^9 J/m^3, as compared with 1.6×10^5 J/m^3 in the atmosphere; the potential-energy density $\frac{1}{2}\varrho_w gH$ is 2×10^7 J/m^3 in the ocean and 4×10^4 J/m^3 in the atmosphere; the densities of available potential energy are comparable: 7×10^2 J/m^3 in the ocean and 5×10^2 J/m^3 in the atmosphere.

EQUATORIAL SURFACE COUNTERCURRENTS

Some of the above-described qualitative features of ocean currents have simple theoretical explanations. For instance, the existence of the equatorial surface countercurrents can be explained, with the aid of the formula for the total flow (14.7), as an effect of the latitudinal nonuniformity of the wind stress τ_0 (Shtokman, 1948). Actually, assuming the x axis to be directed eastward and the y axis northward, let us consider the wind-driven currents in a basin of constant depth, occupying a region $0 \leqslant x \leqslant L$, $0 \leqslant y \leqslant l$ for $l \ll L$. We assume that the wind-driven currents are produced by the zonal frictional stress of the wind $\tau_{x0} = \tau_{x0}(y)$, $\tau_{y0} = 0$. From the continuity equation for the total flow $(\partial S_x/\partial x) + (\partial S_y/\partial y) = 0$ and the boundary condition at the walls $S_n = 0$, it follows that for all x we have $\int_0^1 S_x' \, dy = 0$, so that at certain latitudes y the zonal current must change sign. The quantity τ_{x0}, and thus also $\mathbf{S}$ and $\nabla \zeta$ in the middle sections of the basin around $x = L/2$, will not depend on x. Therefore, from the continuity equation it follows that $S_y = 0$ there, and since $(\partial/\partial x)(\partial \zeta/\partial y) = 0$, it will also be true that $\partial \zeta/\partial x = \text{const}$. Now, if we assume that $\zeta = 0$ for $y = 0, 1$, then, with an accuracy of the order of the Ekman number E, from (14.7) we get

$$\zeta = -\frac{2}{\varrho_w g h_E'} \int_0^y (\tau_{x0} - \bar{\tau}_{x0}) \, dy, \tag{15.1}$$

where $\bar{\tau}_{x0}$ is the mean value of τ_{x0} over the interval $0 \leqslant y \leqslant l$. Consequently, for instance for

$$\tau_{x0} = -\frac{\tau_0}{2}\left(1 + \cos\frac{2\pi y}{l}\right),$$

that is, for a wind of constant direction, we have

$$\zeta = \left(\frac{\tau_0 l}{2\pi \varrho_w g h'_E} \right) \sin \frac{2\pi y}{l} ,$$

indicating that the deviation of the ocean level changes sign at $y = 0.5$. A calculation of the current velocity according to formula (14.6) in this case shows that in the middle zone of the basin the current is opposite to the wind even at the surface of the ocean.

WESTERN BOUNDARY CURRENTS

The east–west asymmetry of the subtropical anticyclonic oceanic gyres, with the intensification of the western boundary currents, can be explained with the aid of this same formula (14.7) of the theory of the Ekman boundary layer, as a consequence of the "β effect", that is, the latitudinal variation $\beta = df/dy$ of the Coriolis parameter f (Stommel, 1948). Actually, if we introduce with the aid of the continuity equation the total-flux function ψ, via the relations $S_x = -\partial\psi/\partial y$ and $S_y = \partial\psi/\partial x$, and if in addition we express $\nabla\zeta$ from (14.7) in terms of it, then the condition curl $(\nabla\zeta) = 0$ can, with an accuracy of the order of the Ekman number, be written as

$$\frac{fh'_E}{2H^2} \Delta\psi + \frac{\partial(\psi, f/H)}{\partial(x, y)} = \text{curl} \frac{\tau_0}{\varrho_w H} . \tag{15.2}$$

The boundary condition at the ocean shore $S_n = 0$ now takes the form $\psi = \text{const}$, where for a simply-connected ocean region this constant is arbitrary, whereas in the opposite case the values of ψ on each closed boundary contour must be found from the condition that function $\zeta(x, y)$ be single-valued.

First let us consider an ocean of constant depth $H = \text{const}$. In the open ocean the first term on the left-hand side of (15.2), describing the effect of friction in the bottom-adjacent Ekman boundary layer, can be neglected. Then formula (15.2) becomes the relation of Sverdrup (1947)

$$\beta \frac{\partial\psi}{\partial x} = \text{curl} \frac{\tau_0}{\varrho_w} , \tag{15.3}$$

which makes it clear that $\tau_0/\varrho_w\beta$ can serve as a typical value of the total flux function ψ. If, in addition, we select a length scale $L = f/2\beta$, then equation (15.2) can be written in terms of dimensionless variables as

$$\text{E}\, \Delta\psi + \partial\psi/\partial x = F(x, y), \tag{15.4}$$

where the right-hand side can be assumed known. Let us look for a solution of (15.4) in the square $0 \leqslant x \leqslant 1$, $0 \leqslant y \leqslant 1$, with zero values of ψ at the boundaries, in the form of an exponential series in the small parameter E. A series of the form $\psi = \sum_{n=1}^{\infty} \psi_n(x, y)\, \text{E}^n$, where functions ψ_n and their derivatives with respect to x and y are of the order of unity, is not suitable for this, since for functions ψ_n we would then get first-order equations $\partial\psi_n/\partial x = F_n - \Delta\psi_{n-1}$, making it impossible to satisfy both of the boundary conditions for x (at $x = 0$ and at $x = 1$). Therefore, in the vicinities of the western shore $x = 0$ and the eastern shore $x = 1$, we have to introduce boundary layers with "fast" transverse coordinates $x' = x/\mathrm{E}$ and $x'' = (1 - x)/\mathrm{E}$ and then seek a olution in the form

$$\psi = \sum_{n=0}^{\infty} [\psi_n(x, y) + \psi'_n(x', y) + \psi''_n(x'', y)\, \mathrm{E}^n, \tag{15.4'}$$

where functions ψ_n, ψ'_n and ψ''_n are quite different from zero, respectfully, in the open ocean and in the western and eastern boundary layers. The last two requirements signify that ψ'_n must damp out rapidly (exponentially, say) with an increase in x', while ψ''_n decreases rapidly with an increase in x''. From (15.4) it follows that

$$\frac{\partial\psi_0}{\partial x} = F_0\ ; \qquad \frac{\partial^2\psi'_0}{\partial x'^2} + \frac{\partial\psi'_0}{\partial x'} = 0; \qquad \frac{\partial^2\psi''_0}{\partial x''^2} - \frac{\partial\psi''_0}{\partial x''} = 0, \tag{15.4''}$$

indicating that no solution ψ''_0 which decreases exponentially with an increase in x'' exists, so that we must set $\psi''_0 = 0$. A solution $\psi'_0 = ce^{-x'}$ is valid, however, so that the flow possesses an east-west asymmetry. The boundary conditions on x now have the form $\psi_0 + \psi'_0 = 0$ for $x = 0$ and $\psi_0 = 0$ for $x = 1$. Consequently, the solution of equation (15.4) outside of the boundary layers at the walls $y = 0$ and $y = 1$ has the form

$$\psi = \begin{cases} -(1 - \mathrm{e}^{-x/E})\displaystyle\int_0^1 F_0(x, y)\, \mathrm{d}x + O(\mathrm{E}) & \text{for } x \leqslant \mathrm{E}, \\ -\displaystyle\int_x^1 F_0(x, y)\, \mathrm{d}x + O(\mathrm{E}) & \text{for } \mathrm{E} \leqslant x \leqslant 1. \end{cases} \tag{15.4'''}$$

As an example, Fig. 15.2 shows isolines of function ψ in the case $\tau_{x0} = -\cos \pi y$, $\tau_{y0} = 0$. Let us note that in the western boundary layer $S_x = O(1)$, $S_y = O(1/\mathrm{E})$, that is, a strong coastal current (boundary current) is present, whereas at the eastern shore there is no such current. It is clear from the foregoing considerations that this asymmetry can be attributed to the second term on the left-hand side of equation (15.4) or (15.2), that is, to the β effect.

The total volume transport of the current through a latitude circle $y = $ const from the western shore $x = x_W(y)$ to the eastern shore $x = x_E(y)$ is $Q = \int_{x_W}^{x_E} S_y\, \mathrm{d}x$, where $S_y = \partial\psi/\partial x$ can be found from (15.3). According to the second row of (15.4'''), this quantity would not be zero and would depend on y, which would contradict the law of conservation of the water mass.

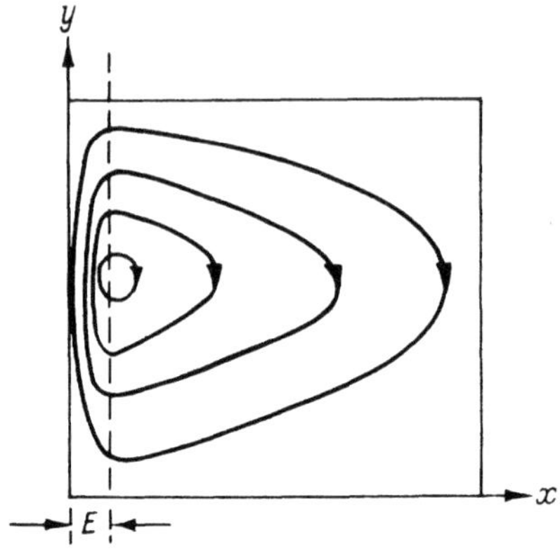

Fig. 15.2. Solution of equation (15.4) for $\tau_{x0} = -\cos \pi y$, $\tau_{y0} = 0$.

Consequently, it must be compensated by the volume transport Q_w of the western boundary current, that is, $Q_w = -Q$, or, taking (15.3) into account,

$$Q_w = -\frac{1}{\beta}\int_{x_W}^{x_E} \operatorname{curl}\frac{\tau_0}{\varrho_w}\,dx \tag{15.5}$$

(Munk, 1950). Calculations made using this formula show a fairly good fit with, for example, empirical estimates of the volume transport of the Gulf Stream south of the point 33°N of its separation from the coast. Note that the volume transport Q_w is independent of the eddy viscosity K. On the other hand, K can be determined from the width EL of the western boundary layer; for instance, for $L = 10^4$ km and $H = 4$ km, assuming EL to be equal to the width of the Gulf Stream (50 km), we get $E = 5 \times 10^{-3}$ and $h'_E = 20$ m, which gives $K = \frac{1}{2}h_E'^2 = 200\ \text{cm}^2/\text{s}$.

In the case of an ocean of varying depth, neglecting the first term on the left-hand side of equation (15.2), we obtain the generalized Sverdrup relation. Now, in the calculation of the total-flux function ψ, the role of the latitude circles $f = \text{const}$ is played by the lines $f/H = \text{const}$. A western boundary current will then exist only on condition that $(\partial/\partial s)(f/H) > 0$, and an eastern current will exist if $(\partial/\partial s)(f/H) < 0$, where d$s$ is directed along the tangent to the northern shoreline. If, for instance, an f/H isoline at some point is tangent to the western shore, then the sign of $(\partial/\partial s)(f/H)$ changes at this point, which indicates that there the boundary current separates from the coast under the influence of the bottom relief.

In order to state the problem more generally, while calculating the ocean currents we must add to the equations of motion additional terms describing the horizontal mixing; for example, on the right-hand sides of equations (14.4) the terms $K_h\Delta_h u$ and $K_h\Delta_h v$ must be introduced. In this case the horizontal Ekman number $E_h = (2K_h/f)^{1/3}L^{-1}$ appears in addition to the vertical Ekman number $E = (2K/f)^{1/2}H^{-1}$. Here it can be demonstrated that the structure of the boundary layers in the general case is the same as in the particular case $E_h \sim E$, namely as follows: (1) in the open ocean u, v, w, and ζ have values of the order of E; (2) a surface Ekman boundary layer of thickness $O(E)$ is formed, in which $u = O(1)$, $v = O(1)$, and $w = O(E)$; (3) a bottom Ekman boundary layer of thickness $O(E)$ is formed, in which $u = O(E)$, $v = O(E)$, and $w = O(E^2)$; (4) at the western shore a Man'kovskii boundary layer of thickness is $O(E^{2/3})$ is formed, in which $u = O(E)$, $v = O(E^{1/3})$, $w = O(E^{2/3})$, and $\zeta = O(E)$ inside this layer at the western shore, and also at the eastern shore, a sublayer of thickness $O(E)$ is formed, in which $u = O(E)$, $v = O(E)$, $\zeta = O(E^2)$, while $w = O(1)$ is not small; (6) boundary layers are also formed at the northern and southern shores.

INERTIAL-VISCOUS BOUNDARY LAYERS

For an even more general statement of the problem, we retain the nonlinear (inertial) terms in the equations of horizontal motion (9.3)–(9.4). If it is the *horizontal* structure of the currents which is of primary interest, then the simplest approach will be to use the *method of total fluxes*, first proposed by Shtokman (1946), that is, to find an approximate average of equations (9.1)–(9.4) over the entire depth of the ocean $-H \leqslant z < 0$.

Such an average (indicated by a bar over the symbol) can be obtained with the aid of the following easily demonstrated hydrodynamic equation:

$$\overline{\frac{\mathrm{d}A}{\mathrm{d}t}} = \frac{\bar{v}_\theta}{a}\frac{\partial \bar{A}}{\partial \theta} + \frac{\bar{v}_\lambda}{a \sin\theta}\frac{\partial \bar{A}}{\partial \lambda} + \frac{A'_s v_{rs}}{H} + \frac{1}{aH\sin\theta}\left(\frac{\partial H\overline{A'v'_\theta}\sin\theta}{\partial\theta} + \frac{\partial H\overline{A'v'_\lambda}}{\partial\lambda}\right), \tag{15.6}$$

where a prime signifies a deviation from the mean value over the depth, while a subscript s indicates the value at $z = 0$. The averaged equations (9.2)–(9.4) have the form

$$\frac{1}{aH\sin\theta}\left(\frac{\partial H\bar{v}_\theta\sin\theta}{\partial\theta} + \frac{\partial H\bar{v}_\lambda}{\partial\lambda}\right) = -\frac{v_{rs}}{H}; \tag{15.7}$$

$$\frac{\bar{v}_\theta}{a}\frac{\partial\bar{v}_\theta}{\partial\theta} + \frac{\bar{v}_\lambda}{a\sin\theta}\frac{\partial\bar{v}_\theta}{\partial\lambda} - \frac{\operatorname{ctg}\theta}{a}\bar{v}_\lambda^2 - 2\Omega\cos\theta\bar{v}_\lambda = -\frac{1}{aH}\frac{\partial}{\partial\theta}H\left(\frac{\bar{p}}{\varrho_0}\right) + \bar{f}_\theta + \varphi_\theta; \tag{15.8}$$

$$\frac{\bar{v}_\theta}{a}\frac{\partial\bar{v}_\lambda}{\partial\theta} + \frac{\bar{v}_\lambda}{a\sin\theta}\frac{\partial\bar{v}_\lambda}{\partial\lambda} - \frac{\operatorname{ctg}\theta}{a}\bar{v}_\theta\bar{v}_\lambda + 2\Omega\cos\theta\bar{v}_\theta = -\frac{1}{aH\sin\theta}\frac{\partial}{\partial\lambda}H\left(\frac{\bar{p}}{\varrho_0}\right) + \bar{f}_\lambda + \varphi_\lambda, \tag{15.9}$$

$$\varphi_\theta = \frac{\mathrm{ctg}\,\theta}{a}\,\overline{v'^2_\lambda} - \frac{\overline{v'_{\theta s} v_{rs}}}{H} - \frac{1}{aH\sin\theta}\times$$

$$\times\left(\frac{\partial H\overline{v'^2_\theta}\sin\theta}{\partial\theta} + \frac{\partial H\overline{v'_\theta v'_\lambda}}{\partial\lambda}\right) + \frac{1}{aH}\left(\frac{p}{\varrho_0}\right)_b\frac{\partial H}{\partial\theta}; \qquad (15.8')$$

$$\varphi_\lambda = -\frac{\mathrm{ctg}\,\theta}{a}\,\overline{v'_\theta v'_\lambda} - \frac{\overline{v'_{\lambda s} v_{rs}}}{H} - \frac{1}{aH\sin\theta}\times \qquad (15.9')$$

$$\times\left(\frac{\partial H\overline{v'_\theta v'_\lambda}\sin\theta}{\partial\theta} + \frac{\partial H\overline{v'^2_\lambda}}{\partial\lambda}\right) + \frac{1}{aH\sin\theta}\left(\frac{p}{\varrho_0}\right)_b\frac{\partial H}{\partial\lambda},$$

after first introducing into (9.3) and (9.4) the Boussinesq approximation, that is, replacing the density ϱ in the multiplier of ∇p by the standard density value ϱ_0 (here taken to be constant); subscript b indicates the value at the ocean bottom (for $z = -H$). To find the mean values $\bar{f}_\theta$ and $\bar{f}_\lambda$, we can use formulas (9.3′) and (9.4′), and then we can express the mean and bottom values of $\tau_{\theta\theta}$, $\tau_{\theta\lambda}$, and $\tau_{\lambda\lambda}$ in terms of $\bar{v}_\theta$ and $\bar{v}_\lambda$, with the aid of some semiempirical formulas. The simplest of these are:

$$\bar{f}_\theta = \frac{\tau_{\theta s} - \tau_{\theta b}}{\varrho_0 H} + K_h\left(\Delta_h\bar{v}_\theta - \frac{\cos 2\theta}{a^2\sin^2\theta}\bar{v}_\theta - \frac{2\cos\theta}{a^2\sin^2\theta}\frac{\partial\bar{v}_\lambda}{\partial\lambda}\right); \qquad (15.8'')$$

$$\bar{f}_\lambda = \frac{\tau_{\lambda s} - \tau_{\lambda b}}{\varrho_0 H} + K_h\left(\Delta_h\bar{v}_\lambda - \frac{\cos 2\theta}{a^2\sin^2\theta}\bar{v}_{\lambda_1} + \frac{2\cos\theta}{a^2\sin^2\theta}\frac{\partial\bar{v}_\theta}{\partial\lambda}\right), \qquad (15.9'')$$

in which, by the way, it can be assumed that the bottom friction is about two orders of magnitude weaker than the surface friction. Quantities v'_θ and v'_λ describe the baroclinic component of the horizontal-velocity field, which varies with depth. Thus, quantities (15.8′) and (15.9′) describe the 'CEBAR' combined effect of baroclinicity and (bottom) relief, which, in general, is not small (the last terms in (15.8′) and (15.9′) can be calculated on the basis of a given relief $H(\theta, \lambda)$ and density field $\varrho_w\ (z, \theta, \lambda)$ using the equation of hydrostatics (9.1), according to which $p_b = p_a + g\varrho_0\zeta + g\bar{\varrho}_w H$; note by the way, that the density field cannot be specified arbitrarily, and that the velocities turn out to be very sensitive to slight variations in this field.

Next let us consider for simplicity just the case H = const, so that terms φ_θ and φ_λ can be neglected. Then, the relative vorticity of the mean horizontal velocity $\omega = \mathrm{curl}\ \bar{v}_h$ is found from (15.8) and (15.9) with the aid of the equation

$$\bar{\mathbf{v}}_h\cdot\nabla(\omega + 2\Omega\cos\theta) \approx K_h\left(\nabla_h\omega + \frac{2\omega}{a^2}\right) + \mathrm{curl}\,\frac{\tau_0}{\varrho_0 H}, \qquad (15.10)$$

where div $\bar{\mathbf{v}}_h \approx 0$, so that a contour integral of the left-hand side is equal to zero. Thus it is evident that, at any rate in regions of atmospheric anticyclones, where curl $\tau_0 < 0$, horizontal mixing must in general be taken into account. This mixing, and also the advection of ω, can be neglected in the open ocean, and there Sverdrup's relation $\beta \bar{v}_\theta = -(1/\varrho_0 H)$ curl τ_0 is obtained from (15.10), giving a southward motion with a volume transport depending on the latitude. The corresponding mass defect should move the nearshore currents northward, the advecton of ω and the horizontal mixing in these currents already being sizable, that is, inertial-viscous boundary layers will form. It is easy to show that in these layers the total advection $\int v\omega\, \mathrm{d}n$ of the relative vorticity $\omega \approx \partial v/\partial n$, as well as the contribution of the term $2K_h\omega/a^2$, are small, so that the advection βv of the vorticity of the Earth's rotation must be compensated by diffusion of the relative vorticity from the wall. This can happen at the western shore (positive vorticity travels northward and diffuses out from the wall). At the eastern shore for a northward current this cannot happen (negative vorticity diffuses out from the wall), whereas for a southward current it can take place only owing to $(1/\varrho_0 H)$ curl τ_0, which can produce only a weak current.

In order to analyze the structure of the inertial-viscous boundary layer, let us go over, in the simplified equations (15.7)–(15.9), to local Cartesian coordinates x, y (measured in terms of a scale L), and let us measure the velocity in terms of a scale $U_0 = \tau_0/\varrho_0 \beta L H$ and the pressure in terms of a scale $2\varrho_0 \Omega U_0 L$. Then, in dimensionless form the inertial terms will have a small multiplier $\varepsilon^2 = (L_i/L)^2 = K_i$, $L_i^2 = U_0\beta$, while the viscous terms will have a multiplier $\varepsilon^3/\mathrm{Re}$, where $\mathrm{Re} = U_0\, L_i/K^h$ is the Reynolds number of the boundary layer. The solution for the dimensionless flux function ψ of the mean current will now be sought in the form (15.4′), with E replaced by ε. Similarly to (15.4‴) we then get

$$\psi = \begin{cases} \dfrac{\mathrm{d}\tau}{\mathrm{d}y} + \psi_0'\left(\dfrac{x}{\varepsilon}, y\right) + O(\varepsilon) & \text{for } x \leqslant \varepsilon, \\[2ex] (1-x)\dfrac{\mathrm{d}\tau}{\mathrm{d}y} + O(\varepsilon) & \text{for } \varepsilon \leqslant x \leqslant 1-\varepsilon, \\[2ex] O(\varepsilon);\quad \dfrac{\partial\psi}{\partial x} = -\left(1 - \exp\left(-\dfrac{1-x}{\varepsilon}\mathrm{Re}^{1/3}\right)\right)\dfrac{\mathrm{d}\tau}{\mathrm{d}y} + O(\varepsilon) & \\[1ex] & \text{for } x \geqslant 1-\varepsilon, \end{cases} \tag{15.11}$$

where function $\psi_0'(x', y)$ is the solution of the following problem:

$$\begin{aligned} &-\frac{1}{\mathrm{Re}}\frac{\partial^3\psi_0'}{\partial x'^3} + \frac{\partial\psi_0'}{\partial x'}\frac{\partial^2\psi_0'}{\partial x'\,\partial y} - \\ &-\left(\frac{\mathrm{d}^2\tau}{\mathrm{d}y^2} + \frac{\partial\psi_0'}{\partial y}\right)\frac{\partial^2\psi_0'}{\partial x'^2} + \psi_0' = 0; \end{aligned} \tag{15.12}$$

Analysis indicates that for large Re (comparatively small K_h) and for $d^2\tau/dy^2 > 0$ (southern half of basin), there exists an inertial boundary layer with a viscous nearshore sublayer (whereas for $d^2\tau/dy^2 < 0$, that is, in the northern half of the basin, such a layer is already absent for Re $\geqslant 10$). For small Re (very large $K_h \geqslant 10^7$ to 10^8 cm²/s), function $\psi_0'(x', y)$ can be sought in the form of a series $\Sigma_{n=0}^{\infty}\psi_{0n}'(\xi, y)\,(\mathrm{Re})^{2n/3}$, where $\xi = x'\,(\mathrm{Re})^{1/3}$, so that the principal term

$$\psi_{00}' = -\frac{2}{\sqrt{3}}\frac{d\tau}{dy}\,e^{\xi/2}\sin\left(\frac{\sqrt{3}}{2}\xi + \frac{\pi}{3}\right) \tag{15.13}$$

describes the viscous boundary layer, in which the inertial terms are negligible.

EQUATORIAL DEEP COUNTERCURRENTS

Now let us consider briefly the theoretical explanation of the equatorial deep countercurrents. With the approach to the equator, $f = 2\Omega\cos\theta \rightarrow 0$, so that the velocities of the geostrophic motions prevailing in the open ocean of the middle and high latitudes, the thicknesses of the Ekman boundary layer h_E and h_E', and the Ekman number E will all increase without restriction. Consequently, in the equatorial zone an inner boundary layer is created. In order to analyze its structure, let us consider in the simplest case H = const the central part of the equatorial zone of the ocean, where the currents are zonally homogeneous, that is, the derivatives with respect to λ of all the quantities except the ocean level ζ are zero. Then, in view of continuity equation (15.7) (for $v_{rs} \approx 0$), the velocity $\bar{v}_\theta = 0$ there, and the equation of zonal motion (15.9) – (15.9′) neglecting horizontal mixing becomes

$$\frac{1}{a}\frac{\partial\overline{v_\theta' v_\lambda'}}{\partial\theta} = -\frac{g}{a}\frac{\partial\zeta}{\partial\lambda} + \frac{\tau_{\lambda s} - \tau_{\lambda b}}{\varrho_w H}\ . \tag{15.14}$$

At the borders between the equatorial boundary layer and the open ocean, just as in the open ocean in general, the nonlinear terms and the bottom friction can be neglected. Accordingly, the longitudinal drop in the ocean level must be compensated largely by the frictional stress of the trade wind:

$$\frac{g}{a}\frac{\partial\zeta}{\partial\lambda} \approx \frac{\tau_{\lambda s}}{\varrho_w H} < 0.$$

This same relation must also be valid at the equator, since otherwise the quantities $(\partial/\partial\theta)(\partial\zeta/\partial\lambda) = (\partial/\partial\lambda)(\partial\zeta/\partial\theta)$ would not be small, which would contradict the foregoing assumption of zonal homogeneity of the currents in the central part of the equatorial zone. Then, from (15.14) the relation

$$\frac{1}{a}\frac{\partial\overline{v_\theta' v_\lambda'}}{\partial\theta} \approx -\frac{\tau_{\lambda B}}{\varrho_w H}$$

remains. Calculations in the Ekman statement of the problem, for a zonal wind, indicate that the left-hand side is negative, so that we have $\tau_{\lambda b} > 0$, that is, near the bottom the zonal current is directed eastward; this current is thus an equatorial deep countercurrent (the "bottom" in this case refers to the lower boundary of the upper mixed layer, that is, the thermocline, which in view of its high vertical stability prevents the turbulent diffusion of momentum in the underlying layers). These calculations in the Ekman statement (Charney, 1960) are based on the use of equations of motions (14.4), supplemented by the nonlinear terms:

$$v\frac{\partial u}{\partial y} + w\frac{\partial u}{\partial z} - \beta y v = -\frac{\tau_{x0}}{\varrho_w H} + K\frac{\partial^2 u}{\partial z^2};$$

$$v\frac{\partial v}{\partial y} + w\frac{\partial v}{\partial z} + \beta y u = -g\frac{\partial \zeta}{\partial y} + K\frac{\partial^2 v}{\partial z^2}, \qquad (15.15)$$

for the boundary conditions $K(\partial u/\partial z) \to \tau_{x0}/\varrho_w$, $K(\partial v/\partial z) \to 0$ as $z \to 0$ and $u, v, w \to 0$ as $z \to -H$. After differentiating the second of these equations with respect to y, replacing $\partial v/y$ by $-\partial w/\partial z$ with the aid of the continuity equation, and then setting $y = v = 0$, from (15.15) we get

$$w\frac{\partial u}{\partial z} = -\frac{\tau_{x0}}{\varrho_w H} + K\frac{\partial^2 u}{\partial z^2},$$

$$\left(\frac{\partial w}{\partial z}\right)^2 - w\frac{\partial^2 w}{\partial z^2} + \beta u = -g\left(\frac{\partial^2 \zeta}{\partial y^2}\right)_{y=0} - K\frac{\partial^3 w}{\partial z^3} \qquad (15.15')$$

for the boundary conditions $K(\partial u/\partial z) \to \tau_{x0}/\varrho_w$, $w \to 0$, $\partial^2 w/\partial z^2 \to 0$ as $z \to 0$ and u, w, $\partial w/\partial z \to 0$ as $z \to -H$ (the "extra" condition, of these six conditions, makes it possible to determine the constant $(\partial^2\zeta/\partial y^2)_{y=0}$). Numerical integration of these *ordinary* differential equations for $H \approx 150$ m and $\tau_{x0} \approx -0.05$ Pa indicated that for $K = 15$ to 20 cm^2/s a deep equatorial countercurrent is obtained, with velocities close to those observed in nature.

It should be stressed that currents at the equator which flow eastward possess a unique stability, since deviations of these to the north or south give rise to a Coriolis force which acts upon them as a restoring force. This effect is inertial in nature and can easily be calculated using a model of inertial motion, described by the equations

$$\frac{du}{dt} - \frac{uv\tan\varphi}{a} - 2\Omega v\sin\varphi = 0;$$

$$\frac{dv}{dt} + \frac{u^2\tan\varphi}{a} + 2\Omega u\sin\varphi = 0 \qquad (15.16)$$

(Monin, 1972), where λ, φ are the longitude and latitude, and $u = a(d\lambda/dt)\cos\varphi$ and $v = a(d\varphi/dt)$ are the corresponding velocity components. Thus,

(15.16) is a system of *ordinary* differential equations in functions $\lambda(t)$, $\varphi(t)$, being of fourth order in time. It has two first integrals: the energy $u^2 + v^2 = u_0^2 + v_0^2$, and the angular momentum $(u + \Omega a \cos \varphi)\, a \cos \varphi = (u_0 + \Omega a \cos \varphi_0)\, a \cos \varphi_0$, where the initial values are denoted by a zero subscript.

Let us assume that $\cos \varphi_0 = 1$ (equator), $u_0 = U \cos \alpha > 0$ (current with eastward component), $v_0 = U \sin \alpha$, and $\varepsilon = U/2a\Omega \ll 1$ (for $U = 1$ m/s we have $\varepsilon \approx 1/900$). Then it is evident that $|u| \leqslant U$, and thus the integral of the angular momentum gives $\cos \varphi \geqslant (1 + 2\varepsilon \cos \alpha + \varepsilon^2)^{1/2} - \varepsilon$, or approximately $|\sin (\varphi/2| \leqslant \varepsilon^{1/2} |\sin (\alpha/2)|$, so that a latitude zone containing current meanders may have only a small width. Consequently, in it the approximation of a β plane is valid, $y = a\varphi$ being small, so that $v = dy/dt$ is also small, and therefore $u \approx U$, making it possible to simplify the second equation (15.16) to $d^2y/dt^2 + 2\Omega U/ay = 0$. Hence the period of oscillation of coordinate y will be $\tau = 2\pi(a/2\Omega U)^{1/2}$, and the wavelength of the meanders will be $l \approx U_\tau = 2\pi(aU/2\Omega)^{1/2}$. For $U = 1$ m/s we get $\tau \approx 15$ days and $l \approx 1200$ km. Meanders of equatorial deep countercurrents, detected by special observations carried out during recent years, turned out to have periods and wavelengths of precisely this order of magnitude.

The most detailed numerical model of an equatorial deep countercurrent was constructed by Semtner and Holland (1980), who integrated equations (9.1)–(9.4) with the linearized equation of state (12.3′), but unfortunately without the dependence of ϱ_w on the salinity, and with the heat-conduction equation for a constant coefficient of vertical exchange $K = 1.5$. A "convective adjustment" for the temperature was used for $\partial t/dz < 0$. A biharmonic horizontal exchange was assumed, with $K_h = 8 \times 10^{18}$ and $K_{hT} = 3 \times 10^{19}$ in an ocean of constant depth $H = 4$ km (with 14 irregularly arranged levels and a quadratic bottom friction for $c_\tau = 1.3 \times 10^{-3}$), with a length 30° in longitude (with $\delta x = 0.5°$) and with latitudes from 9.875°S to 10.125°N (with $\delta y = 0.25°$). The frictional stress of the wind at the ocean surface was assumed to be constant ($\tau_{x0} = 0.05$ Pa), and the heat flux was taken to be 2.5 MJ/m² per day $(25 - T_s)$. Mass and momentum fluxes were assumed to be absent at the latitudinal walls, and the heat flux there was specified as 15×10^4 MJ/m² per day $(T_0 - T)$. Fluxes of mass, momentum, and heat were assumed to be absent at the longitudinal walls. The integration with respect to time was from a state of rest at $T_0 = 4 + 20 \exp(z/500\text{m})$ in intervals $\delta t = 30$ min for 5 years (the first two years of which corresponded to acceleration). As a result, a gradually and slightly meandering equatorial countercurrent was obtained, with a maximum of 1 m/s, lifted toward the east; on both sides of the equator the baroclinic instability is assumed to produce equatorially trapped westward Rossby waves $m = n = 1$ with periods of 33 days and lengths of 800 km, as well as shorter inertial-gravity waves in the thermocline.

MODELING OF GLOBAL CIRCULATION

Many numerical models have been constructed of the circulation in individual regions of the World Ocean. These models are of methodological and regional interest (let us cite, in particular, the excellent modeling by Cox (1970) of the seasonal variability of currents in the Indian Ocean, with its very highly developed monsoon effects). However, all the various regions of the World Ocean are parts of an interconnected whole, and the theory of climate requires numerical models of the circulation in the entire World Ocean, with realistic delineations of its shores and bottom relief. Only a few such models have been constructed so far.

First of all, we should mention the barotropic models (mean annual total fluxes) of Il'in *et al.* (1969), Takano (1969), Liakhtman *et al.* (1971), and Kagan *et al.* (1972); the latter team of modelers calculated the seasonal variations of the total fluxes. Secondly, we cite the two-layer models of Veronis (1973b with wind drive, 1976 with thermal drive) and Andreev *et al.* (1976). In the last of these the variations of the thickness of the upper layer (equal to 700 m, on the average) were linearized. As an example, Fig. 15.3 shows the mean annual total fluxes in the upper and lower layers obtained by Andreev *et al.* These demonstrate clearly the sizable variations in the circulation pattern with depth.

Thirdly, let us mention a number of multilevel models: the six-level steady-state model of Marchuk and Zalesnyi (1974); the model of the circulation of the atmosphere and ocean of Marchuk *et al.* (1975); the three models of Cox (1975) : (1) with constant density and wind drive, (2) with the density found from observations (a so-called *diagnostic* model), (3) with the *initial* T and S fields found from observations; and the five-level model of Takano (1975), with the density (a function only of T) adjusted to the zonal boundary condition of heat exchange, proportional to $T_a(\Theta) - T_w$ at the ocean surface. Note, too, the six-level steady-state model of Kochergin and Shcherbakov (1977), with an unsteady-state equation of "density diffusion", integrated over 30-day intervals for 5 years and 8 months with an initial density field satisfying the steady-state diffusion equation and specified at the ocean surface, like the wind stress, on the basis of winter observations. Finally, let us mention the 12-level model of Bryan, Manabe, and Pacanowski (1975), intended to be an element in a global model of the climate system (Manabe, Bryan, and Spelman, 1979), but also useful in the calculation (for 1150 years) of the state of the ocean for specified atmospheric effects with a seasonal variation (Bryan and Lewis, 1979).

The only questionable thing about all these models, perhaps, is the traditional parametrization of the horizontal mixing (produced primarily by synoptic eddies) with the help of operators of the form $K_h \Delta_h$ having large positive constant coefficients K_h, since "negative viscosity" effects are not allowed for thereby, and such effects are necessary to create the observed *narrow-stream nature* of the major ocean currents. Until computers can handle calculations based on *global eddy resolving models*, one means of solving this problem may be to supplement equations (9.3), (9.4), (9.3′), and (9.4′) with the Friedman-Keller equations for the single-point second moments of the synoptic fluctuations of the velocity field (first of all, for $\tau_{\theta\theta}$, $\tau_{\theta\lambda}$, and $\tau_{\lambda\lambda}$), simplified by some means or other.

MEANDERS AND RINGS

Even before the Second World War it was established that the Gulf Stream meanders. In the part from Cape Hatteras to the Grand Newfoundland Bank transverse waves 300–400 km long form on its main steam. These waves, which move eastward at velocities of 6–10 cm/s, have a tendency to intensify and change into meanders with an amplitude from south to north of as much as 500 km. The Kuroshio and many other currents also meander in a similar manner. The main causes of the meanders are, apparently, the baroclinic instability of the jet streams and the effects of flow around irregularities of the bottom relief. Let us present an approximate linear theory of such meandering in the case where the bottom relief $z = -H + Sy$ is independent of coordinate x (that is, the isobaths are straight lines parallel to the x axis). We assume in addition that the main flow is along the isobaths with a velocity $u = U(z)$, which is a function of the depth z, the fields of pressure and temperature being in this principal state determined from the geostrophicity condition $p = -\varrho_0 fyU(z) + \varrho_0 \alpha g \int^z \bar{T}(z)\, dz$ and the equation of hydrostatics, written in the form $T = \bar{T}(z) - (fy/\alpha g)(dU/dz)$. Let us assume that the meanders superimposed onto this current are quasi-geostrophic, and that the functions u', v', w', p', and T' describing them can be expressed as exponential series of the Kibel' number Ki in the form $a' = a_0(x, z, t) + \text{Ki}\, a_1(x, y)$. From the equations of hydrodynamics it follows that $u_0 = w_0 = 0$. The equation for the vertical component of the vorticity in the zero approximation takes the form

$$\left(\frac{\partial}{\partial t} + U\frac{\partial}{\partial x}\right)\frac{\partial v_0}{\partial x} - f\frac{\partial w_1}{\partial z} = 0,$$

or, after integration over the entire depth of the ocean taking into account the smallness of the bottom inclination $S = O(\text{Ki})$ and the kinematic boundary condition at the bottom $w_1 = Sv_0$ for $z = -H$, it takes the form

$$\int_{-H}^{0}\left(\frac{\partial}{\partial t} + U\frac{\partial}{\partial x}\right)\frac{\partial v_0}{\partial x}\, dz + fSv_0(x, -H, t) = 0. \qquad (15.17)$$

In addition, we make use of the equation of temperature shift in the zero approximation

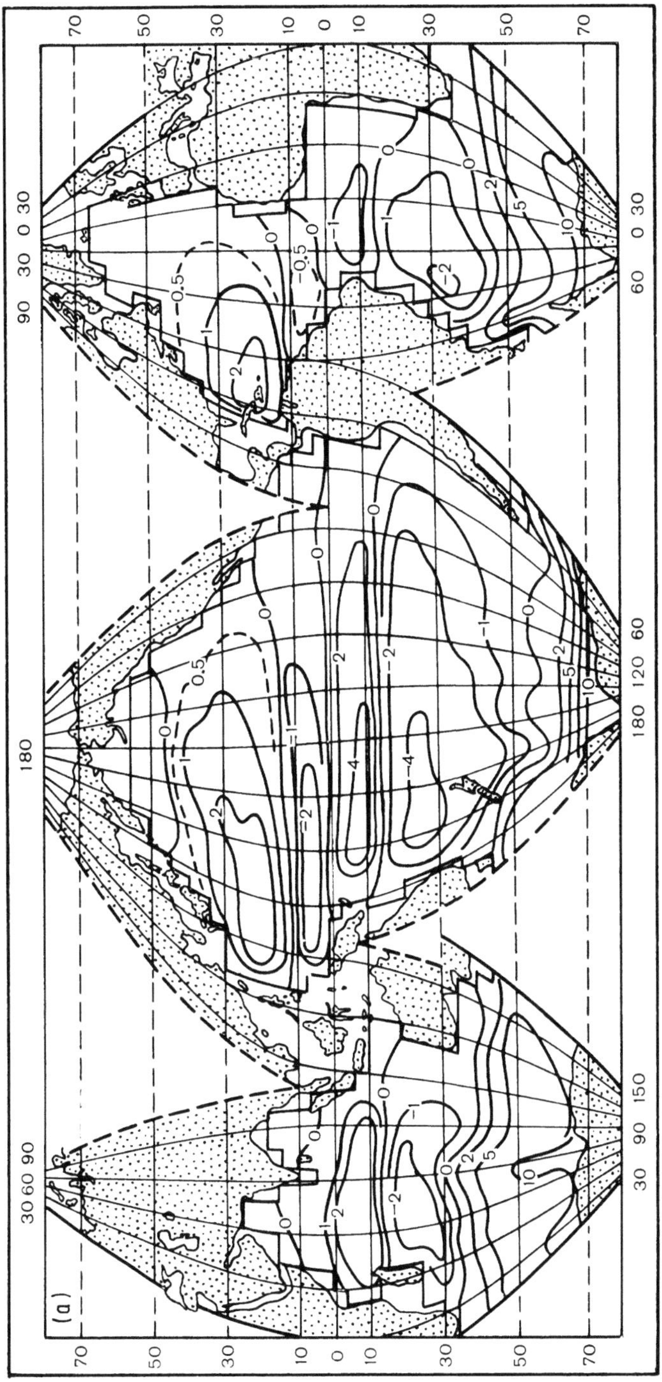

Fig. 15.3(a). Mean annual fluxes in 6.37 × 10^6 m^3/s in upper (a) and lower (b) layers of World Ocean, according to model of Andreev *et al.* (1976).

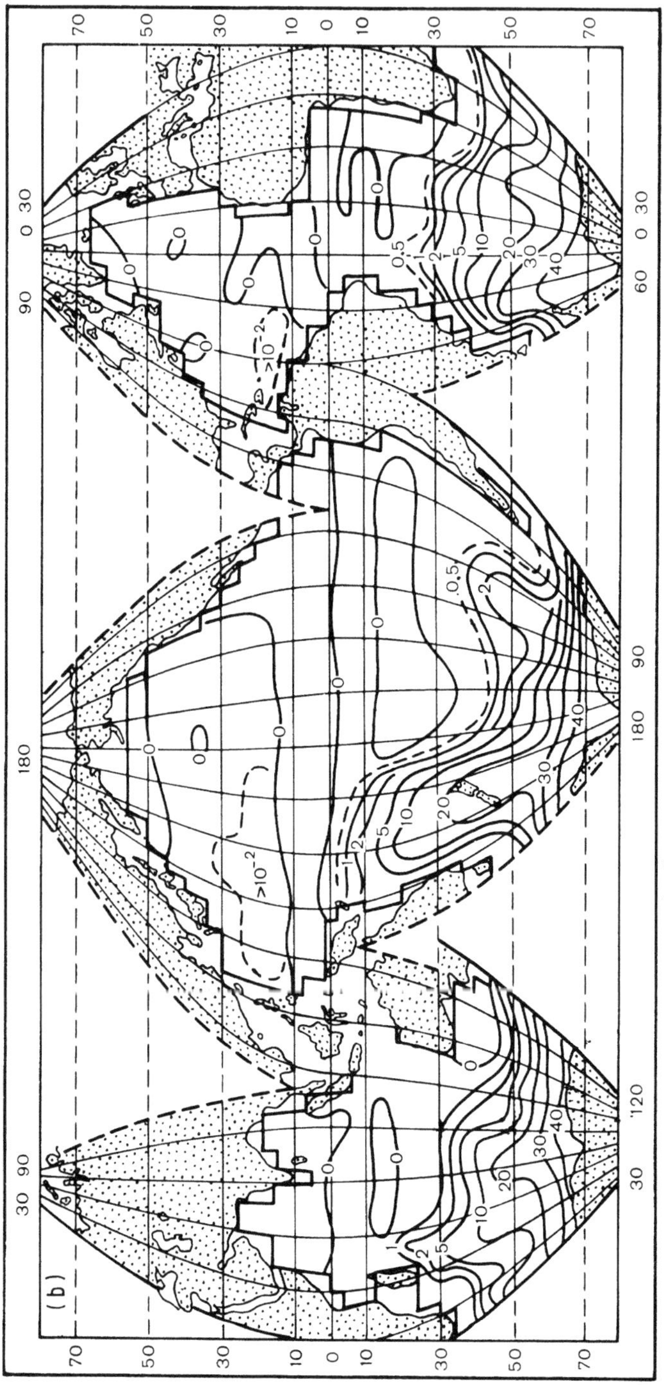

Fig. 15.3. Mean annual flows in 6.37 × 10^6 m^3/s in upper (a) and lower (b) layers of World Ocean, according to model of Andreev *et al.* (1976).

$$\left(\frac{\partial}{\partial t} + U\frac{\partial}{\partial x}\right)T_0 - \frac{v_0 f}{\alpha g}\frac{\partial U}{\partial z} = 0,$$

which after differentiation with respect to x, taking into account the formula $\partial T_0/\partial x = f/\alpha g\ (\partial v_0/\partial z)$ (an analog of the thermal-wind formula), becomes

$$\left(\frac{\partial}{\partial t} + U\frac{\partial}{\partial x}\right)\frac{\partial v_0}{\partial z} - \frac{\mathrm{d}U}{\mathrm{d}z}\frac{\partial v_0}{\partial x} = 0. \tag{15.18}$$

Equations (15.17) and (15.18) have a wave solution of the form $v_0 = A(z)$ exp $[i(kx - \omega t)]$. From (15.18) we get $A(z) = U(z) - \omega/k$, and then (15.17) gives the dispersion relation

$$\omega^2 + \left(\frac{fS}{kH} - 2k\langle U\rangle\right)\omega + k^2\langle U^2\rangle - \frac{fS}{H}U_b = 0, \tag{15.19}$$

where the angular brackets signify averaging over the depth, while the subscript b denotes a value at the bottom. It follows that for $S \neq 0$ stationary waves ($\omega = 0$) are possible: these are topographic meanders with a wave number

$$k = \left[\frac{fSU_b}{H\langle U^2\rangle}\right]^{1/2}.$$

However, in general from (15.19) we have

$$\omega = k\langle U\rangle - \frac{fS}{2kH} \pm \left\{k^2[\langle U\rangle^2 - \langle U^2\rangle] + \right.$$

$$\left. + \frac{fS}{H}[U_b - \langle U\rangle] + \frac{f^2S^2}{4k^2H^2}\right\}^{1/2}. \tag{15.19'}$$

In the barotropic case (U = const) all the frequencies ω are real, that is, the meanders are always stable. In the baroclinic case the meanders can grow in size: since $\langle U\rangle^2 < \langle U^2\rangle$, for sufficiently small fS/H frequencies ω turn out to be complex. In the Gulf Stream meanders were detected before the Second World War (Church, 1937); they have lengths of 300–400 km and they shift eastward at a velocity of 10 km/day, or 8 cm/s.

The meanders which have grown in size may separate from the main flow, forming closed annular currents, or *rings*. The rings of the Gulf Stream were described by Fuglister and Worthington (1951). South of the Gulf Stream they have a cyclonic (counterclockwise) rotation and they contain northern (cold) water; north of the Gulf Stream, on the other hand, they have an anticyclonic (clockwise) rotation and they contain southern (warm) water.

Newly formed cyclonic rings have diameters of about 200 km, horizontal temperature differences as great as 10–12°C (corresponding to depth differences of isothermal surfaces of up to 600–700 m), and linear velocities of rotation in the upper ocean layers of 2 m/s or even more; they penetrate to depths of at least 3 km, and possibly even to the bottom. Their "wakes" on the ocean surface (cold spots) are detectable even from satellites. As the rings age, mixing with the surrounding water causes their diameters, rotation velocities, and horizontal temperature differences to decrease gradually, so that after several months their "wakes" on the ocean surface are no longer visible.

The cyclonic rings of the Gulf Stream usually move in the Sargasso Sea according to the laws of their dynamics and under the influence of the total flow of the large-scale current, toward the west or southwest with velocities of about 3 cm/s, on the average, in the upper 700–1000 meter layer of the ocean, carrying their water with them (whereas at greater depths the rings may undulate through the water). These rings disappear, either being absorbed by the Florida Current or being attenuated completely. Lai and Richardson (1974) have estimated their mean lifetime to be 2 or 3 years; since in a year an average of 5 or 6 pairs of northern and southern rings are formed in the part of the Gulf Stream between Cape Hatteras and the Grand Banks, it follows that 10 to 15 cyclonic rings may be present in the Sargasso Sea at the same time. As an example, Fig. 15.4 shows an array of these rings, according to bathythermographic measurements made in November 1971. They are depicted here to actual scale (with diameters of about 200 km), and they account for about a third of the area of the water region in which they are located. Clearly, the rings must contribute appreciably to the heat exchange between the subtropical and subpolar zones of the ocean.

The warm northern anticyclonic rings of the Gulf Stream have considerably smaller horizontal and vertical dimensions, energies, and lifetimes (half a year, on the average, so that 2 or 3 such rings may exist simultaneously); they move westward or southwestward at a speed of around 5 cm/s and then disappear, being absorbed by the Gulf Stream near Cape Hatteras. The Kuroshio has rings similar to those of the Gulf Stream, the sole difference being that north of the Kuroshio they do not form individual eddies in a quiet ocean. Rather they constitute elements of a complicated eddy field, formed from the branches and meanders of the Kuroshio, the Oyashio, and the North Pacific Current. Actual data on the rings of the West Wind Drift and the other currents are very scanty.

SYNOPTIC EDDIES OF THE OPEN OCEAN

Forty-five years ago V. B. Shtokman, a Soviet oceanographer, already suspected the existence of a synoptic variability of ocean currents. In order to detect this variability, he organized long-term current measurements on moored buoy stations in limited regions of the ocean ("polygons", or "test

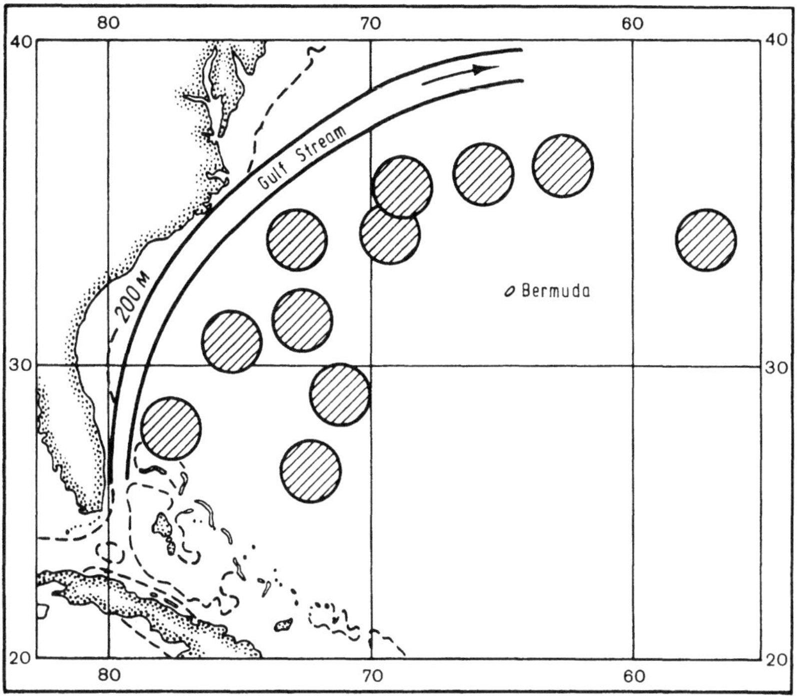

Fig 15.4. Cyclonic rings of Gulf Stream in November 1971, according to Lai and Richardson.

areas"); measurements were made in 1935 in the Caspian, in 1956 in the Black Sea, and in 1958 in the North Atlantic. Later Shtokman's suspicions were confirmed by measurements of deep currents carried out by Swallow with the aid of neutral-bouyancy floats. During these measurements, made in 1958 west of Portugal, and especially in 1959–1960 near the Bermudas (around 31°N, 68°W), strong unsteady currents close in direction were detected at depths of 2 to 4 km (with velocities of up to 41 cm/s at 4 km). Although there were not sufficient data available to interpret their spatial structure, rough estimates indicated space and time scales of the order of 100 km and 50 days, respectively. Let us also mention the temperature spectra obtained by American oceanographers, with a maximum for a period of about 100 days, based on measurements near the Bermudas from 1954 to 1969, and the current spectra, with a maximum for a period of about 40 days, based on measurements made just north of the Gulf Stream from 1965 to 1967.

The first special experiment (known as *Poligon 67*) designed to ascertain the spatial structure of the synoptic variability of ocean currents was performed, on the initiative of Shtokman, by the Institute of Oceanology of the Soviet Academy in 1967 in the Arabian Sea (10–15°N, 63–67°E). For two months current measurements were carried out at seven buoy stations. These measurements revealed considerable unsteadiness of the currents at all depths. Two hydrological surveys were also completed (with an array of 8 × 11 hydrological stations), and the resulting maps of geostrophic currents led to the first discovery of synoptic eddies in the open ocean (albeit by indirect means, rather than by direct current measurements).

The second special experiment on the synoptic variability of ocean currents (*Poligon 70*) was conducted by Soviet oceanographers, under the direction of the Institute of Oceanology, in February–September 1970 in the northern tradewind region of the Atlantic. Here, for six months the currents were measured continually at 10 depths from 25 to 1500 m at 17 moored buoy stations, forming a cross 200 × 200 km in size centered at 16°31′N, 33°30′W; a number of hydrological surveys were also carried out. The (northeasterly) Trades in this region were exceptionally stable, and the mean equatorial current (over the entire measurement period) in the upper layer of the ocean was found to be directed west-southwest with a velocity of about 3 cm/s. However, the instantaneous currents were found to be quite variable, and in the short-period part of the spectrum semidiurnal tidal, and especially inertial, oscillations were detected (with a period of about 42 hours at the latitudes of the observations); for periods of 3 or 4 days there was also a deep minimum in the oscillation spectrum, which made it possible to filter out the short-period oscillations and to distinguish reliably the long-period synoptic variability of the currents.

Maps of the synoptic component of the currents revealed that during the observation period three anticyclonic eddies and two cyclonic eddies passed through the test area. A large anticyclone observed from the end of March to the middle of July, which passed right through the center of the test area, was measured particularly well (Fig. 15.5). According to the maps, "dense packing" of the eddies prevailed, they were about 100 km in size (measured from the center to the region of maximum velocity), and the current velocities in them ranged from 25 cm/s at depths of 200–300 m to 35 cm/s at 400–600 m, 20 cm/s at 1000 m, and 10 cm/s at 1500 m. The vortices moved westward and southwestward at average speeds of 5–6 cm/s. The vertical axis of the May anticyclone was tilted downward with respect to the direction of its motion; this is typical of baroclinically unstable Rossby waves, and a number of the other characteristics of the eddies of *Poligon 70* can be successfully made to fit the theory describing such waves.

Consequently, it can be said that in *Poligon 70* direct measurements of currents led to the discovery of synoptic eddies of the open ocean, perhaps the major oceanographic discovery of the years after the Second World War. *Poligon 70* was repeated on a somewhat smaller scale, and its findings were verified by the American MODE (Mid-Ocean Dynamics Experiment). The main phase of the latter was completed in 1973 in the Sargasso Sea near 28°N, 69°40′W, during an average time of three months. Current measurements were made at four levels deeper than 400 m, since, in an attempt to eliminate

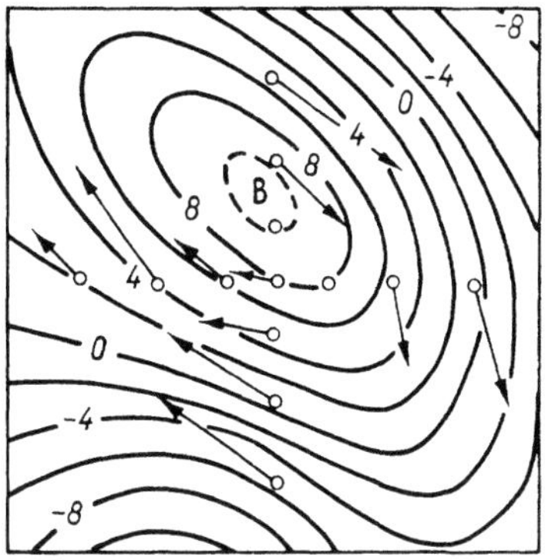

Fig. 15.5. Map of synoptic currents at depth of 300 m in *Poligon 70* on 24 May, according to Grachev and Koshlyakov. Arrows indicate measured currents, streamlines constructed using method of optimum interpolation.

the high-frequency bumping typical of surface buoys (even though it would not have affected the synoptic component of the currents which was the main subject of the experiment), deeply submerged buoys were employed (this resulted in a loss of information about the upper layer of the ocean, which is dynamically the most important layer). A methodological novelty of MODE was the trial launching at a depth of 1500 m of neutral buoyancy floats, with hydroacoustic tracking by a SOFAR system, enabling a quasi-Lagrangian measurement of deep currents.

The next major step in the study of synoptic eddies of the open ocean was the 13-month joint Soviet-American POLYMODE experiment of 1977–1978 in the Sargasso Sea, some brief information about which (and spectra of the currents measured by Soviet participants in the experiment) was presented at the beginning of this section. The Soviet network of 19 POLYMODE buoy stations succeeded in recording the passage of 21 synoptic eddies with diameters mostly of around 100 km (in individual cases up to 300 km), with velocities of rotation in the upper ocean layers of 30–35 cm/s, on the average (sometimes reaching 70–80 cm/s), and velocities of forward motion of 3–10 km/day, mostly toward the west. A number of smaller eddies were also recorded. At depths of 100, 400, 700, and 1400 m the velocities of the mean flow were, respectively, 1.2, 3.2, 2.4, and 1.3 cm/s, while the corresponding rms velocities of the synoptic currents were 23, 24, 18, and 13 cm/s. This means that the kinetic energy of the latter was two orders of magnitude greater than that of the mean flow (by the way, assuming that at a depth of 1500 m the barotropic component of the synoptic currents, that is, the mean component over the depth, is predominant, the data cited indicate that the barotropic and baroclinic components are comparable in the POLYMODE region).

Fig. 15.6 shows typical maps of synoptic currents at four depths, for 28 April 1978. In the NW and SE corners of the test area two anticyclonic eddies are clearly evident (these later approached each other and merged in part, after which the NW eddy intensified sharply, literally "snatching" energy from the adjacent regions). In the NE corner there is a cyclonic eddy, which is especially pronounced at a depth of 400 m but which is poorly defined at depths of 100 and 1400 m. At the latter depth the currents have weakened greatly, indicating the appreciable predominance of their baroclinic component in this case. Another instance of such high baroclinicity is the very strong but small cyclonic eddy observed from December 1977 to February 1978. This eddy, almost a soliton, occupied at least the entire principal thermocline; it moved northward and then northwestward, accelerating from 4–5 km/day to 15 km/day, decreasing in diameter from 150–130 km to 100–80 km, with an increase in its rotation velocities from 50–60 cm/s to 80–90 cm/s and with a velocity minimum at depths of 200 to 500 m (because of the opposite dipping downward of the main and seasonal thermoclines, so that the layer of 18-degree Sargasso water is nearly absent

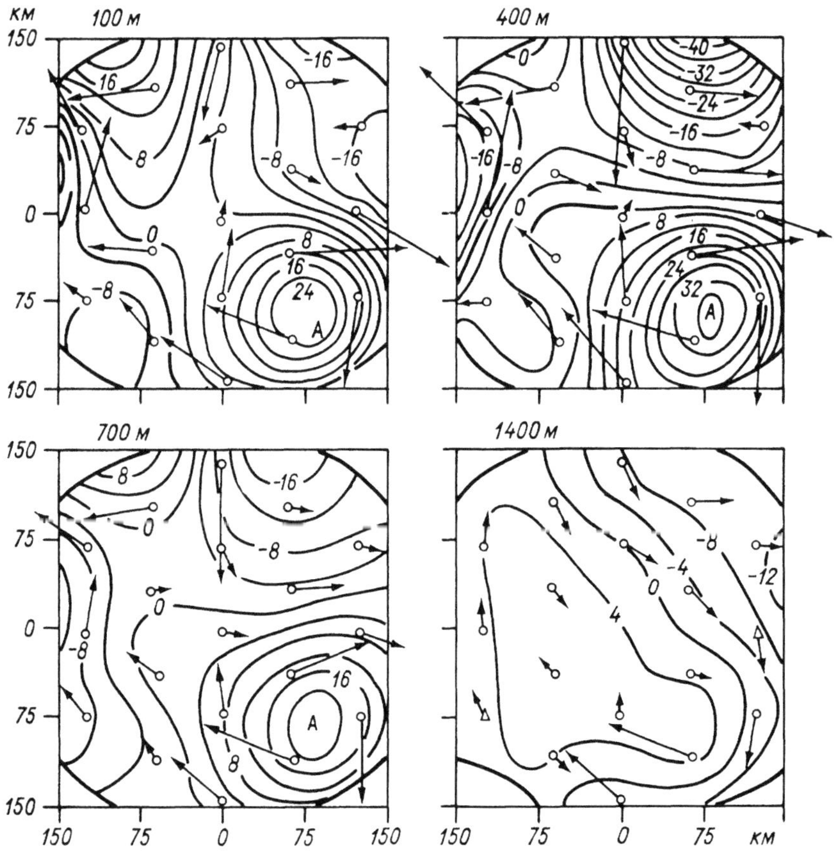

Fig. 15.6 Maps of synoptic POLYMODE currents at four depths on 28 April 1978, according to Koshlyakov, Grachev, and Enikeev.

here). In the upper 600 m this eddy carries with it water which, judging by the T–S and T–O_2 ratios, differs from the surrounding water of the Sargasso Sea. We should also mention here the small (diameter 50–70 km) anticyclonic eddy with a velocity maximum (15–20 cm/s) at a depth of 1600 m, detected during the American local-dynamics POLYMODE experiment (31°N, 70°W) in May–June 1978. This eddy was well-defined only in the layer from 1000 to 2500 m.

During the evolution of the POLYMODE synoptic currents, the formation of narrow intense jets or fronts in the velocity field was repeatedly observed. These jets or fronts are typical of two-dimensional turbulence, in which they are produced by a spectral transfer of enstrophy (in the ocean, of potential enstrophy) from large scales to small scales (see §10). During the winter-spring period abrupt temperature fronts were observed in the upper mixed layer and the seasonal thermocline. These fronts were located, as a rule, at the margins of intense eddies and they moved along with them, demonstrating the advective, rather than wave, nature of the movement of such eddies. For example, at the end of February and beginning of March 1978, along the eastern periphery of an intense anticyclonic eddy in the POLYMODE region a quasi-meridional front 10–15 km wide was observed near the ocean surface. This front, which divided warm water on the west from cool water on the east, with a temperature differential of 1.5–2°C, moved toward the west with the anticyclone at a speed of 3 to 3.5 km/day.

Consequently, the POLYMODE findings and other existing data go to show that the field of synoptic motions in the ocean contains eddies, forming a quasi-two-dimensional turbulence, as well as Rossby waves, just like the field of synoptic motions in the atmosphere, described in §10. Thus its statistical dynamics should be governed by the laws presented in §10: the spectral transfer of potential enstrophy from large scales to small scales (mentioned above); the spectral transfer of kinetic energy from small scales to large scales, leading to an increase in the average size of the eddies, to their subsequent transformation into Rossby waves (which move westward and become stretched out in the zonal direction), and to their barotropization. Oceanic synoptic processes differ from their atmospheric counterparts mainly in that their spatial scales are much smaller, their time scales are much greater, and their generation mechanisms are more diverse.

Actually, the typical horizontal size of the synoptic processes, of the baroclinic Rossby waves of the first mode, say, can be determined as the Rossby deformation radius $L_R = \bar{N}H/f$, where H is the effective depth of the atmosphere or ocean, f is the Coriolis parameter, and $\bar{N}$ is the mean Väisälä–Brunt frequency over this depth. For Rossby waves on an atmospheric jet stream, assuming for its vicinity in the upper troposphere and lower stratosphere that $\bar{N} = 2 \times 10^{-2}\ s^{-1}$ (in the troposphere $\bar{N} = [gT_0^{-1}\ (\gamma_a - \gamma)]^{1/2} \approx 10^{-2}\ s^{-1}$), $f = 10^{-4}\ s^{-1}$, and $H = 10$ km, we get $L_R = 2000$ km (if this is the size of the vortex, then the wavelength will be twice as great, or 4000 km). The mean stratification of the ocean, measured in terms of the quantity $\bar{N}$, is *considerably less stable than the stratification of the atmosphere*. This is because the layer of density discontinuity with large local values of N (approximately 3×10^{-2} s^{-1}), which forms in regions where the ocean is heated from above, is thin, so that its stratification has little effect on the mean value $\bar{N}$. In regions where the ocean is cooled from above mixing occurs, making the stratification close to adiabatic; then $\bar{N} \approx 10^{-3}$ s^{-1}, which is only 1/20 of the atmospheric value assumed above. For $H = 5$ km we get $L_R = 50$ km for the ocean, which is in very good agreement with the empirical values presented above.

Furthermore, the period of a zonal Rossby wave of the first baroclinic mode $\tau = 2\ (\beta L_R)^{-1}$, where β is the derivative of f with respect to the arc of the meridian. For $\beta = 2 \times 10^{-8}$ km^{-1} s^{-1} for the atmosphere, we obtain $\tau = 5 \times 10^4$ s $\approx$ 14 hours, whereas for the ocean $\tau = 2 \times 10^6$ s = 560 hours $\simeq$ 23 days. The phase velocity of the oceanic vortices $c = 2L_R\tau^{-1}$ turns out to be of the order of 5 cm/s. Therefore, because the mean stratification of the ocean is only 1/10 as stable as that of the atmosphere (and its effective thickness is only half as great), synoptic eddies in the ocean turn out to be tens of times smaller than, and hundreds of times slower (and more long-lived) than, synoptic eddies in the atmosphere.

In both the atmosphere and ocean synoptic eddies and Rossby waves may be generated due to baroclinic instability of the large-scale currents. Another mechanism of their generation may be flow around irregularities of the bottom relief, this mechanism being of much more importance in the ocean than in the atmosphere. Actually, the role of the relief is determined by the ratio h/H of the typical height h of an irregularity to the thickness H of the layer; this role becomes comparable to the β effect when $h/H \sim L/a$, where L is the wavelength and a is the Earth's radius. The critical value $h = LH/a$ in the ocean is tens of times less than in the atmosphere, making topographic eddies and Rossby waves much more abundant in the ocean.

In the ocean there are two other possible mechanisms by which synoptic motions can be generated, and these have no analogs in the atmosphere. The first is generation by direct atmospheric action on the ocean surface, that is, by inhomogeneities with the appropriate horizontal scales and periods in the fields of frictional-stress vorticity and atmospheric pressure at this surface. The second mechanism is generation by reflection of Rossby waves from the ocean shore, which transforms large-scale motions into small-scale motions. It should be noted with respect to this that at the western shores the enstrophy increases, whereas at the eastern shores it decreases; since small-scale vortices move more slowly, it follows that both energy and enstrophy accumulate at the western shores, while at the eastern shores they are dispersed.

Consequently, synoptic motions in the ocean can be classified as follows according to the mechanisms of their formation: frontal motions, or meanders in stream currents and rings (I), and free motions or eddies of the open ocean, caused by internal baroclinic instability (II), topographic effects at the bottom and shores (III), and atmospheric effects (IV). Depending on the sign of the rotation, these are divided into cyclonic (C) and anticyclonic (A) motions. Moreover, they can also be divided into eddies moving together with the water in them (E) and waves traveling along the water (W); in the latter case special notice should be taken of solitary waves, or solitons (S). Finally, all these formations can be conveniently classified according to age. In contrast to the atmosphere, where Rossby waves are encountered only in the form of isolated, quite rapidly attenuated, trains, in the ocean the diversity and abundance of generation mechanisms and the weak damping of the synoptic motions can set up a much more developed wave and eddy field, which must play a major role in the mixing of the water and in a number of other oceanic processes.

EDDY-RESOLVING MODELS

The numerical modeling of synoptic motions in the ocean can be achieved with the aid of the equations of dynamics of large-scale motions presented in §12. Here it is convenient to use the quasi-solenoidal approximation, whereby both surface and internal gravity waves are filtered out, or, as long as the scales are not too small, even the quasi-geostrophic approximation (10.7), (10.7′). In the latter case, however, a space grid with sufficiently small horizontal intervals must be employed, at any rate not larger than the radius of the Rossby deformation, that is, not more than several tens of kilometers, in order to ensure the resolution of individual synoptic eddies. Accordingly, numerical models with such small horizontal grid intervals are known as eddies-resolving models. Present-day computers are not yet able to carry out the necessary calculations according to a given model for a long time period and for the entire ocean. Therefore, so far numerical experiments based on such models have been performed only for limited regions of the ocean, squares or rectangles 1000 to 2000 km in size, and only assuming certain artificial boundary conditions (although it is now possible to carry out calculations based on the real boundary conditions of POLYMODE).

The first eddy-resolving model was that of Holland and Lin (1975). It was a two-layer model with primitive equations involving the horizontal viscosity (of the order of $K_h \sim 3 \times 10^6$ cm^2/s). These equations were integrated over a square 1000 × 1000 km on a grid of 51 × 51 points, only a wind drive being present (no differential heating). These conditions produced in the square one large-scale gyre with an intensified westward current to the north, becoming at the northern wall a strong eastward current and then turning into a return current in the eastern and southern regions. At the northern wall and in the return current synoptic eddies appear. These are barotropic, and sometimes in the return current baroclinic as well. The kinetic energy of these eddies is drawn from the potential energy, and at times also the kinetic energy, of the return current. One possible quantitative shortcoming of the results obtained is the smallness of the deviations of the interface between the two layers from the equilibrium position, in comparison with the fluctuations of the isopycnic surfaces in real oceanic synoptic vortices.

From 1977 to 1980 several other eddy-resolving models were constructed. In the USSR the first such models were devised by D. G. Seidov (1978 and 1980). He took into account in the equations the horizontal and vertical viscosity and heat conduction (assuming ϱ to be a function just of T), and he specified at the ocean surface the zonal frictional stress of the wind (which produced one or two large-scale gyres in a rectangular region) and the zonal heat flux, proportional to the difference between the water and air temperatures (a linear function of the latitude). For the integration Seidov used the Euler-Lagrange method of splitting of advection and diffusion operators, five levels along the vertical, and a horizontal interval of 50 km, with an integration time of up to 10 years. One of Seidov's conclusions was that inertial effects play a considerable role in synoptic eddies (in contrast to large-scale gyres). He also pointed out the spatial inhomogeneity of the energy-conversion processes, and in particular the important role of the negative-viscosity effect in zones of oceanic stream currents (which will have to be taken into account when attempting to parametrize the statistical effect of synoptic eddies on large-scale currents).

CHAPTER 6

THE LAND

16. The Hydrology of the Land

The "land" has to be an element of any mathematical model of the climate system. This is because, firstly, in the boundary conditions at the land surface for the fluxes of momentum, heat, and moisture certain characteristics of the land must be introduced, in addition to the atmospheric parameters and quantities, determinable from the equations of dynamics of the atmosphere. These characteristics of the land are specified or determined from some additional relations or equations. It should be noted, by the way, that the extremely fine scale and the great diversity of the characteristics of the land make it necessary to average them (sometimes with the appropriate weighting factors) over the squares of the space grid being used.

The first characteristic of the land which must be specified is, of course, the surface relief $z = h(\theta, \lambda)$. In order to describe the surface runoff of water on a land surface $z = h(\theta, \lambda)$, we have to construct extremum lines: *watersheds* along the ridge axes and *riverbeds* along the trough axes (the areas between these can be called *slopes*). Obviously, the precision of this procedure will depend on the scales of the space grid, over the squares of which the relief heights $h(\theta, \lambda)$ were averaged. In addition to the relief, smoothed over the grid squares, and the extremum lines drawn on it, it is convenient to specify in each square the rms values of the surface slope, which are needed in order to describe the diffusion of the downhill runoff.

The momentum flux, that is, the wind friction on the land surface, is calculated by specifying in addition the "roughness height" z_0 of this surface (which varies from fractions of a millimeter for a snow cover to centimeters for meadows and fields and decimeters for forests), the means of intrasquare averaging being important here.

In order to calculate the heat fluxes at the land surface, which enter into the heat budget equation (9.10), we must first give the albedo A of this surface (which varies from 0.1 for coniferous forests to 0.20–0.25 for meadows and fields, 0.4–0.65 for ice, and 0.8–0.9 for fresh snow) and its emissivity f (which varies from 0.85 to 1). Since these quantities are to a large degree determined by the vegetation and by the presence of a snow cover, they may well be subject to appreciable seasonal and some interannual fluctuations. The classification of regional climates according to the type of landscape, devised by L. S. Berg, can serve as a basis for a tentative mapping of these.

The second step is to calculate the heat flux into the soil H_s and the temperature of the land surface T_s, the latter being needed to find the long-wave self-radiation of the Earth's surface $B(T_s)$ as well as the turbulent heat flux H. The specific humidity at the land surface q_s also has to be calculated, in order to determine the heat of vaporization $\mathscr{L}E$, as will be discussed below. This requires solution of the equation of heat conduction in the soil, having specified the distributions of the soil conductivity and soil heat capacity, together with their dependence on the moisture content of the soil (and on the degree of soil freezing). Since the solution of the heat-conduction equation is a relatively easy task, we will not consider it further here.

Finally, in order to compute the moisture budget at the land surface, described by equation (9.11), it is necessary to determine at each point (θ, λ) the moisture flux into the soil I_s, and also the surface runoff or ground-water runoff, that is, to solve the fundamental problems of the hydrology of the land, which constitutes the main topic of the present section. To do this, we have to specify the relief $z=h_2\ (\theta, \lambda) \leqslant h\ (\theta, \lambda)$ of the subsoil surface of the water-resistant layer and the characteristics of the layer $h_2 \leqslant z \leqslant h$, considered to be a porous medium through which moisture can percolate; the distribution of *porosity* in it (that is, the pore fraction of the soil volume) and the *permeability* of the soil have to be known. Here it is assumed that the soil contains both tiny *capillary* pores (capillary porosity m_0), which give it a water-retaining capacity, and large *noncapillary* pores (noncapillary porosity m_1), permitting a rapid penetration of water from the soil surface $z=h\ (\theta, \lambda)$ down to the free surface of the ground water $z=h_1\ (\theta, \lambda, t)$, which can be calculated (where $h_2 \leqslant h_1 \leqslant h$).

The entire soil layer $h_2 \leqslant z \leqslant h$ in which water can move is divided, in general, into the following five layers (from top to bottom): (1) a layer of surface moistening of the soil, which includes the root systems of plants, so that the transpiration of moisture proceeds from it; (2) a layer of capillary-suspended moisture; (3) an air-saturated layer of soil aeration; (4) a layer of capillary rise of ground water; (5) a ground-water layer $h_2 \leqslant z \leqslant h_1$, the horizontal flow in which is known as the *subsurface runoff*.

The capillary seepage of water in layers (1)–(4), when they are incompletely saturated, that is, when the volume fraction σ of liquid water in the soil (the moisture content of the soil, or *soil humidity*) is less than the capillary porosity m_0, is calculated with the aid of the equation of conservation of moisture $\partial\sigma/\partial t + \partial V_z/\partial z = 0$, where V_z is the vertical component of the volume transport of water through a unit area, measured in cm/s. This quantity, known as the *seepage rate*, is defined by *Darcy's law*

$$V_z = -k_0 \frac{\partial Z}{\partial z}; \qquad Z = z + \frac{p}{g\varrho_w}, \tag{16.1}$$

where Z is the so-called *head* ($p = p_a + p_c$ being the water pressure in the capillaries, if p_a is the atmospheric pressure and p_c is the pressure created by the surface tension of the water on the curved meniscus surfaces in the capillaries), and k_0 is the coefficient of capillary permeability. The permeability is a function of a number of soil characteristics (primarily the capillary porosity m_0, the size of the soil particles or of the pores in the soil, and, what is very important, the soil humidity σ), as well as of the properties of the seeping liquid (primarily its viscosity). Sometimes the dependences of k_0 and p_c on the soil humidity σ are described by simple empirical formulas, such as the following ones of S. F. Aver'yanov:

$$k_0 = k_m \left(\frac{\sigma - \sigma_0}{m_0 - \sigma_0} \right)^{7/2} ; \qquad p_c = - p_0 \frac{\sigma_0 \sigma_1^3 - \sigma^3}{\sigma \sigma_1^3 - \sigma_0^3} , \tag{16.2}$$

where σ_0 is the volume fraction of bound water in the soil, σ_1 is the total moisture content of the soil (for $p = p_a$), k_m is the value of k_0 for complete saturation of the soil with water (that is, for $\sigma = m_0$), and p_0 is the value $(-p_c)$ when only bound water is present in the soil (that is, for $\sigma = \sigma_0$). Using Darcy's law for the seepage rate, we can rewrite the equation of conservation of moisture as

$$\frac{\partial \sigma}{\partial t} - W \frac{\partial \sigma}{\partial z} = \frac{\partial}{\partial z} D \frac{\partial \sigma}{\partial z} ; \quad W = \frac{\partial k_0}{\partial \sigma} ; \quad D = \frac{k_0}{g \varrho_w} \frac{\partial p_c}{\partial \sigma} \tag{16.3}$$

Mathematically speaking, this equation is a nonlinear equation of convective diffusion in which both the settling rate W and the diffusion coefficient D depend on the unknown function σ. The boundary condition for the unknown function σ at the ground surface $z = h$ when liquid water is present on it has the form $\sigma = m_0$ (here, for the air humidity at $z = h$ we can use the saturation condition $q = q_m (T_s, p_a)$ at a temperature T_s instead of the moisture-budget equation (9.11)). In the absence of liquid water on the ground surface it should be assumed that $\varrho_w V_z = E$. The second boundary condition can be put into the form $\sigma = m_0$ at the surface of the ground water $z = h_1$ (which has to be determined, together with the subsurface runoff, from equations to be given below). The moisture flux into the soil I_s, which enters into the moisture-budget equation (9.11), is the sum of the capillary flow $\varrho_w V_z$ and the so-called *infiltration*, that is, the noncapillary flow $-\varrho_w k_1$ which rapidly penetrates to the level of the ground water; the latter flow is obtained from a formula analogous to (16.1) for $Z = z$, where k_1 is here the noncapillary permeability.

Now let us calculate the subsurface runoff, produced by the horizontal motion of ground water in the layer $h_2 \leqslant z \leqslant h_1$. This calculation can be carried out in the so-called *hydraulic approximation*, that is, by averaging over the thickness of the layer $h_1 - h_2$. Summed up over the layer $h_1 - h_2$, the horizontal components of the subsurface runoff per unit length of the parallel and meridian will be given by the following formulas, similar to (16.1):

$$\varrho_w (h_1 - h_2)V_\theta = -\frac{\varrho_w k\,(h_1 - h_2)}{a}\frac{\partial h_1}{\partial \theta};$$
$$\varrho_w (h_1 - h_2)V_\lambda = -\frac{\varrho_w k\,(h_1 - h_2)}{a \sin \theta}\frac{\partial h_1}{\partial \lambda}, \tag{16.4}$$

where k is the permeability of the ground-water layer. If it is integrated over the layer $h_1 - h_2$, then the equation of moisture conservation, taking formulas (16.4) into account, takes the form of the following *Boussinesq equation* for the height $z=h_1\,(\theta, \lambda, t)$ of the free surface of the ground water:

$$\frac{\partial m_1 h_1}{\partial t} = \frac{1}{a^2 \sin \theta}\left(\frac{\partial}{\partial \theta} D_1 \sin \theta \frac{\partial h_1}{\partial \theta} + \frac{1}{\sin \theta}\frac{\partial}{\partial \lambda} D_1 \frac{\partial h_1}{\partial \lambda}\right) +$$
$$+\frac{I_1 - E_1}{\varrho_w}; \qquad D_1 = k_1\,(h_1 - h_2,) \tag{16.5}$$

where m_1 and k_1 are, respectively, the porosity and permeability of the ground-water layer (D_1 is called the *conductivity of the stratum*), while I_1 and E_1 are the infiltration and evaporation at its upper surface (E_1 is sometimes given as a decreasing linear function of the depth $h - h_1$).

Now let us note that, according to the moisture-budget equation (9.11), the precipitation P, the evaporation E, and the moisture flux into the soil I_s (plus a source of liquid water for melting of ice or snow or a loss of liquid water for freezing) at each point on the Earth's surface do not, in general, compensate one another. Their algebraic sum E_s turns out to be the source of the formation of the surface runoff. The latter is the sum of the downhill runoff and the channel (river) runoff, this division being dependent on the scales of the space grid used in the calculations: if very large scales are employed, then only the largest rivers will figure into the channel runoff, and smaller rivers and streams will be described just on the average, rather than individually (that is, as a part of the downhill runoff); if, on the other hand, a very fine grid is used, then even small rivers and streams can be described individually (that is, as a part of the channel runoff).

The downhill runoff can be described as a *two-dimensional* flow of a compressible liquid with a surface density $\varrho_s = \varrho_w h_w$. Here h_w is the effective thickness of the layer of downhill flow, determinable from the continuity equation with sources E_s; the latter are found from the moisture-budget equation (9.11):

$$\frac{\partial h_w}{\partial t} + \frac{1}{a \sin \theta}\left(\frac{\partial h_w v_\theta \sin \theta}{\partial \theta} + \frac{\partial h_w v_\lambda}{\partial \lambda}\right) = \frac{E_s}{\varrho_w}, \tag{16.6}$$

valid in regions (θ, λ, t) where $h_w > 0$. The horizontal components of the downhill runoff per unit length of the parallel and meridian are $\varrho_w h_w v_\theta$ and

$\varrho_w h_w v_\lambda$, where velocity components $v_\theta(\theta, \lambda, t)$ and $v_\lambda(\theta, \lambda, t)$ must be found from the equations of motion:

$$\begin{aligned}
&\frac{dv_\theta}{dt} - \frac{\operatorname{ctg}\theta}{a} v_\lambda^2 - 2\Omega \cos\theta v_\lambda \\
&\quad = -\frac{g}{a}\frac{\partial h}{\partial \theta} - \frac{E_s v_\theta}{\varrho_w h_w} - \frac{v v_\theta}{l_w} + f_\theta, \\
&\frac{dv_\lambda}{dt} + \frac{\operatorname{ctg}\theta}{a} v_\theta v_\lambda + 2\Omega \cos\theta v_\theta \\
&\quad = -\frac{g}{a\sin\theta}\frac{\partial h}{\partial \lambda} - \frac{E_s v_\lambda}{\varrho_w h_w} - \frac{v v_\lambda}{l_w} + f_\lambda.
\end{aligned} \tag{16.7}$$

The second terms on the right-hand sides of (16.7) appear because we have passed from the equations for the momentum ($\varrho_s v_\theta$, $\varrho_s v_\lambda$) to the equations for the velocities (v_θ, v_λ), using the continuity equation (16.6), which has on its right-hand side the quantity E_s/ϱ_w, describing the sources or losses of surface runoff. The third terms on the right-hand sides, in which $v = (v_\theta^2 + v_\lambda^2)^{1/2}$ appears, as well as l_w, a coefficient with units of length, describe the v^2 frictional resistance at the land surface (in hydrology such a description of the frictional resistance is known as the *Chezy formula*). The fourth terms describe the diffusion of the downhill runoff due to the presence of small-scale inclinations of the relief. The quantities $\varrho_s f_\theta$ and $\varrho_s f_\lambda$ should be expressed in terms of the frictional stresses τ_{ij}, with the aid of formulas (9.3′) and (9.4′), taken for $\tau_{\theta z} = \tau_{\lambda z} = 0$. Here stresses τ_{ij} can be expressed in terms of the tensor of the deformation rates, as in the case of a two-dimensional viscous liquid, it being assumed that the viscosity is a function of both the rms size of the small-scale inclinations of the relief and the large-scale curvature of the slope, the sign of which must be the same as that of this coefficient (so that for a positive slope curvature the downhill flow must diverge, and for a negative curvature it must converge).

The two-dimensional hydrodynamics described by equations (16.6) and (16.7) is complicated by the presence of water sources or sinks E_s, which may give the downhill flow irregular margins, and may cause it to have "holes" in it and to "drip"; the evolution of the boundaries of this water layer, where h_w goes to zero, has to be calculated. The second terms on the right-hand sides of equations (16.7), containing h_w in their denominators, may in the vicinity of the flow boundaries create difficulties during the numerical calculations.

Channel flow can be described as the *one-dimensional* flow of a compressible liquid with a linear density $\varrho_l = \varrho_w \Sigma$, where Σ is the area of the active section of the (water-filled) river, determinable from the continuity equation with sources present. If the effects of channel curvature are neglected, this equation can be written as

$$\frac{\partial \Sigma}{\partial t} + \frac{\partial \Sigma U}{\partial x} = E_l, \tag{16.8}$$

where x is the coordinate along the river (increasing downstream), U is the mean flow rate over the river section, and E_l is the lateral influx of water (the volume of inflowing water per unit length of the river per unit time). The lateral influx is the sum of the components (normal to the river) of the downhill runoff $h_w v_n$ and the subsurface runoff $(h_1 - h_2)\, V_n$ from the right and left slopes; it should be noted that, in regions with a stable snow cover, only the subsurface runoff will remain. The channel runoff, that is, the volume transport of water in the river, is equal to ΣU, the rate U being found from the equation of motion

$$\frac{\partial U}{\partial t} + U\frac{\partial U}{\partial x} = - g\frac{\partial\,(h + h_r)}{\partial x} - \frac{E_l U}{\Sigma} - \frac{U^2}{l_r} + \nu_r\frac{\partial^2 U}{\partial x^2}, \tag{16.9}$$

where $(-\,\partial h/\partial x)$ is the gradient of the river bottom, and h_r is the river depth, so that $[-\,\partial(h + h_r)/\partial x]$ is the gradient of the river surface. The second, third, and fourth terms on the right-hand side of (16.9) are analogous to the corresponding terms in equations (16.7). In (16.9) l_r is a parameter with units of length ($1/l_r$ is the Chezy coefficient), and ν_r is the coefficient of longitudinal (Taylor) diffusion. In hydrology equations (16.8) and (16.9) are known as the *St. Venant equations*. In them it is still necessary to specify the dependence of the cross-sectional area of the river Σ on its depth h_r (an analog of the equation of state, expressing the density as a function of the pressure). For instance, if h_r is the maximum depth, then for a rectangular section of constant width we have $\Sigma = ah_r$, for a triangular section $\Sigma = bh_r^2$, and for a trapezoidal section $\Sigma = ah_r + bh_r^2$. Similarly, we can set $l_r = ch_r$.

Using the water-budget equation (9.11) and calculating the lateral influx of water due to the channel, slope, and subsurface runoff, we can now find the regime of the continental water bodies (lakes and reservoirs) formed in the depressions of the relief (bounded by closed contour lines).

It should be noted that these methods for describing hydrological processes with the aid of equations of the hydrodynamical type have not yet been applied with any degree of completeness, not in global models nor even in regional models, although the equations of the theory of flow through porous media have, on the other hand, been widely used to solve local problems. According to hydrologists, this is because of our imperfect knowledge about the parameters entering into these equations, which makes the material in the present section rather a kind of program for future studies. For the present, however, in global models of the climate system these equations have been replaced by very simple empirical relations, which provide neither a qualitative explanation of the processes nor a quantitatively accurate description of them. Examples of such empirical relations will be given in later chapters.

17. Cryology

Now let us consider how to describe forms of frozen water in the climate system: snow cover, sea ice (making the seas similar to the dry land in a number of respects), continental ice sheets, and permafrost. These forms are of definite climatic significance, firstly as indicators of the state of the climate (such a role, which is especially important for monitoring climatic fluctuations, is played, in particular, by the regions and areas of spreading of the winter snow cover, as well as by the sea ice in the Arctic and Antarctic) and as factors active in shaping the climate (as is the case with the continental ice sheets, which constitute an inertial element in the climate system that is second only to the oceans; this is what gives the system its auto-oscillating nature).

Many of the characteristics of the land which were listed at the beginning of the last section are affected by the presence of a snow cover. These include: the roughness height z_0; the albedo of the Earth's surface A; the emissivity f; and the porosity, heat conductivity, and heat capacity of the upper layer of the ground. With regard to the albedo, it should be noted that the albedo of the Earth's surface may differ considerably from that of the Earth-atmosphere system, primarily because of clouds, including high-level clouds, which mask the effect of the higher albedo of a snow cover. For instance, the mean amount of cloud cover in the Arctic is estimated to be 0.50 from December to March, 0.55 in April, 0.70 in May, 0.75 in June and July, 0.80 in August and September, 0.70 in October, and 0.60 in November. This means that the masking effect of the clouds is greatest in summer, when the amount of solar radiation is also at its peak. The dynamics of a snow cover can be described by the following equation:

$$\frac{\partial \bar{h}_s}{\partial t} = P_s - E + E_{wi} \tag{17.1}$$

(valid for periods of time when $\bar{h}_s > 0$), where $\bar{h}_s$ is the thickness of the snow cover, expressed as a water equivalent (that is, $\bar{h}_s = \varrho_s h_s$, where h_s is the geometrical thickness and ϱ_s is the density of the snow); P_s is the rate of fall of precipitation, found from the equations of dynamics of the atmosphere; E is the evaporation rate; and the quantity $E_{wi} = H_{wi}/\mathscr{L}_{wi}$, differing from zero only for a temperature T_s equal to the freezing point of water T_{wi}, is found from the heat-budget equation (9.10) for $T_s = T_{wi}$, in which H_s is calculated using the equation of heat conduction in the ground (once again taking into account the snow cover).

Now, passing to a marine ice cover, let us consider first the freezing and melting of ice (taking into account, in particular, the fall of precipitation onto its surface) and then the drifting of ice and the formation of leads and hummocks. The freezing and melting of an ice layer (of thickness h_i), on which there may be a layer of snow (of thickness h_s), are calculated with the aid of the heat-conduction equation

$$c\varrho \frac{\partial T}{\partial t} = \frac{\partial}{\partial z}\left(c\varrho k \frac{\partial T}{\partial z} + F_0^{\downarrow} \exp\left(\int_0^z \varkappa \, dz\right)\right) \tag{17.2}$$

where $F_0^{\downarrow}$ is the flux of short-wave solar radiation penetrating into these layers through their upper surface $z = 0$, and $\varkappa = \varkappa(z)$ is the coefficient of absorption of this radiation; note that coefficients $c\varrho$, k, and $\varkappa$ are different for snow and ice (for ice $c\varrho$ and k depend on the ice temperature and salinity). The heat-budget equation (9.10) can serve as a boundary condition for equation (17.2), provided the left-hand side of (9.10) is replaced by $(c\varrho k \, \partial T/\partial z)_{z=0} + F_0^{\downarrow}$. For temperatures below the freezing point of seawater (often taken to be 271.2 K) we should set $H_{wi} = 0$, and at the freezing point this boundary condition defines H_{wi}, and thus the rate of freezing and melting E_{wi} as well. For a known E_{wi} variations in the thickness of the snow layer are calculated with the aid of equation (17.1).

At the boundary between the snow and ice $z = -h_s$ the temperature and the heat flux must be continuous. Finally, at the lower boundary of the ice $z = -(h_s + h_i)$ the difference of the heat fluxes in the ice and in the water (the heat flux in the water being determined with the aid of the equations of dynamics of the ocean) defines H_{wi}, and thus E_{wi} as well, giving as a water equivalent the rate of freezing or melting of the ice from below. It should be noted that, for a precise calculation of this heat flux in the water (directed upward, from the water to the ice), the salinity of the water, variations in the thickness and the other characteristics of the upper mixed layer, and its interactions with deeper layers of the ocean all must be taken into account; for instance, when the ice freezes from below, the liberation of salt increases the density of the water in the mixed layer and makes the development of convection possible. Sometimes, for simplicity the heat flux from the water to the ice is assigned a standard value (2 W/m^2 in the Arctic and 25 W/m^2 in the Antarctic).

A very detailed one-dimensional model describing the evolution of the sea-ice thickness of type (17.2) was constructed by Maykut and Untersteiner (1969, 1971), using a numerical scheme involving layers 10 cm thick.

The large-scale horizontal motion of sea ice, assuming variations in its thickness due to hummocking and the formation of leads, can be described as the motion of a two-dimensional compressible film, satisfying, first of all, the general equation of continuity with sources present

$$\frac{\partial \Phi}{\partial t} + \frac{1}{a \sin \theta}\left(\frac{\partial \Phi v_\theta \sin \theta}{\partial \theta} + \frac{\partial \Phi v_\lambda}{\partial \lambda}\right) + \frac{\partial \Phi \varepsilon_i}{\partial h_i} = \Psi, \tag{17.3}$$

where Φ $(h_i;\ \theta, \lambda;\ t)$ is the probability density for a thickness h_i of the ice cover, so normalized that $n = \int_{+0}^{\infty} \Phi \, dh_i$ is the *compactness* of the ice cover, that is, the portion of the area covered by ice; $\varepsilon_i = \partial h_i/\partial t$ is the rate of freezing or melting of the ice, determined using equation (17.2); Ψ is a redistribution function, describing hummocking and the formation of leads, which must be specified on the basis of models of these processes. Velocity components v_θ and v_λ have to be found from equations of motion of the type of (9.3), (9.4), which must involve, generally speaking, all the forces affecting the horizontal motion: the sums of the wind-friction force on the upper surface of the ice cover and the force of the relative water flow at its lower surface, the force due to inclinations of sea level (the dynamic topography of the sea surface), the Coriolis force, and the internal forces of interaction between ice floes (that is, of components f_θ and f_λ of the divergence of the tensor of the horizontal stresses, specified with the aid of the rheological laws of a viscous fluid or of a viscoelastic or plastic medium).

A detailed model of the dynamics of a marine ice cover was worked out in 1973–1976 in connection with AIDJEX (Arctic Ice Dynamics Joint Experiment). This model included the distribution function of ice thickness Φ (h_i; θ, λ; t), equations of motion involving the five above-mentioned forces, and an elasticoplastic rheological law (Coon *et al.*, 1974; Pritchard, 1975, 1976; Hall *et al.*, 1976). The model was tested for periods of about 24 hours in the vicinity of the Beaufort Sea, comprising about 1/20 of the area of the Arctic.

In a model devised at the National Center for Atmospheric Research (USA) (Parkinson and Washington, 1979), a thermodynamic calculation of the evolution of the thickness of a marine ice and snow cover follows a somewhat more simplified scheme than that of Maykut and Untersteiner. However, approximate account is taken of variations in time of the fraction A of the area taken up by leads in each cell of the space grid, because of the melting of ice or the freezing over of leads, these variations depending on the total heat flux in the leads and the water temperature in them (this is important, since at the freezing point of water the vertical fluxes of heat and moisture over the leads may be two orders of magnitude higher than over the water, so that even for $A = 1\%$ leads may play a decisive role in interactions between the atmosphere and ocean).

The velocities of horizontal motion of an ice cover in this model are found from the condition of equilibrium between the stresses of wind friction and the ocean current, the gradient of dynamic topography, and the Coriolis force, neglecting accelerations and, for the time being, without taking into account the internal resistance. If, even during only a single time interval, these velocities lead to an intersection with a shore boundary, then the corresponding velocity component is replaced by zero. If, in some square of the space grid, they lead to a compactness of the ice greater than $n = 1 - A_{\min}$ (where $A_{\min} = 0.005$ in the Arctic and 0.02 in the Antarctic), then the velocities of the ice flows into the square are reduced in proportion to $(1 - A_{\min} - R)(n - R)^{-1}$, where R is the fraction of the square area covered by ice which remains in the square, and iterations are used to make all the squares match. This model gives a fairly good representation of the seasonal fluctuations of the amount of ice in the Arctic and Antarctic, the greatest compactness of ice being in the Arctic and the obtained pattern of ice drift being a quite realistic one. In the Arctic the ice minimum was found to be in September, when the edge of the ice cover moves away almost completely from the shore, and the maximum was in March, when ice covers the Bering Sea, blocks the north of Iceland, and spreads south of Greenland, with a maximum $h_i \sim 4$ m. In the Antarctic the ice minimum was found to be in March, when the ice cover is confined to the continent, and the maximum was at the end of August, when it spreads north of 60°S, with a maximum $h_i = 1.4$ m.

The evolution of the continental ice sheets constitutes a specific problem of the theory of long-period climatic fluctuations, involving periods of thousands or tens of thousands of years. These sheets can be considered to be plastic bodies which flow as a result of their own weight. Large-scale motions in them can be described with the aid of equations (9.1)–(9.4), (9.3′), (9.4′), plus a rheological law relating the deformations of the ice to the stresses in it. On the basis of data from laboratory and field measurements, it can be said that ice flows like a viscous, but non-Newtonian, fluid. Thus it has been called a *pseudoplastic* fluid, in which an increase in deformation rate is attended by a lessening growth of the viscous stresses, so that the viscosity is reduced (in addition, the viscosity decreases with a rise in temperature). The viscosity of such a fluid is given, for example, by a formula such as the following:

$$\nu = A\nu_i f\left(\frac{\varrho_i \nu_i \Phi}{\tau_i}\right) \exp\left[B(T_{wi} - T)T_{wi}^{-1}\right], \tag{17.4}$$

where Φ is the rms deformation rate, ν_i is a typical value of the kinematic viscosity of ice at low deformation rates (apparently, $\nu_i \sim 10^{14}$ cm²/s), τ_i is a typical value of the viscous stresses in kg/cm², ϱ_i is the density of the ice, T_{wi} is its melting point (which decreases slightly with a rise in pressure), $f(\xi)$ is a decreasing function (equal, let us say, to unity for $\xi = 0$, and apparently approaching ξ^{-1} asymptotically for higher ξ), and A and B are numerical constants. As Φ^2 we can here take the second invariant of the tensor of the deformation rates, which in spherical coordinates has the form

$$\begin{aligned}
\Phi^2 = {} & 4\left(\frac{\partial v_r}{\partial z}\right)^2 + 4\left(\frac{1}{a}\frac{\partial v_\theta}{\partial \theta} + \frac{v_r}{a}\right)^2 + \\
& + 4\left(\frac{1}{a\sin\theta}\frac{\partial v_\lambda}{\partial \lambda} + \frac{v_r}{a} + \frac{v_\theta \operatorname{ctg}\theta}{a}\right)^2 + \\
& + 2\left(\frac{1}{a}\frac{\partial v_r}{\partial \theta} + \frac{\partial v_\theta}{\partial z} - \frac{v_\theta}{a}\right)^2 + \\
& + 2\left(\frac{1}{a\sin\theta}\frac{\partial v_\theta}{\partial \lambda} + \frac{1}{a}\frac{\partial v_\lambda}{\partial \theta} - \frac{v_\lambda \operatorname{ctg}\theta}{a}\right)^2 + \\
& + 2\left(\frac{\partial v_\lambda}{\partial z} + \frac{1}{a\sin\theta}\frac{\partial v_r}{\partial \lambda} - \frac{v_\lambda}{a}\right)^2.
\end{aligned} \tag{17.5}$$

In the foregoing equations describing the motion of ice a number of small terms are usually dropped. The appropriate boundary conditions must, of course, be appended to these equations. One such is the kinematic boundary condition at the surface of the ice sheet $z = \zeta(\theta, \lambda, t)$:

$$v_z = \frac{\partial \xi}{\partial t} + \frac{v_\theta}{a}\frac{\partial \zeta}{\partial \theta} + \frac{v_\lambda}{a \sin\theta}\frac{\partial \zeta}{\partial \lambda} - \frac{1}{\varrho_i}(P_s - E + E_{wi}). \quad (17.6)$$

Since this condition can be used to calculate the level and shape of the surface of the ice sheet, it constitutes one of the fundamental equations of dynamics of the ice sheet. The quantity $\alpha = (1/\varrho_i)\,(P_s - E + E_{wi})$ on the right-hand side of (17.6), describing the ice-mass budget at the surface of the ice sheet, is of great significance for ice-sheet dynamics. Calculations of Verbitskii (1981), made using a simplified form of equation (17.6), indicate that even slight changes in α may cause radical variations in the size of an ice sheet. For instance, a precipitation deficit $\alpha \sim 150$ mm/yr caused the edges of the Laurentian and Scandinavian Pleistocene ice sheets to recede at rates of the order of 100 m/yr.

At the bottom of an ice sheet, that is, at the surface of the solid Earth $z = h(\theta, \lambda, t)$ (in very precise calculations the slight subsidence caused by the weight of the ice must be taken into account), all three components of the velocity must go to zero $v_z = v_\theta = v_\lambda = 0$. Note, however, that, if in the thin near-bottom layer of ice its viscosity decreases abruptly or liquid water even forms, then at the upper boundary of this near-bottom layer only the normal velocity component will go to zero; the formulation of boundary conditions for ice shelves, floating on the water, requires separate consideration.

The ice density ϱ_i depends very slightly on the pressure and temperature. Except in the upper layer of a glacier several meters or tens of meters thick, where snow becomes firn and then ice, with a subsequent compaction of the latter, ϱ_i can be assumed to be approximately constant (about 0.9 g/cm^3; pure ice at atmospheric pressure and zero temperature has a density $\varrho_i = 0.91676$ g/cm^3). However, the temperature field in a glacier nevertheless affects the motion of the ice, albeit via the viscosity (17.4) rather than the density ϱ_i, and perhaps (in the case of bottom melting) via the boundary condition at the bottom. The temperature field can be calculated with the aid of a heat-conduction equation like (17.2), provided to its right-hand side is added the heat influx due to the viscous dissipation of kinetic energy $\frac{1}{2}\varrho_i \nu \Phi^2$, with a boundary condition of the heat budget of form (9.10) at the glacier surface $z = \zeta$. At the bottom of the glacier it is apparently necessary to allow for the geothermal heat flux and the possibility that the ice will melt. Similarly, at the lower surface of an ice shelf the heat flux from water to ice must be taken into account, and the possibility that ice will melt or water will freeze. The formation of water runoff from a glacier (and its contribution to the heat transfer) can be considered to be a separate problem.

Recently, Verbitskii and Chalikov (1980) constructed a numerical model of the continental ice sheets, in accordance with the principles outlined above, that is, including equations of motion with the kinematic boundary condition (17.6) at the ice-sheet surface $z = \zeta$ and the heat-conduction equation. They expressed rheological law (17.4), in terms of the relation between the tensors of the deformation rates $\varepsilon_{ij} = 1/2\ (\partial v_i/\partial x_j + \partial v_j/\partial x_i)$ and the stresses τ_{ij}, as follows:

$$\varepsilon_{ij} = K\tau^{n-1}\tau_{ij} \exp\left[-\varkappa\left(\frac{T_{wi}}{T} - 1\right)\right], \qquad (17.4')$$

where $\varkappa \approx 21.1$. For actual small stresses τ, not exceeding 10^5 Pa, it was assumed that $n = 1$, that is, the ice was considered to be a linearly viscous substance. The stresses were found from the equations of motion, neglecting accelerations, expressed as conditions of equilibrium between the pressure forces and the viscous stresses, which gave

$$\tau_{xz} = -g\varrho_i(\zeta - z)\frac{\partial\zeta}{\partial x}; \qquad \tau_{yz} = -g\varrho_i(\zeta - z)\frac{\partial\zeta}{\partial y}, \qquad (17.7)$$

so that rheological relations (17.4′) became

$$\frac{1}{2}\frac{\partial u}{\partial z} = -Kg\varrho_i(\zeta - z)\frac{\partial\zeta}{\partial x}\exp\left[-\varkappa\left(\frac{T_{wi}}{T} - 1\right)\right];$$

$$\frac{1}{2}\frac{\partial v}{\partial z} = -Kg\varrho_i(\zeta - z)\frac{\partial\zeta}{\partial y}\exp\left[-\varkappa\left(\frac{T_{wi}}{T} - 1\right)\right]. \qquad (17.4'')$$

After integrating these equations over z (from 0 to z) and substituting the results into the continuity equation (nondivergence of velocities), also integrated over z (from 0 to z), Verbitskii and Chalikov took into account boundary condition (17.6) and obtained for ζ a nonlinear elliptic equation with coefficients depending on T. The latter was solved numerically together with the steady-state equation describing the heat influx (allowing for the viscous dissipation of kinetic energy into heat), integrated over the thickness of the glacier, the vertical temperature profile being approximated as a quadratic polynomial in z/ζ. These equations were used first of all to calculate the present-day ice sheets of Antarctica, Greenland, and Severnaya Zemlya; the horizontal boundaries, the distributions of T and $P_s - E + E_{wi}$ over the surface, and the geothermal heat flux at the bottom were all assumed to be given for these ice sheets.

The results, including those for the surface relief of the ice sheets and the temperature distribution in them, showed a good fit with empirical data. For instance, the volume obtained for the Eastern Antarctic Ice Sheet was 24.3×10^6 km^3, which compares favorably with the empirical range of 23×10^6 to 30×10^6 km^3. A large region of melting was detected at the bottom of the sheet, with a maximum (up to 3 mm/yr) at the bottom center and with increasing melting again at the very edge, where the dissipation of kinetic energy is a maximum (and comparable with the geothermal heat). The volume of bottom melting obtained was 16.8 km^3/yr, as compared with the empirical value of 20 km^3/yr. The velocities of motion obtained were about 1 m/yr at the center of the sheet and 10 m/yr at the margins.

A volume of 2.76×10^6 km^3 was obtained for the Greenland Ice Sheet; its bottom temperature was found to be –5°C at the sheet center, dropping to –10°C toward the margins and then rising again at the very edges. The volume of 590 km^3 obtained for the ice cap on Severnaya Zemlya agrees with empirical data. These successes prompted Verbitskii to calculate the Phanerozoic evolution of the antarctic glaciation as well, and to make some predictions about the future.

Another possible problem in climate theory is calculation of the permafrost regime by solving the equation of heat conduction in the upper layer of the continental Earth's crust, which is up to 1.5 km thick, taking into account the vertical inhomogeneity of the thermal properties and moisture content of this layer, with sources due to phase transitions of moisture and with a specified heat budget (9.10) at the ground surface as well as a geothermal heat flux at the lower boundary of the permafrost layer.

CHAPTER 7

INTEGRAL-PARAMETER CLIMATE MODELS

18. A Similarity Theory for the Circulation of Planetary Atmospheres

Parts of the atmosphere-ocean-land climate system possess considerable spatial inhomogeneity, vertical as well as horizontal, which can moreover vary with time (for instance, with the seasons). Any description of these using a small number of parameters (for example, the mean air temperature at the Earth's surface, the mean seawater temperature in the mixed layer, the mean atmospheric pressure at the Earth's surface, the mean content of moisture and CO_2 in the atmosphere, the mean cloud cover, etc.) would have to be a very incomplete one. Consequently, *integral-parameter models* of the climate and one-dimensional models, describing only the vertical, or only the latitudinal, inhomogeneity of a climate system, in terms of its overall structure, are bound to be inadequate, and their quantitative results should not, in general, be taken too seriously. On the other hand, such models may at times provide a qualitative picture of the possible role of certain climatic factors and feedbacks. The most well-known models of this type will be described in this chapter and in Chapter 8.

The simplest model of all involves a single parameter, the mean temperature of the outgoing radiation of the Earth T_r. This model is constructed by equating the mean daily uptake of solar radiation by the Earth's surface $q = \frac{1}{4} I_0 (1 - A) \approx 244$ W/m² (where I_0 is the solar constant, see §3, and A is the mean planetary albedo of the Earth, equal to about 0.28, according to satellite data) to the flux of outgoing radiation $f\sigma T_r^4$ (where f is the emissivity, equal to about unity, and $\sigma \approx 5.67 \times 10^{-8}$ W/(m² K⁴) is the Stefan-Boltzmann constant), that is,

$$f\sigma T_r^4 = q \quad [= \tfrac{1}{4} I_0 (1 - A)]. \tag{18.1}$$

This gives $T_r \approx 256$ K. Such a temperature exists in the atmosphere at a height of about 5 km. The ratio of it $\alpha_r = T_r/T_s \approx 0.89$ to the mean temperature of the surface layer of air $T_s \approx 288$ K characterizes the vertical temperature distribution in the atmosphere.

However, parameter T_r (or T_s) is not sufficient; in order to estimate the circulation of the atmosphere, we would have to add at least the mean difference δT between the surface air temperatures at the equator and the poles. This quantity enters into the definition of the specific rate of generation of kinetic energy in the atmosphere (Golitsyn, 1968):

$$\varepsilon = \eta \frac{q}{M}; \qquad \eta = k \frac{\delta T}{T_m}, \tag{18.2}$$

where M is the mass of a unit air column, η is the efficiency of the atmospheric "heat engine", T_m is the highest temperature of the surface layer of air (in the equatorial zone), and k is a utilization factor ($k < 1$, since the atmospheric heat engine is far from ideal; according to empirical data, which give $\delta T \approx 50$ K and $\varepsilon \approx 4$ cm^2/s^3, it can be assumed that factor $k \approx 0.1$).

Next, from T_r we find the velocity of sound $c_r = (\varkappa\, R/\mu\, T_r)^{1/2} \approx 320$ m/s, where $\varkappa = c_p/c_v \approx 1.4$ is the ratio of specific heats, $R \approx 8.314$ J/(K mole) is the universal gas constant, and $\mu = 29$ is the relative molecular mass of atmospheric air, and we introduce the scale of the synoptic processes $L = c_r/2\omega \approx 3000$ km, where ω is the angular velocity of the Earth's rotation. This gives us estimates of the typical wind velocity $U \sim (\varepsilon\, L)^{1/3} \approx 10$ m/s, the typical time of synoptic processes $\tau \sim \varepsilon^{-1/3}\, L^{2/3} \approx 3$ days, the typical coefficient of horizontal mixing $K_h \sim 0.1\ \varepsilon^{1/3}\ L^{4/3} \approx 3 \times 10^6$ m^2/s, and several other derivative quantities. Note that the value of δT need not be specified, but rather is found from some additional relation. This may be, for instance, the condition of equilibrium between the divergence of the total (over the depth of the atmosphere) meridional heat flux produced by air currents, on the one hand, and the heat losses to radiation. This condition can be written in simplified form as

$$c_p M U \frac{\delta T}{\pi a/2} = f \sigma T_r^4, \tag{18.3}$$

where a is the radius of the planet. Using the previous definitions of U, ε, and T_r, we now get

$$\delta T = (f\sigma)^{-1/16}\, c_p^{-3/4}\, q^{9/16} \left(\frac{\pi a}{2M} \right)^{1/2} \left(\frac{\pi}{2k_1} \frac{a}{L} \right)^{1/4}, \tag{18.4}$$

where $k_1 = kT_r/T_m$. The multiplier of a/L describes the effect of the rotational velocity of the planet. For rapid rotation $L \sim c_r/\omega$, making the synoptic vortices in the atmosphere relatively small ($a/L \approx a\omega/c_r < 1$); in this case the vortices do not vigorously transfer heat from the equator to the poles, and the temperature difference between the equator and the poles $\delta T \sim \omega^{1/4}$ is quite large. If the planet's rotation is slow, then $L \sim 2a$, and the rotation does not affect δT. The following dependences of some parameters of the atmospheric circulation on the mass of a unit air column M should also be noted:

$$\eta \sim M^{-1/2}; \qquad \varepsilon \sim M^{-3/2}; \qquad U \sim M^{-1/2}; \qquad \delta T \sim M^{-1/2};$$
$$\tau \sim M^{1/2}.$$

The relation $U \sim M^{-1/2}$ indicates, in particular, that the total kinetic energy of the atmospheric motions $E \sim MU^2$ is independent of M.

It should be mentioned that all the formulas cited in this section have contained only six genuinely "external" dimensional parameters: the assimilated flux (uptake) of solar radiation q, the radius of the planet a, the angular velocity of its rotation ω, the mass of a unit air column M, the specific heat of atmospheric air c_p, and the Stefan–Boltzmann constant σ (or, more precisely, $\sigma_1 = f\sigma$). There is also a dimensionless external parameter which appears in the formulas: the relative molecular mass of atmospheric air μ (entering into the formulas only via $R/\mu = c_p - c_v$). Finally, henceforth our seventh dimensional external parameter will be the typical acceleration of gravity in the planetary atmosphere, determined, on the one hand, by the mass and radius of the planet, or, on the other hand, by the mass of a unit air column and the pressure p_0 at its lower boundary ($g = p_0/M$). On the basis of his hypothesis that these seven dimensional parameters and one dimensionless parameter completely define the atmospheric circulation, Golitsyn (1970, 1973) formulated the corresponding similarity theory for the circulation of a planetary atmosphere.

Since four independent dimensions are involved here (length, time, mass, and temperature), the seven dimensional external parameters can be used to construct three independent dimensionless combinations, or similarity criteria. First it will be convenient to calculate the radiation temperature T_r, the quantities $c_v = c_p - R/\mu$ and $\varkappa = c_p/c_v$, and the velocity of sound

$$c_r = \left(\varkappa \frac{R}{\mu} T_r\right)^{1/2} = (\varkappa - 1)^{1/2}\sigma_1^{-1/8}c^{1/2}q^{1/8}. \tag{18.5}$$

Then the three similarity criteria (dimensionless numbers) can be defined as follows:

$$\begin{aligned}
&\Pi_\omega = a/2L = \omega a/c_r \ (L = c_r/2\omega);\\
&\Pi_g = H/a = \frac{c_r^2}{\varkappa g a}\left(H = \frac{c_r^2}{\varkappa g}\right);\\
&\Pi_M = \tau_p/\tau_r = \sigma_1^{3/8}c_p^{-3/2}q^{5/8}a/M\\
&(\tau_p = a/c_p;\ \tau_r = c_p M T_r/q).
\end{aligned} \tag{18.6}$$

Criterion Π_ω can be called the rotational Mach number. Let us stipulate that the angular velocity of rotation of the planet ω enters into all the formulas only via Π_ω (where, if ω is so low that $L > 2a$, we must set $L = 2a$ and $\Pi_\omega = 1/4$). In the expression for Π_g the quantity H is the so-called height (or thickness) of the homogeneous atmosphere. Let us also stipulate that the acceleration of gravity g enters into all the formulas only via Π_g. Dimensionless number Π_M is the ratio of the relaxation time τ_p for perturbations of the pressure or density to the thermal relaxation time τ_r (that is, to a state of local radiation equilibrium). It is usually small (for Venus $\Pi_M = 1.3 \times 10^{-5}$; for Earth $\Pi_M = 1.1 \times 10^{-3}$; for Mars $\Pi_M = 3.4 \times 10^{-2}$). Now, according to Golitsyn's similarity theory, any characteristic of the atmospheric circulation F can be written as

$$F = F_0 (\sigma_1, c_p, q, a)\, \Psi_F (\Pi_\omega, \Pi_g, \Pi_M), \tag{18.7}$$

where F_0 is a combination of parameters σ_1, c_p, q, and a having the dimensions of F, while Ψ_F is some dimensionless universal function of the dimensionless numbers (for low Π_M this function varies with Π_M, probably, as $(\Pi_M)n_F$). For instance, the total energy E takes the form

$$E = \sigma_1^{1/8} c_p^{-1/2} q^{7/8} a^3 \Psi_E (\Pi_\omega, \Pi_g). \tag{18.8}$$

If the rms velocity of atmospheric motions U_0 is defined in terms of the total kinetic energy of the atmosphere $E = \frac{1}{2}(4\pi\, a^2 M)\, U_0^2$, then Π_M will differ from the square of the Mach number $(\mathrm{Ma})^2 = (U_0/c_r)^2$ only by a factor $2\pi(\varkappa - 1)\Psi_E^{-1}$. Now, if the time scale of the atmospheric processes is defined as $\tau_0 = L/U_0$ and the rate of generation of kinetic energy is defined as $\varepsilon = E/4\pi a^2 \tau_0$, then a comparison of the latter expression with formula (18.2) for the efficiency η gives an explicit expression (proportional to Ma) from which δT can be found using the second formula of (18.2). A comparison with (18.4) then gives function Ψ_E as

$$\Psi_E = (2\pi)^{4/3} \left(k \frac{T_r}{T_M} \right)^{1/2} \left(\frac{a}{L} \right)^{-1/2} \quad \text{for } L < a. \tag{18.9}$$

The fundamental parameters of the general circulation of the atmosphere can now be written in the following form:

$$\eta = \frac{(2\pi)^{1/2}}{2} k_1^{3/4} \sigma_1^{3/16} c_p^{-3/4} q^{5/16} a^{1/2} M^{-1/2} \left(\frac{a}{L} \right)^{1/4};$$

$$\varepsilon = \frac{(2\pi)^{1/2}}{2} k_1^{3/4} \sigma_1^{3/16} c_p^{-3/4} q^{21/16} a^{1/2} M^{-3/2} \left(\frac{a}{L} \right)^{1/4};$$

$$E = (2\pi)^{4/3} k_1^{1/2} \sigma_1^{1/8} c_p^{-1/2} q^{7/8} a^3 \left(\frac{a}{L} \right)^{-1/2};$$

$$U = (2\pi)^{1/6} k_1^{1/4} \sigma_1^{1/16} c_p^{-1/4} q^{7/16} a^{1/2} M^{-1/2} \left(\frac{a}{L} \right)^{-1/4};$$

$$\delta T = \frac{(2\pi)^{1/2}}{2} k_1^{-1/4} \sigma_1^{-1/16} c_p^{-3/4} q^{9/16} a^{1/2} M^{-1/2} \left(\frac{a}{L} \right)^{1/4}. \tag{18.10}$$

In like manner, Golitsyn and his followers have also computed several other parameters of the climate system, including the parameters of turbulence in the atmospheric boundary layer.

In 1973 at the Oceanology Institute of the Soviet Academy of Sciences a further development of the above similarity theory was proposed, with a view toward evaluating the variations of the climatic characteristics accompanying changes in the assimilated heat flux q (heat uptake by the Earth's surface) and in the meridional distribution of this flux. In this case equation (18.1) remains unchanged, but in (18.2) q is replaced by the typical difference δq between the assimilated heat fluxes at the equator and at the pole. Then, accurate to some fixed multiplier, equation (18.2) can be written as $\varepsilon \sim (\delta T/T)\, \delta q$. Equation (18.3) takes the form $c_p M(LU/a)\,(L\,\delta\, T/a) \sim \delta q$, where $L \sim c_r/\omega \sim T^{1/2}$, so that, accurate to some fixed multiplier, we have $TU\,\delta\, T \sim \delta q$. Finally, an estimate for a specific rate of dissipation of kinetic energy $\varepsilon \sim U\tau$ is added, where $\tau \sim \varrho U^2$ is the typical frictional stress in the atmospheric boundary layer and $\varrho \sim 1/T$ is the air density, so that $\varepsilon \sim U^3/T$. From the above four equations we now get

$$T \sim q^{1/4}; \qquad \delta T \sim q^{-3/16}\,(\delta q)^{1/2};$$

$$U \sim q^{-1/16}\,(\delta q)^{1/2}; \qquad \varepsilon \sim q^{-7/16}(\delta q)^{3/2}. \tag{18.11}$$

These formulas will now be applied to evaluate the seasonal fluctuations of quantities (18.11), keeping in mind that in the summer hemisphere q is $(1 + \alpha)$ times greater than the mean annual value (according to astronomical data, on the day of the summer solstice $\alpha = 0.33$), the additional flux αq_{av} being incident mainly on the polar region, giving δq a multiplier of $1 - \alpha$. For the winter hemisphere α has to be replaced by $-\alpha_1 = -0.39$. The results of the calculations are given in Table 18.1.

TABLE 18.1
Seasonal fluctuations

	T K	δT°C	U m/s	ε cm²/s³
Annual mean	288	60	20	5
Summer extremum	309.3	46.6	16.0	2.4
Winter extremum	254.6	77.6	24.4	10.2
Multiplier for summer	1.07	0.78	0.80	0.48
Multiplier for winter	0.88	1.29	1.22	2.04

The seasonal fluctuations of the mean temperature T turned out to be much too high (since the real atmosphere, during the course of its seasonal fluctuations, deviates considerably from the state of radiation equilibrium described by formula (18.1)). At the same time, the seasonal fluctuations obtained for δT and U were quite satisfactory. For instance, in the Southern Hemisphere empirical data gave for summer (January) $\delta T = 40$°C and for winter (July) 74°C; the kinetic energy $\pi a^2 M U^2$ was found to be 3.9 J in summer and 7.1×10^{20} J in winter (as compared with 3.4 and 7.8 J, according to the data of Table 18.1).

The success of the results obtained with formulas (18.11) suggested that they be used to evaluate artificial situations. For instance, if the solar constant were to increase by a factor of $(1 + \alpha)$ or if artificial heat sources were placed along the equator (with a source strength of αq, calculated per unit area of the entire Earth), then we would have $q \sim 1 + \alpha$ and $\delta q \sim 1 + \alpha$. If these sources were distributed uniformly over the entire Earth's surface, then we would have $q \sim 1 + \alpha$ but δq would not change. If they were placed at one of the poles, then in this hemisphere we would have $q \sim 1 + \alpha$ and $\delta q \sim 1 - \alpha$. Such an effect might be produced if the albedo of the arctic region were reduced, but it would be very slight; a drop in albedo of, say, from 0.8 to 0.3 would increase the local heat flux in the Arctic by a factor of $(1–0.3)/(1–0.8) = 3.5$, or $2.5q$. However, figured per unit surface area of the entire Earth, equal to 40 times the area of the Arctic, this corresponds to the very low value $\alpha = 2.5/40 \approx 0.06$, only 1/6 of that for the seasonal fluctuations. Finally, if sources of strength $\alpha q/2$ were placed at both poles, we would have $q \sim 1 + \alpha$ and $\delta q \sim 1 - \alpha/2$. Table 18.2 gives the results of calculations using formulas (18.11) in the indicated artificial situations for $\alpha = 0.1$. All these quantities, except the rise in the mean temperature T, are small, so that from the point of view of the simplified theory being considered here it can be said that the minimum strength of artificial heat sources needed to produce appreciable changes in the global climate has to create a heat flux of at least $0.1q$, that is, it must be of the order of 10^{13} kW, which is equivalent to the burning of 2 Mtons of TNT per second.

TABLE 18.2
Effects of additional heat sources

Source distribution	T K	δT°C	U m/s	ε cm^2/s^3
At equator	7.2	+1.9	+0.9	+0.5
Uniform	7.2	−1.1	−0.1	−0.2
At one pole	7.2	−4.1	−1.1	−1.0
At both poles	7.2	−2.6	−0.6	−0.6

In 1976 at the Oceanology Institute of the Soviet Academy of Sciences Golitsyn's similarity theory was developed in another direction as well: the effects of the global interaction between the ocean and the atmosphere were included. To accomplish this, the following modification of equation (18.3) was used:

$$Q_a(1 + \xi) = q; \qquad Q_a = c_p M \frac{\overline{U'T'}}{(\pi a/2)} = \frac{m}{\Pi_M \Pi_\omega^2} \frac{U\delta T}{c_r T_r} q;$$

$$\xi = \frac{4}{3}\frac{C}{Q_a}, \tag{18.3'}$$

where Q_a is the meridional heat flux in the atmosphere, and Q is a typical value of the vertical heat flux from the atmosphere to the ocean at low latitudes and from the ocean to the atmosphere at high latitudes, so that the meridional heat flux in the ocean is equal to $2Q$ (in (18.3′) this quantity is multiplied by the relative area of the World Ocean $\approx 2/3$). In the definition of Q_a, in contrast to (18.3), in which the product $U\delta T$ appears, the mean value of the product of the synoptic perturbations of the wind velocity and the air temperature $\overline{U'T'} \sim \sigma_v\sigma_\tau \sim (LU/a)(L\delta T/a)$ was used. Thus, taking formula (18.6) into account, the second formula of (18.3′) was obtained, in which $m \approx 1.2$ is a numerical multiplier. Here, instead of the previous definition of $U = (\varepsilon L)^{1/3}$, the following difference analog of the thermal-wind formula was used:

$$U = \frac{\varkappa - 1}{\varkappa} \frac{c_p \delta T}{\sqrt{2}\omega\, \pi a/2}, \tag{18.12}$$

this being derived from the formula of the geostrophic wind (with an effective Coriolis parameter of $\omega\sqrt{2}$) with the aid of the equation of hydrostatics (on condition that $\delta p/p \approx \delta T/T$) and the equation of state of an ideal gas.

The vertical exchange in the ocean was described by the equations

$$\frac{Q}{c_w\varrho_w} = K\frac{\delta T_w}{h}; \qquad u_*^2 = \frac{K}{\alpha_H}\frac{\delta U_w}{h};$$

$$\mathrm{Rf} = \frac{\beta Q}{c_w\varrho_w}\left(u_*^2\frac{\delta U_w}{h}\right)^{-1}, \tag{18.13}$$

where c_w and ϱ_w are the specific heat and density of seawater, $\beta \approx 0.2$ cm/(s^2 K) is its buoyancy parameter, $u_* = (c_f\varrho_a/\varrho_w U^2)^{1/2}$ is the friction velocity in the ocean (where $c_f \sim 10^{-3}$ is the friction coefficient of the sea surface), K and K/α_H are the virtual temperature conductivity and viscosity in the ocean, h is the thickness of the upper active layer of the ocean, $\delta T_w \sim \delta T$ and δU_w are typical vertical differences of the temperature and current velocity in this layer, and Rf is the dynamic Richardson number. Both Rf and α_H are small: $\alpha_H \mathrm{Rf} = \mathrm{Ro} \approx 5 \times 10^{-4}$. Assuming this condition, for $u_* = 1$ cm/s, $\delta T_w = 27°$C, and h = 300 m, we get $K = u_*^2(\mathrm{Ro}\ h)^{1/2}(\beta\delta T_w)^{-1/2} - 5/3$ cm^2/s. Accordingly, the vertical mass flux at the ocean surface $M_* = \beta Q(gc_w)^{-1} = 9.6$ g/(cm^{2}yr), which agrees fairly well with empirical estimates. However, it is desirable, rather than specifying h, to determine it as a result of the ocean-atmosphere interaction, keeping in mind that during the cold half year h increases with time according to the law $\partial h/\partial t = bK/h$, where b is a numerical constant and the right-hand side represents the rate of entrainment of water into the active layer of the ocean via its lower boundary. Then a typical value of h will be $(bKt_0)^{1/2}$, where t_0 is the annual period, and in order to fit previous estimates we must be set $b = 17$. Using this formula for h, we now get from (18.13) the following relations:

$$K = \left(\frac{\text{Ro}\, u_*^4}{\beta \delta T_w} \right)^{2/3} (bt_0)^{1/3}; \qquad h = \left(\frac{\text{Ro}\, u_*^4}{\beta \delta T_w} \right) (bt_0)^{2/3};$$

$$\frac{Q}{c_w \varrho_w} = \left(\frac{\text{Ro}\, u_*^4}{\beta b t_0} \right)^{1/3} (\delta T_w)^{2/3}. \tag{18.14}$$

For fixed u_* the fact that the dependence of Q on δT_w is slower than a direct proportion can be attributed to the hindered vertical heat exchange of the stable stratification of the ocean. However, climatically u_* is not fixed, being according to (18.12) proportional to δT. If we do not distinguish between δT_w and δT, then for climatic calculations we have $Q \sim K \sim (\delta T)^2$ and $h \sim \delta T$. Now, substituting Q from (18.14) into relations (18.3′) and (18.12) and solving these formulas for δT and ξ, we obtain

$$\frac{\delta T}{T_r} = \left(\frac{2.2}{m} \; \frac{\Pi_M \Pi_Q^3}{1 + \xi} \right)^{1/2}; \qquad \xi = \Pi_Q \left(\frac{\delta T_w}{\delta T} \right)^{2/3};$$

$$\Pi_Q = \frac{4}{3m} \left(\frac{\text{Ro}}{2.2b} \right)^{1/3} \left(\frac{\varrho_a c_f}{\varrho_w} \right)^{2/3} \frac{c_w}{c_p} \; \frac{\varrho_w a^{8/3} \omega^{5/3}}{M c_r^{4/3} (t_0 \beta T_r)^{1/3}} \approx 0.42. \tag{18.15}$$

The dimensionless multiplier Π_Q introduced here has a meaning of upper limit of the ratio ξ of the oceanic meridional heat flux to the atmospheric flux. If we know δT and the mean temperature of the surface layer of air $T_s = T_r/\alpha_r$, then we can find the temperatures at the equator $T_E = T_s + 1/3\ \delta T$ and at the pole $T_p = T_s - 2/3\ \delta T$ and then set $\delta T_w = T_E - \max\,(T_p, T_c)$, where T_c is the freezing point of water. The above theory can be used to calculate from the external parameters the climatic characteristics of the atmosphere (T_s, δT, T_E, T_p, U, Q_a) and of the ocean (δT_w, K, h), as well as their small-scale interactions (τ and Q).

For instance, for the present-day climate, when Antarctica is covered with ice while the Arctic is semi-isolated from the World Ocean, we can set $\xi = 0$ in formula (18.15) for δT. Then, given the present values $T_r = 255$ K, $\Pi_\omega = 1.7$, and $\Pi_M = 2.17 \times 10^{-3}$, from (18.15) we have $\delta T = 36°$C and thus $T_s = 15°$C, $T_E = 27°$C, $T_p = -9°$C, $U = 10$ m/s, $\tau = 0.1$ Pa, $Q_a = 244$ W/m^2, 4/3 $Q = 84$ W/m^2, $\delta T = 27°$C, $K = 1.7$ cm^2/s, and $h = 300$ m. On the whole, these values show a good fit with empirical data.

During the Mesozoic both poles were apparently located in the open ocean and there was no perennial ice in the polar regions, so that the total albedo of the Earth was probably somewhat lower than the present value. If it is assumed that q was 4% higher than at present, then T_r and T_s turn out to have been 3°C higher, while $\xi = 0.42$, $\delta T = 31°$C, $T_E = 301$ K (just 1 K higher than at present), and $T_p = 270$ K (6 K higher); hence the above-mentioned absence of perennial ice.

During the Permo-Carboniferous the South Pole was in Gondwanaland

and the North Pole was apparently in the Pacific. If, as a first approximation, we evaluate the Permo-Carboniferous climates of the Southern and Northern Hemispheres separately, assuming a value of $\xi = 0$ in the Southern Hemisphere and, say, a mean albedo of 0.37 (corresponding to 20% of the area ice-covered with an albedo of 0.8 and 80% of it ice-free with an albedo of 0.26), then for this hemisphere we obtain T_r and T_s values 7°C lower than at present, and $\delta T = 35°C$, $T_E = 293$ K, and $T_p = 258$ K. For the Northern Hemisphere, on the other hand, assuming the Mesozoic value $\delta T = 31°C$, we get a value close to the present temperature of the Arctic $T_p = (293–31)$ K = 262 K.

The above theory can also be used to evaluate the climate of a rapidly or slowly rotating Earth, etc. Let us consider, for example, the climate of a dense carbon-dioxide atmosphere with an adiabatic vertical temperature gradient $\gamma_a = g/c_p = 11$ K/km. The height of formation of the outgoing radiation $|\partial T/\partial Z| = (T_s - T_r)|\partial T/\partial Z|^{-1}$ in this atmosphere will be higher than the present height, and for $|\partial T/\partial Z| = 6.5$ K/km it will be 5 km. The condition $(T_s - T_r)\ \gamma_a^{-1} > 5$ km for a carbon-dioxide atmosphere leads to the very striking result $T_s > 37°C$.

19. A Model of Glacial Periods

In a series of studies carried out by two brothers, V. Ya. Sergin and S. Ya. Sergin, which were summarized in their 1978 book, an important model with concentrated parameters was constructed. This model, designed to describe climatic fluctuations during the glacial periods of the Pleistocene, was based on semiempirical expressions for the heat budget at the outer limit of the atmosphere $Q_a = F_a + \Phi_a$ and at the Earth's surface $Q_s = F_s + \Phi_s + \Psi_s$, where F_a, F_s, and Φ_a, Φ_s are the total fluxes of short-wave and long-wave radiation, respectively (recall that Φ_s is called the *effective radiation* of the Earth's surface), while Ψ_s is the nonradiative heat flux between the Earth's surface and the atmosphere (see equation (9.10)). Assuming that the heat capacity of the atmosphere is much lower than that of the ocean or the land, the Sergins set $Q_a \approx Q_s = Q$. They represented the deviations of this quantity from the mean during the climatic fluctuations of the Pleistocene (all such fluctuations will henceforth be denoted by the same letters as the initial quantities) as a linear function of the deviations of the temperature of the Earth's surface Θ, the cloud amount N, the ocean area S, the ice-cover area S_i, and the evaporation rate E:

$$Q = \left(\frac{\partial \Phi_s}{\partial B} \frac{\partial B}{\partial \Theta} + \frac{\partial \Phi_s}{\partial \Theta} \right) \Theta + \left(\frac{\partial F_s}{\partial N} + \frac{\partial \Phi_s}{\partial N} \right) N + \left(\frac{\partial \Phi_s}{\partial S} \right) S + \left(\frac{\partial \Phi_s}{\partial S_i} \right) S_i + k_\Theta \left(\frac{\partial \Phi_s}{\partial E} \right) E, \qquad (19.1)$$

where B is the absolute humidity of the air (for an unchanged relative humidity depending only on the temperature); and k_Θ is an empirical coefficient taking into account the presence of a turbulent flux of sensible heat in addition to evaporation. It should be noted that in the derivation of (19.1) the Sergins neglected the effects of Pleistocene fluctuations of other factors, such as the CO_2 content, the aerosol content, the relative air humidity, the snow cover, and the thermal effects of the melting of ice.

Next they assumed the fluctuations of the temperature Θ to have a single-valued relationship with Q, expressible in terms of operators as

$$\Theta = P_\Theta[Q] \approx \frac{k}{Tp} \quad Q, \tag{19.2}$$

where the approximate form of operator P_Θ is determined after solving the equation of turbulent heat conduction in the ocean with a specified function $Q(t)$ at the ocean surface. In (19.2) p is the operator d/dt, k = 1 cm^3 yr K/kcal = 0.24 cm^3 yr K/J is the turbulent temperature conductivity in the ocean, and T = 400 yr is the time constant of the ocean. The fluctuations of the cloud amount were taken to be a function just of the fluctuations of the evaporation rate:

$$N = N(E) \approx \varkappa E \approx \varkappa (\eta_\Theta \Theta + \eta_u u), \tag{19.3}$$

where $\varkappa$ = 0.0046 yr/cm, u represents the wind-velocity fluctuations, η_Θ = 7.5 cm/(yr K), and η_u = 30 cm/yr m s^{-1}. Instead of E, this formula introduces into (19.1) the wind velocity u, the fluctuations of which were described by the formula

$$u = c_1 \Gamma - u_0; \qquad \Gamma = \Gamma_0 + \gamma S_{ic}; \qquad S_{ic} = S_{ci} + S_{mi}, \tag{19.4}$$

where c_1 = 0.14 m/(s K), u_0 = 6 m/s, Γ is the temperature difference between the equatorial and polar zones (this is, as noted above, a necessary characteristic of the general circulation of the atmosphere), Γ_0 = 29°C, $S_{ic} = S_{ci} + S_{mi}$ is the total ice-cover area, the sum of the continental ice (S_{ci}) and the marine ice (S_{mi}), and γ = 0.5°C/10^{6}km^2. Here it was assumed that the mean annual boundary of the sea ice lies along the $T_s = T_{\varphi i}$ isotherm, so that S_{mi} is a function of Θ and Γ:

$$S_{mi} = \lambda_0 + \lambda_1 \frac{\Theta - T_{\varphi i}}{\Gamma}, \tag{19.5}$$

where $\lambda_0 = 45$ and $\lambda_1 = -49$ if S_{mi} is measured in 10^6 km^2. Taken together with (19.4), formula (19.5) gives a quadratic equation in S_{mi}, for which a nonnegative root must be selected. Moreover, the fluctuations of the ocean area S were assumed to be a function of the fluctuations of the sea level h, giving $S = S(h) \approx K_1 h$. The corresponding coefficient in (19.1) was calculated with the aid of the formula $\Phi_s \approx K_2\, \alpha \approx K_2 K_3 S$, where for each latitude zone coefficient K_2 is an empirical function of the cloud amount and $\alpha = K_3 S$

represents the albedo fluctuations (coefficient $K_3 = (\alpha_0 - \alpha_1)/S_1$, where α_0 and α_1 are the albedos of the ocean and land, and S_1 is the area of the latitude zone). Consequently,

$$(\partial \Phi_s/\partial S)\, S \approx K_s h, \tag{19.6}$$

where $K_s = K_1 K_2 K_3 = 0.01$ kcal/(cm^2 yr m) $= 0.4$ MJ/(m^3 yr) for the Northern Hemisphere and 0.006 kcal/(cm^2 yr m) $= 0.024$ kJ/(m^3 yr) for the Southern Hemisphere. Now the height of sea level h, instead of the ocean area S will enter in (19.1). Its fluctuations are related as follows to the fluctuations of the water volume in the continental ice V_{ci}:

$$h = P_h[V_{ci}] \approx \left(-K_1 + \frac{K_2}{1 + T_{1p}} \right) V_{ci} \quad , \tag{19.7}$$

where in the first term of the definition of operator P_h coefficient $K_1 = 2.8 \times 10^{-9}$km^{-2} is the reciprocal of the area of the World Ocean, while the second term takes into account the isostatic compensation in the asthenosphere beneath a melting ice sheet; in it $K_2 = (\varrho_w/\varrho_A)\, K_1 = 0.8 \times 10^{-9}$km^{-2} (where ϱ_w and ϱ_A are the densities of water and of the material of the asthenosphere) and $T_1 = 2000$ yr is the relaxation time of the asthenosphere. The variation rate dV_{ci}/dt of the quantity V_{ci} in (19.7) is equal to the difference Ac − Ab between the rates of accumulation and ablation of ice. According to the so-called transport model of the ice flow in a continental ice sheet (from center to periphery along the deepening flow lines, with the corresponding lag), Ab can be expressed approximately in terms of Ac if we multiply Ac by the operator conversion function $(1 + T_i p)^{-1}$, where T_i is the time constant of the ice sheet, which depends slightly on the size of the sheet (for Antarctica, apparently, $T_i \approx 9000$ yr, while for Greenland $T_i \approx 6000$ yr). Consequently,

$$V_{ci} = P_{ci}\,[\mathrm{Ac}] \approx \frac{T_i}{1 + T_i p}\,\mathrm{Ac}. \tag{19.8}$$

The accumulation Ac is defined as AS_{ci}, where A is the mean specific accumulation, taken to be a specified function of the rate of fall of precipitation P (assumed to have a single-valued relationship with the evaporation rate E), the area of continental ice S_{ci} (equal to its volume V_{ci} divided by its mean thickness $H \approx 2$ km), the temperature Θ, and the temperature differential Γ. The above relations form a closed system of equations, provided that the missing coefficients in formula (19.1) are supplied. The following empirical values of these were assumed: $K_\Theta = \partial Q/\partial\Theta = 0.53$ kcal/(cm^3 yr K) $= 22.21$ MJ/(m^2 yr K); $K_N = \partial Q/\partial N = -26$ kcal/(cm^2 yr) $= -1086.4$ MJ/(m^2 yr); $K_{S_i} = \partial Q/\partial S_i = -0.15$ kcal/(cm^2 yr 10^6 km^2) $= -6.28$ MJ/(m^2 yr 10^6 km^2); $K_E = \partial Q/\partial E = 0.2$ kcal/g $= 0.84$ kJ/g.

This system of equations can be depicted with the aid of the operator diagram in Fig. 19.1. The block representing continental ice S_{ci} in the diagram is interpreted differently for the Northern and Southern Hemispheres of

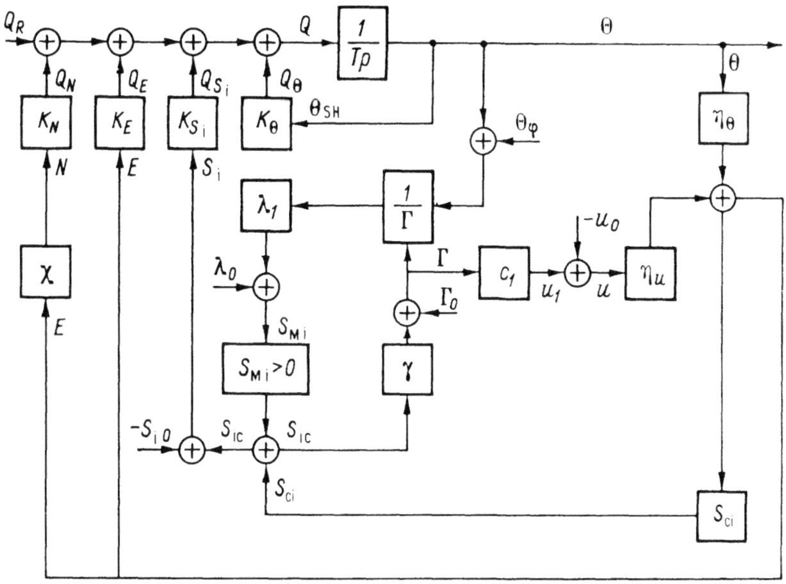

Fig. 19.1. Operator diagram of model of glacial periods devised by V. Ya. Sergin and S. Ya. Sergin.

the Earth. In the Northern Hemisphere the total area of continental ice $S_{ci}^{NH} = nS_{ci}$ may vary considerably with time, and its fluctuations have to be carefully computed. Moreover, the number n of ice sheets there can also vary, and the Sergins proposed the following tentative formula for it:

$$n = 1 + \mu S_{ci}^{NH} \tag{19.9}$$

where $\mu \leqslant 0.15 \times 10^{-6}$ km^{-2}. Figure 19.2 shows an operator diagram for the continental ice of the Northern Hemisphere. The diagrams in Figs. 19.1 and 19.2 were realized with an analog computer, zero (that is, present-day) initial conditions having been selected for all the variables except $\Theta_{NH}(0) = +2$°C.

Because of the negative feedback between the ocean and the continental ice (rise in $T_w \rightarrow$ increased E and $P \rightarrow$ increase in $S_{ci} \rightarrow$ drop in T_w) in the Northern Hemisphere self-oscillations occurred, with a period of 35,000 years, a 16° range of Θ values (the variations toward cooling being twice as great as the variations toward heating), and increases in V_{ci}^{NH} of up to 17×10^6 km^3. An increase in ocean depth led to an abrupt increase in the amplitude of these self-oscillations, whereas a rise in T_i increased the amplitude slightly and the period considerably. An enhancement of the greenhouse effect (K_Θ) eliminated the self-oscillations, that is, it stabilized the climate; variations in K_N, and particularly in K_E, greatly affected both the period and amplitude of the self-oscillations. Twofold variations in K_{S_i} affected the self-oscillations only comparatively slightly, which militates against the suggestions in the literature that the feedback on K_{S_i} may cause the Earth to reach a state of

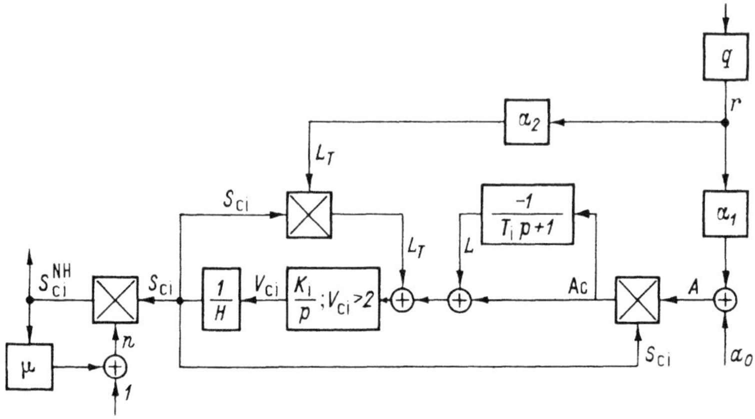

Fig. 19.2. Operator diagram of continental ice of Northern Hemisphere.

complete glaciation. An increase in the mean level of precipitation in the region of glaciation increased appreciably both the period and the amplitude of the self-oscillations. The effect of the initial conditions, and also of the mean thickness of the ice sheets, was found to be negligible.

In the Earth's Southern Hemisphere the area of the continental ice S_{ci}^{SH} is determined mainly by the configuration of Antarctica, and it does not vary much in time. The fluctuations of this area depend just on the fluctuations of the ocean level h, approximately linearly, that is, $S_h = K_h h$, where according to the hypsographic curve $K_h = 0.007/m^{-1} \times 10^6$ km^2. Figure 19.3 gives the operator diagram for the continental ice of the Southern Hemisphere. The combination of diagrams 19.2 and 19.3 describes an aperiodic regime. Its stabilization time (time constant) is 40,000 to 50,000 years, according to experiments on an analog computer.

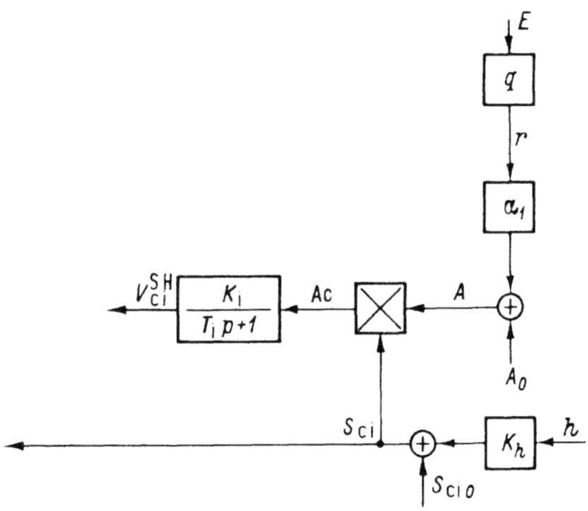

Fig. 19.3. Operator diagram of continental ice of Southern Hemisphere.

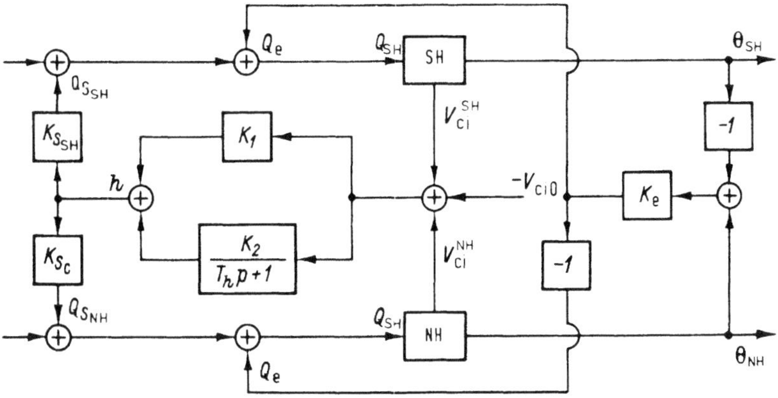

Fig. 19.4. Operator diagram of interaction between Southern and Northern Hemispheres.

The interaction between the hemispheres occurs, firstly, because of the transfer of energy across the equator, the fluctuations of which (per cm^2 of surface of the hemisphere) were taken to be proportional to the difference between the fluctuations of the mean hemispheric temperatures: $Q_e = K_e$ $(\Theta_{SH} - \Theta_{SH})$, where $K_e = 0.4$ kcal/(cm^2/ yr K) = 16.76 MJ/(m^2 K) per year. The second reason for this interaction is the establishment of a single ocean level according to formula (19.7).

Figure 19.4 is an operator diagram of the interaction between the Northern and Southern Hemispheres. The overall diagram (see Figs. 19.1–19.4) was arrived at using an analog computer, the fluctuations in time of quantities Θ_{NH}, Θ_{SH}, V_{ci}^{NH}, V_{ci}^{SH}, h, and Γ_{NH} being recorded on a loop oscillograph. This diagram generated self-oscillations, for nominal values of the parameters, with a period of about 60,000 years, with a very large range of Θ_{NH} values (about 20°C, the swing toward cooling being twice as great as the swing toward warming), with variations of Θ_{SH} of about 7°C, of V_{ci}^{NH} from 2.4×10^6 km^3 to 27×10^6 km^3, of V_{ci}^{SH} from 19×10^6 km^3 to 30×10^6 km^3, and, finally, with h variations of more than 60 m. With an increase in K_e, the fluctuations in the Southern Hemisphere were enhanced (that is, they are induced fluctuations); the fluctuations in the Northern Hemisphere were enhanced as well thereby. Variations in the other parameters had qualitative effects which were the same as those in an isolated Northern Hemisphere (see above).

The combined operator diagram (see Figs. 19.1–19.4) may lead to a more usual form of the differential equations. It comprises two first-order equations for temperatures Θ_{NH} and Θ_{SH}, second-order equations for h and for V_{ci}^{NH}, and the finite relations

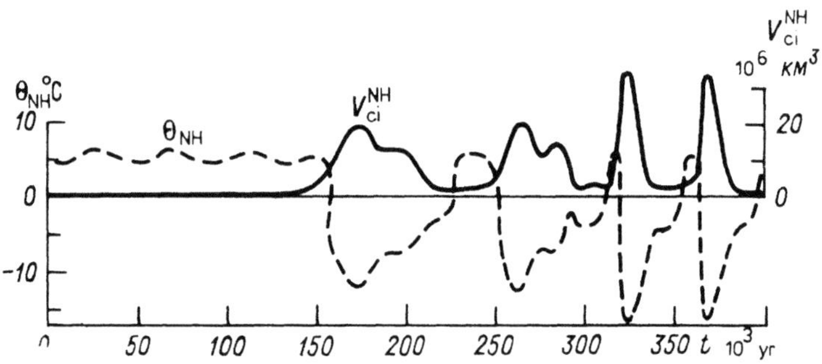

Fig. 19.5. Results of modeling transition from Pliocene to Pleistocene.

$$\gamma(S_{ic}^{NH,SH})^2 + [(\Gamma_0 + \Gamma_{ex}^{NH,SH}) - \gamma S_{ci}^{NH,SH} - \lambda_0\gamma]\, S_{ic}^{NH,SH} =$$
$$(\Gamma_0 + \Gamma_{ex}^{NH,SH})\, S_{ci}^{NH,SH} + \lambda_1\, (\Theta_{NH,SH} + \Theta_{\varphi i}) +$$
$$+ \lambda_0\, (\Gamma_0 + \Gamma_{ex}^{NH,SH});\quad S_{ci}^{SH} = S_{ci0} + K_h h \qquad (19.10)$$

together with the obvious inequalities $S_{ic}^{NH} \geqslant V_{ci}^{NH}/H$; $S_{ic}^{SH} \geqslant S_{ci}^{SH}$; $V_{ci}^{NH} \geqslant 2H/(H - 2\mu)$. These equations were integrated using a digital computer for a time interval of 10 years; here, in addition to the self-oscillations, the response of the system to various external disturbances was calculated, such as astronomical oscillations of the Earth's orbital parameters, expressed as fluctuations of the temperature differences between equator and poles Γ_{ex}^{NH} and Γ_{ex}^{SH}. It was found, in particular, that the modulating influence of external disturbances is augmented with a decrease in the period of the self-oscillations. The amplitudes of the forced oscillations generally showed a satisfactory qualitative, and sometimes even quantitative, fit with available empirical estimates (although the temperature fluctuations given by the model were too great, while the ice-volume fluctuations were too small). In accordance with the empirical data, the calculated temperature fluctuations lagged in phase behind the fluctuations in glaciation (by about 20°), indicating that the glaciations are not controlled by the temperature (but rather, apparently, by the precipitation). A calculation of the forced oscillations for a gradual increase (from 0 to 1 during a time between 100,000 and 300,000 years) in coefficient Π, introduced into the diagram after the block of glaciation formation, enabled a reconstruction of the transition of climatic variations from the aperiodic regime of the Pliocene to the oscillatory regime of the Pleistocene (Fig. 19.5).

CHAPTER 8

ONE-DIMENSIONAL MODELS

20. Models of the Vertical Structure

The vertical structure of the atmosphere is known to vary quite appreciably from place to place, especially from latitude to latitude, and also in time, especially from season to season. In spite of this, however, the mean vertical structure is such an expressive climatic characteristic of the state of the atmosphere as a whole that meteorologists have coined the special term "standard atmosphere" and have worked out for it a whole series of increasingly accurate empirical models.

For example, according to a model published in 1962 by the (American) Committee on Extension to the Standard Atmosphere (COESA), the following sea-level values are assumed: atmospheric pressure p_0 = 1013.25 mb, air temperature T_0 = 288.15 K, air density $\varrho_0 = 1.225 \times 10^{-3}$ g cm^{-3}, and relative molecular mass μ_0 = 28.9644 (assumed to be constant up to a height of 90 km and then to decrease). In the troposphere up to a height of 11 km the temperature is assumed to drop to 216.65 K (and the density to 3.6480×10^{-4} g/cm^3). In the stratosphere the temperature is assumed to be constant to 20 km, after which it rises to 270.65 K at a height of 47 km; in the mesosphere it is assumed to be invariable up to 52 km and then to drop to 180.65 K at a height of 79 km; finally, in the thermosphere the temperature is assumed to be unchanged up to 90 km and then to drop abruptly (Fig. 20.1). In view of the above-mentioned variability of the vertical structure of the atmosphere, COESA also worked out additional standard atmospheres for January and July at 15, 30, 45, 60, and 75° N (for a given humidity in the troposphere).

The standard atmosphere evidently requires both a qualitative physical explanation and a quantitative theoretical formulation with the aid of simplified models of the climate. For instance, the temperature drop with height in

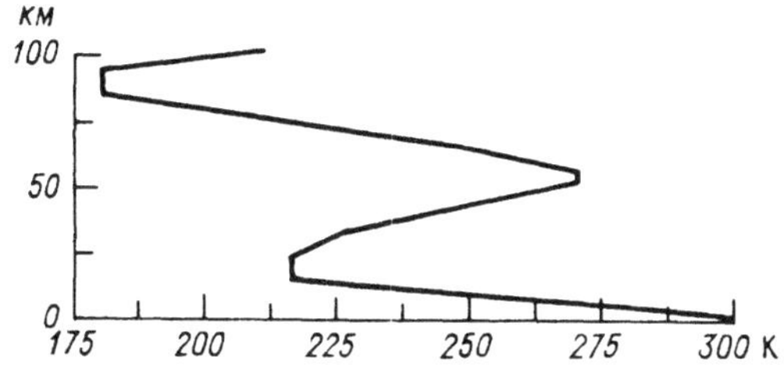

Fig. 20.1. Vertical distribution of air temperature according to COESA-62 standard atmosphere.

the troposphere is attributed to the greenhouse effect, whereby the atmosphere is heated from below but more gradually than would be the case for radiation equilibrium, because of the smoothing effect of convective (turbulent) heat exchange. The isothermy and the subsequent temperature rise with height in the stratosphere are attributed to heating as a result of the absorption of ultraviolet solar radiation by ozone.

A theoretical model of the vertical thermal structure of the troposphere and lower stratosphere, which has by now become classical, was constructed back in 1943 by I. A Kibel' (it was mentioned above, in §5). A fundamental equation of this model is the equation of heat influx, written as a time average in steady-state form and taking into account only the vertical heat exchange. Here the divergence of the heat fluxes caused by large-scale atmospheric motions is neglected, so that this equation portrays the equilibrium between the heat influxes due to the turbulent and radiative vertical heat exchange:

$$\frac{\mathrm{d}}{\mathrm{d}z} \lambda \frac{\mathrm{d}T}{\mathrm{d}z} + \alpha \varrho_{\mathrm{ab}} (F^{\downarrow} + F^{\uparrow} - 2fB + \gamma S) = 0, \tag{20.1}$$

where λ is here the vertical turbulent heat conductivity, α is the coefficient of absorption of "graybody" long-wave radiation, ϱ_{ab} is the density of absorbing substances, $F^{\downarrow}$ is the downward flux of graybody long-wave radiation of the atmosphere, $F^{\uparrow}$ is the upward flux of graybody long-wave radiation of the Earth's surface and atmosphere, $B = \sigma T^4$ is the integral radiation flux of a perfect black body in the hemisphere, f is a correction factor taking into account the difference of the emissivity from that of a perfect black body, S is the downward flux of graybody short-wave solar radiation, and γ is the ratio of the absorption coefficients for graybody short-wave and long-wave radiation.

Formula (20.1) is joined by the following equations for the fluxes of radiative energy:

$$\frac{\mathrm{d}F^{\downarrow}}{\mathrm{d}z} = \alpha \varrho_{\mathrm{ab}} (F^{\downarrow} - fB); \qquad \frac{\mathrm{d}F^{\uparrow}}{\mathrm{d}z} = \alpha \varrho_{\mathrm{ab}} (fB - F^{\uparrow}); \tag{20.2}$$

$$\frac{\mathrm{d}S}{\mathrm{d}z} = \alpha \gamma \varrho_{\mathrm{ab}} S.$$

These are analogous to equations (5.12) of §5 from Emden's theory of radiation equilibrium (only the notation is different). However, Emden's theory assumed that $F^{\uparrow} - F^{\downarrow} - S = \text{const}$, which would have made it possible to obtain directly from equations (20.2) and the boundary conditions an expression for the temperature; now, since the above expression cannot be assumed constant, equation (20.1) has to be used to relate the fluxes of radiative energy and the temperature. The boundary conditions for equations (20.1)–(20.2) have the form:

$$z = 0\colon -\lambda \frac{dT}{dz} = F^{\downarrow} + S - F^{\uparrow}; \qquad F^{\uparrow} = f_* B; \tag{20.3}$$

$$z = \infty\colon F^{\downarrow} = 0; \qquad F^{\uparrow} = S = q = \frac{1}{4} I_0(1 - A). \tag{20.4}$$

The first of conditions (20.3) is the simplified equation (9.10) of the heat budget at the Earth's surface, in which, in particular, the heat flux H_s into the depths of the ocean or the active layer of the land, averaged over the year, is neglected. In the second condition f_* is a correction factor, indicating the difference between the emissivity of the Earth's surface and that of a perfect black body. In conditions (20.4) q is the mean daily assimilated flux of solar radiation (A is the mean planetary albedo of the Earth). Moreover, as a vertical coordinate, instead of z it is convenient to use the optical thickness for "graybody" radiation $\tau = \int_z^{\infty} \alpha\varrho_{ab}\, dz$. Likewise, in equations (20.1)–(20.4) temperature T can everywhere be expressed in terms of the function $B = \sigma T^4$, so that $dT/dz = -(\alpha\varrho_{ab}/4\sigma T^3)(dB/d\tau)$. It will be useful to introduce the notation:

$$F^{\uparrow} + F^{\downarrow} = q\xi; \qquad F^{\uparrow} - F^{\downarrow} = q\eta; \qquad B = \frac{q}{f}\,\zeta;$$

$$\frac{\alpha\varrho_{ab}\lambda}{4f\sigma T^3} = \frac{2}{m^2 - 1}, \tag{20.5}$$

with the aid of which equations (20.1)–(20.4) can be written as

$$\frac{d}{d\tau}\,\frac{2}{m^2 - 1}\,\frac{d\zeta}{d\tau} + \xi - 2\zeta + \gamma\, e^{-\gamma\tau} = 0; \tag{20.1'}$$

$$d\xi/d\tau = \eta; \qquad d\eta/d\tau = \xi - 2\zeta; \tag{20.2'}$$

$$\tau = \tau_0\colon \frac{2}{m^2 - 1}\,\frac{d\zeta}{d\tau} = e^{-\gamma\tau_0} - \eta; \qquad \xi + \eta = 2\frac{f_*}{f}\,\zeta; \tag{20.3'}$$

$$\tau = 0\colon \xi = \eta = 1, \tag{20.4'}$$

where τ_0 is the total optical thickness of the atmosphere. With the aid of (20.2′) and (20.4′), quantities ξ and ζ can be expressed in terms of η:

$$\xi = 1 + \int_0^{\tau} \eta\, d\tau; \qquad \zeta = \frac{1}{2}\left(1 + \int_0^{\tau} \eta\, d\tau - \frac{d\eta}{d\tau}\right). \tag{20.2''}$$

Next, taking into account the second equation (20.2′) and the first boundary condition (20.3′), equation (20.1′) is integrated over τ, after which, having eliminated ζ using the second formula (20.2″), we obtain for η the simple equation

$$\mathrm{d}^2\eta/\mathrm{d}\tau^2 - m^2\eta = -(m^2 - 1)\,\mathrm{e}^{-\gamma\tau}, \tag{20.6}$$

$$\tau = 0:\ \eta = 1; \qquad \tau = \tau_0:\ \eta + \frac{f_*}{f}\frac{\mathrm{d}\eta}{\mathrm{d}\tau} = \left(\frac{f_*}{f} - 1\right)\left(1 + \int_0^{\tau} \eta\,\mathrm{d}\tau\right). \tag{20.7}$$

Boundary conditions (20.7) here serve as the second of conditions (20.3′) and (20.4′). As a theoretical model of the vertical structure of the troposphere and lower stratosphere, Kibel' proposed an analytical solution of the described equations, with the bold simplifying assumption that m = const. This solution has the form:

$$\eta = \frac{m^2 - 1}{m^2 - \gamma^2}\,(C_1\,\mathrm{e}^{m\tau} + C_2\,\mathrm{e}^{-m\tau} + \mathrm{e}^{-\gamma\tau});$$

$$\zeta = \frac{1}{2} + \frac{1}{2}\frac{m^2 - 1}{m^2 - \gamma^2}\left\{\frac{C_1}{m}\,[(1 - m^2)\,\mathrm{e}^{m\tau} - 1] - \frac{C_2}{m}\,[(1 - m^2)\,\mathrm{e}^{-m\tau} - 1] - \frac{1}{\gamma}\,[(1 - \gamma^2)\,\mathrm{e}^{-\gamma\tau} - 1]\right\}, \tag{20.8}$$

where constants C_1 and C_2 are found from conditions (20.7).

Numerical calculations were carried out with these formulas for 42° latitude, where $q = 0.138$ cal/(cm^2 min) = 96.32 W/m^2. Here it was assumed that $\alpha = 7.25$ cm^2/g, $\lambda = 1$ cal/(cm s K) = 4.19×10^2 W/(m K), m = 1.75, $f_*/f = 1.15$, $\varrho_{ab} = \varrho_{abo}\,(T_0/T)\,\mathrm{e}^{-z/L}$ g/cm^3 for $\varrho_{abo} = 6.2 \times 10^{-6}$ g/cm^2, L = 8 km, $T = T_\infty(1 + \tau)^{1/4}$, and $T_0/T_\infty = 280/220$. This formula for ϱ_{ab} was used to calculate dependence $\tau(z)$ with the aid of the relation

$$\frac{4}{5}\,[(1 + \tau)^{5/4} - 1] = \alpha\varrho_{abo}\,\frac{T_0}{T_\infty}\,\frac{L}{\ln 10}\,\mathrm{e}^{-z/L}, \tag{20.9}$$

but the value obtained for the total optical thickness of the atmosphere $\tau_0 = 12.6$ was apparently too high (so that the fluxes of radiative energy in this model are not realistic). The following formula was obtained for the temperature:

$$f\sigma T^4 = 0.047\,[6.8 - 4.8\,\mathrm{e}^{-0.2\tau} + 0.57\,\mathrm{e}^{-1.75\tau} - 0.36\,\mathrm{e}^{1.75(\tau - 12.6)}], \tag{20.10}$$

which, together with (20.9), gives a fairly good fit with the empirical data: temperatures of 8°C at a height of 300 m, 9°C at heights from 0.5 to 1 km, 4.5°C at a height of 2 km, −13°C at 5 km, −40°C at 9 km, −49°C at 11 km, and −53°C at 13 km.

It may be that a different choice of parameters (and in particular of relation (20.9) between τ and z) would yield values of τ_0 and fluxes of radiative energy which are closer to the actual ones, without disturbing the good fit between the temperature structure and the empirical data. It is more likely, however, that neglecting the transfer of heat by large-scale atmospheric movements, which is unavoidable in a one-dimensional model, leads us so far away from the true relationships between the local vertical distributions of temperature and radiation that a correct simultaneous portrayal of both distributions in this model cannot be achieved.

21. Models of the Latitudinal Structure

As stated in §3, insolation is the primary energy source of the climate system, but the daily totals of the insolation are independent of longitude, that is, their distribution always possesses latitudinal zonality. Consequently, even though the latitudinal zonality of the distributions of reflected and absorbed radiation is disturbed because of the horizontal inhomogeneity of the Earth's surface and atmosphere (and, in particular, of the cloud cover), still the insolation does ensure an approximate latitudinal zonality of the climate system. Therefore, the mean characteristics of the climate system over the longitude, that is, the mean meridional sections of the system, are the most important characteristics of the state of the climate system.

These sections are two-dimensional, that is, they have a height structure and a latitudinal structure. The mean vertical structure of the atmosphere over all latitudes was considered in the preceding section. However, the latitudinal structure of the atmosphere can be described either in terms of its "base" (in the ground-adjacent layer of the atmosphere) or as an average over its thickness (similarly to what was done for the ocean in §13, and in particular in Fig. 13.1). The first of these characteristics, the latitudinal structure of the atmospheric layer near the ground, is obviously the most interesting one, and it will be the subject of the present section.

Figure 21.1 shows the mean annual distributions of the zonal temperatures T of the ground-adjacent layer of the atmosphere in the Northern and Southern Hemispheres. These distributions are plotted as a function of the annual insolation W_0, given by formulas (3.10)–(3.11). The latitudes corresponding to the W_0 values were tabulated in §3 and can be found from the lower horizontal scale of Fig. 21.1. This figure provides a simplified qualitative picture of the latitudinal distributions of the mean annual zonal temperatures T of the atmospheric layer near the ground: the greater the annual insolation W_0, the higher T is. The curve of $T(W_0)$ in the middle and low latitudes $|\varphi| < 60°$ is approximately linear [$T°\text{C} \approx 27/11\,(W_0 - 15)$], whereas at high latitudes T is lower than the linear dependence would give (especially in the Southern Hemisphere). This is because the albedo of the Earth's surface is

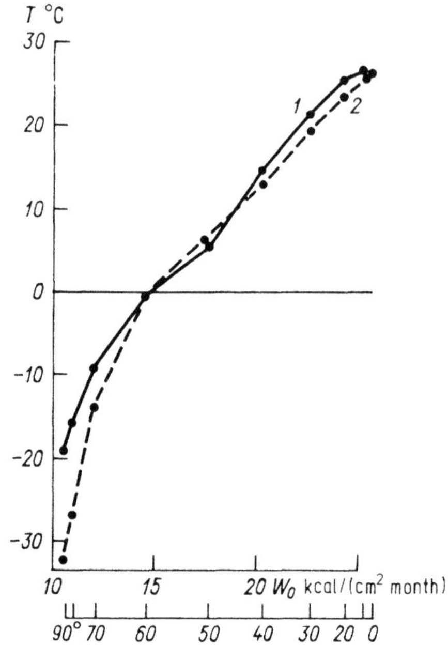

Fig. 21.1. Mean annual zonal temperatures of ground-adjacent layer of atmosphere in Northern (1) and Southern (2) Hemispheres. 1 : kcal/(cm^2 month) = 41.9 MJ/m^2.

higher there, in view of the presence of a winter snow cover or, in polar land regions (which are especially large in the Southern Hemisphere), a permanent snow cover.

It is not easy to come up with a convincing quantitative explanation of distribution $T(\varphi)$, since it is physically obvious that the $T(\varphi)$ curve for fixed φ values will depend on more than just the local values of the insolation and albedo. The state of the surface layer of the atmosphere must depend on the state of the other parts of the climate system. Nevertheless, several attempts have been made to calculate the zonal distribution of the surface temperature with the aid of just the equation of the mean annual heat budget for the latitudinal zones of the atmosphere as a whole. This heat-budget equation is written as

$$QS(x)\,[1 - \alpha(x)] - I(x) = D(x), \tag{21.1}$$

where $x = \sin \varphi$, $Q = 1/4\, I_0$ is a quarter of the solar constant, $QS(x) = W_0/T_0$ is the mean annual insolation, defined by the astronomical formulas (3.10) and (3.11), $\alpha(x)$ is the planetary albedo, so that $QS(1 - \alpha)$ is the mean annual assimilated solar radiation, $I(x)$ is the flux of outgoing long-wave radiation of the Earth, so that the left-hand side of (21.1) represents the total mean annual flux of radiative energy at the outer limit of atmosphere, and, finally, $D(x)$ is the divergence of the integral (over the entire thickness of the atmosphere and ocean) mean annual meridional heat flux produced by large-scale motions.

For $D \equiv 0$ and $I = f\sigma T^4$ equation (21.1) becomes (18.1). For a more precise approach, when determining $T(x)$ from equation (21.1), we must express functions $I(x)$, $D(x)$, and $\alpha(x)$ in terms of $T(x)$, with the aid of some approximate semiempirical formulas (although, as noted above, a closed equation for $T(x)$ can hardly correspond to reality). The first to do this was Budyko (1968), who set

$$I(x) = a + bT(x) - [a_1 + b_1 T(x)]\, n = A + BT(x); \tag{21.2}$$

$$D(x) = \beta[T(x - \bar{T}]; \tag{21.3}$$

$$\alpha(x) = \begin{cases} \alpha_i & \text{for } T(x) < T_c, \\ \alpha_w < \alpha_i & \text{for } T(x) > T_c \end{cases} \tag{21.4}$$

where in (21.2) $a = 14.0$ and $a_1 = 3.0$ kcal/(cm^2 month) = 125.7 MJ/m^2 per month; $b = 0.14$ and $b_1 = 0.10$ kcal/(cm^2 month K) = 4.19 MJ/(m^2 K) per month; n is the cloud amount, assumed to have a mean value of 0.5, so that $A = 12.5$ kcal/(cm^2 month) = 523.75 MJ/m^2 per month and $B = 0.09$ kcal/(cm^2 month K) = 3.77 MJ/(m^2 K) per month. In Newton's law (21.3) $\bar{T}$ is the mean planetary temperature of the surface layer of air, and $\beta = 0.235$ kcal/(cm^2 month K) = 9.85 MJ/(m^2 K) per month. In formula (21.4) for the feedback between the albedo and the temperature, $\alpha_i = 0.62$ and $\alpha_w = 0.32$ are the mean albedos of the ice-covered and ice-free latitude zones, and $T_c = -10$°C is the mean annual temperature at the boundary between these zones (which is at present in the Arctic, at 72°N). At about the same time, Sellers (1969) proposed a similar "parametrization" of the quantities I, D, and α (that is, he expressed them in terms of T):

$$I(x) = \sigma T^4(x) \;\{1 - \tfrac{1}{2} \tanh [19 \times 10^{-16} T^6(x)]\} \;; \tag{21.2'}$$

$$D(x) = - \frac{d}{dx} (1 - x^2)\, D \;\frac{dT(x)}{dx} \;; \tag{21.3'}$$

$$\alpha(x) = \begin{cases} 0.4860 - 0.092T\text{°C} & \text{for } \alpha \leqslant 0.85, \quad T < 10\text{°C}, \\ \text{const in other cases} \end{cases} \tag{21.4'}$$

Instead of the linear formula (21.2) for $I(x)$, Sellers used the more cumbersome empirical formula (21.2′). On the other hand, he described the feedback between the albedo and the temperature with the aid of the linear law (21.4′), where T is the temperature reduced to sea level, with replacement of this law by α = const in cases where it leads to values of $\alpha \geqslant 0.85$ or where $T \geqslant +10$°C. Sellers represented the total meridional heat flux (sensible, latent, and oceanic) as the sum of the flux produced by the mean meridional circulation (the rate of which was determined by the meridional temperature gradient) and the diffusive flux, which is proportional to the meridional temperature gradient. Sellers expressed the meridional gradients in terms of differences, and later Faegre (1972) and North (1975) introduced the differential expression (21.3′), where $D \approx 0.45$ W/(m^2 K).

Steady-state models of the latitudinal structure of the ground-adjacent layer of the atmosphere, such as those of Budyko and Sellers, have been analyzed in a number of works. The solutions obtained with these models are typically nonunique (possibly indicating intransitiveness of the climate, in Lorenz's sense of the word; see §2), and most of the solutions are unstable with respect to small variations of the parameters of the models. For instance, Budyko (1968) established that the solution of equations (21.1)–(21.4) corresponding to the present-day climate with polar ice is unstable relative to a slight decrease in the insolation Q. If the insolation were reduced by only 1.6%, then the limit of the polar ice cover (where $T = T_c$) would reach to a latitude of 50°. For a further reduction of Q, the limit of the ice would jumpwise shift toward the equator, that is, eventually the whole planet would become covered with ice, a state which would presumably be stable relative to small variations of Q. Budyko also noted that for the present value of Q a climatic regime without polar ice could exist, which would presumably be unstable relative to small variations of the parameters, and of parameter β in particular. These statements will be defined below.

The dependences on Q were derived analytically by Chýlek and Coakley (1975), who obtained from equations (21.1) – (21.4) at the critical latitude $x = x_c$, where $T = T_c$, the relation

$$F(x_c) \equiv \frac{I_c}{Q} \frac{1+\beta_1}{\beta_1} \frac{1-\alpha_c}{\beta_1} S(x_c) + \bar{\alpha}(x_c) - 1 = 0, \tag{21.5}$$

where $I_c = A + BT_c$, $\beta_1 = \beta/B = 2.61$, α_c is the value of the albedo at $T = T_c$, intermediate between α_i and α_w, and $\bar{\alpha} = \int_0^1 \alpha(x)S(x)\,dx$ is the weighted-mean planetary albedo. If $S(x)$ is approximated by a piecewise-linear function, then $\bar{\alpha}$ will be a quadratic function of x_c and (21.5) will be a quadratic equation in x_c, having for present-day values of the parameters two real roots (see curve 1 of function $F(x_c)$ in Fig. 21.2), the larger one corresponding to the climate at present. If Q is reduced, these roots converge in value (curve 2 in the figure corresponds to a 1% reduction of Q). If Q is reduced by 1.6%, they merge (curve 3), and for even lower Q the roots become complex (curve 4 corresponds to a 2% reduction of Q).

It should be stressed, however, that this instability is inherent not so much to the dependence on Q itself as to the relation between the parameter $I_c Q^{-1} \beta_1^{-2}(1 + \beta_1)(1 - \alpha_c)$ and the values adopted for $\alpha(x, x_c)$, that is, to the model as a whole, and it may well be that it is due to certain features of the model rather than to physical factors. With regard to this, the calculations of

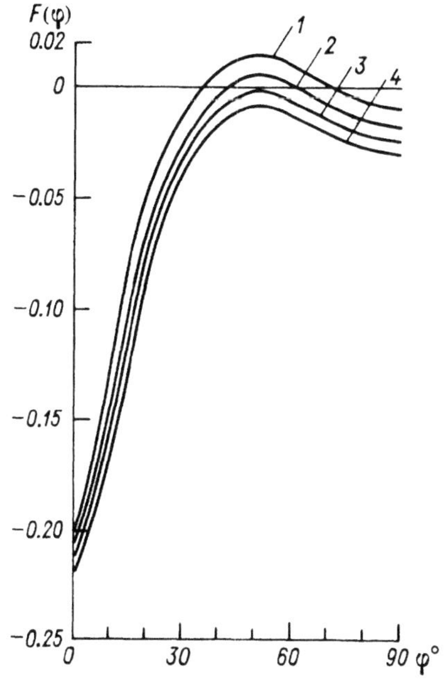

Fig. 21.2. Curves of functions $F(x_c)$ of formula (21.5) for four values of Q (indicated in text).

Golitsyn and Mokhov (1978) should be mentioned. These showed that the rate of entropy exchange between the climate system and the surroundings reaches a minimum (ensuring stability of the state of the system) in the given model for a meridional heat exchange considerably lower (as indicated by β, for instance) than that assumed above, and thus for a state with a more contrasting temperature difference between equator and poles than exists in reality. In other words, with the model in question the parameters can be so selected as to give either a temperature contrast which is correct in value but unstable, as was done above, or else a minimum exchange of entropy plus stability, but a temperature contrast which is too high, as was obtained by Golitsyn and Mokhov.

A convenient means of checking the stability of a solution of the equations of the climate model is to construct the functional $\Phi\{T(x)\}$, for which the equations of the model serve as the Lagrange equations (obtained by equating to zero its first variation with respect to $T(x)$, that is, $\delta\Phi = 0$). Then the condition for stability of the solution is that the second variation be nonnegative $\delta^2\Phi \geqslant 0$. Golitsyn and Mokhov (1978) constructed such functionals for model (21.1)–(21.3), but with formula (21.4) for the albedo replaced by a linear dependence of α on T of the type of Sellers' formula (21.4′): $\alpha = \alpha_0 - (\mathrm{d}\alpha/\mathrm{d}T)\,T$ in the temperature interval $T_2 = T(x_2) \leqslant T \leqslant T(x_1) = T_1$, outside of which α is independent of T (here α_0 and $\mathrm{d}\alpha/\mathrm{d}T$ can be specified functions of x). In the so-called Fourier representation (in which the heat flux is represented in the form $-\lambda\nabla T$ for λ = const) the indicated functional has the form

$$\Phi = -\int_0^1 \left[QS(x)\left(1 - \alpha_0 + \frac{1}{2}\frac{d\alpha}{dT}T\right) - A + BT_0 - \frac{B}{2}T - \right.$$

$$\left. - \frac{\beta}{2}\left(T - \int_0^1 T\,dx\right)\right] T\,dx, \qquad (21.6)$$

where T_0 is the freezing point of water. This gives the stability condition

$$\delta^2\Phi = (B + \beta)\int_0^1 (\delta T)^2\,dx - \beta\left(\int_0^1 \delta T\,dx\right)^2 -$$

$$- Q\int_{x_1}^{x_2} S\frac{d\alpha}{dT}(\delta T)^2\,dx \geqslant 0. \qquad (21.7)$$

Let us make this inequality greater, by discarding the second term and introducing the quantities $c_1(x_2, 1) = \max_{x_2 \leqslant x \leqslant 1}(\delta T)^2$ and $c_2(0, 1) = \min_{0 \leqslant x \leqslant 1}(\delta T)^2$. This gives us the sufficient condition for stability:

$$(B + \beta)c_2(0, 1) \geqslant Q\frac{d\alpha}{dT}c_1(0, 1)\int_{x_1}^1 S(x)\,dx, \qquad (21.7')$$

where the feedback factor $d\alpha/dT$ is assumed to be constant and where we set $x_2 = 1$ (that is, T_2 is not higher than at the pole). Now, if $S \sim \cos\varphi = (1 - x^2)^{1/2}$ and if there is an equals sign in (21.7′), then for the critical latitude of the ice boundary φ_c, below which complete glaciation of the Earth takes place, we obtain the expression

$$\frac{\pi}{4} - \frac{\varphi_c}{2} - \frac{\sin 2\varphi_c}{4} = \frac{c_2(0, 1)}{c_1(0, 1)}\frac{B + \beta}{Q\,d\alpha/dT}, \qquad (21.8)$$

in which for min φ_c we must assume a constant latitudinal distribution of the temperature fluctuations δT, that is, in (21.8) we must set $c_2(0, 1) = c_1(0, 1)$. If δT is specified as an increasing function of φ, then the critical latitude φ_c obtained will be appreciably higher, meaning that the climate will be more unstable. Thus, min φ corresponds to climates with an ice-boundary latitude higher than φ_c (the present climate, for instance), and these are stable. Note that the negative $I(T)$ feedback is greater than the positive $\alpha(T)$ feedback, so that for $d\alpha/dT = 0$ all climates are stable, including, in particular, a climate with complete glaciation of the Earth. On the other hand, max φ corresponds to intermediate climates with an ice-boundary latitude between φ_c and 0, and

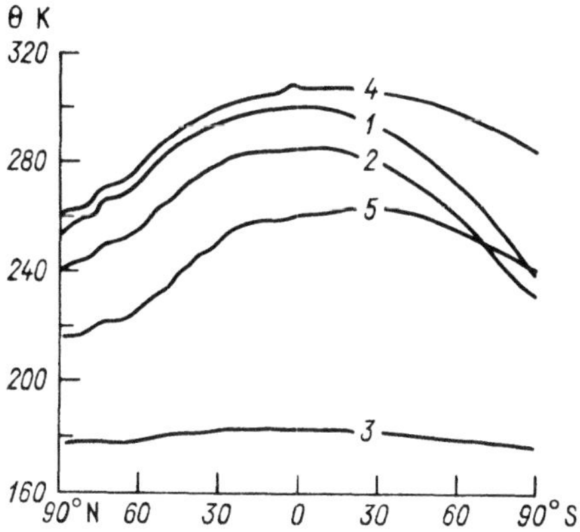

Fig. 21.3. Solutions of Faegre (1972) for model of Sellers: (1) present-day climate, (2) climate of glacial epoch, (3) climate of complete glaciation of Earth, (4) and (5) asymmetrical climates with warmer Southern Hemisphere.

these are unstable. If the following latitude-independent sufficient condition of stability:

$$\frac{d\alpha}{dT} \leqslant \frac{B + \beta}{Q} \tag{21.8'}$$

is satisfied, then the critical latitude disappears, and the transition from partial to total glaciation of the Earth (and vice versa) with a gradual change in parameters Q and $B + \beta$ takes place smoothly.

The observed sensitivity of model (21.1)–(21.4) to small variations of parameter Q (whereby the boundary of polar ice reaches a critical latitude φ_c below which the Earth becomes completely glaciated, for a reduction of Q by only 1.6%, and we have a high value of $\varphi_c \approx 50°$) is unrealistic, and it constitutes a shortcoming of the model. This shortcoming can be obviated to a large extent, bringing the model closer to reality, if we analyze the model parameters in more detail. For instance, in (21.4) the dependence of the albedo on the amount of cloud cover n can be taken into account, by setting

$$\alpha_{i,\,w} = n\alpha_c + (1 - n)\alpha'_{i,\,w}, \tag{21.9}$$

where α_c is the cloud albedo, and $\alpha'_{i,\,w}$ is the albedos of cloudless ice-covered and ice-free latitudes. If, in addition, we specify the dependence of all four parameters n, α_c, and $\alpha'_{i,\,w}$ on the latitude (the last three in terms of their dependence on the mean zenith angle of the Sun at each latitude), then, according to the calculations of Golitsyn and Mokhov, the critical latitude φ_c is lowered to 20° and is reached only for a much larger reduction of Q, by 8%, which agrees well with the values obtained using much more complete three-dimensional models of the climate system. However, the need to specify four functions n, α_c, and $\alpha'_{i,\,w}$ of the latitude makes the model of practically no predictive value whatsoever.

Similar calculations were carried out for a model employing diffusion formula (21.3′) instead of Newton's law (21.3). Faegre solved numerically the equations of Sellers' model (21.1), (21.2′) – (21.4′), but with formula (21.2′) replaced by the simpler expression $I = K\sigma T^4$, where K is some function of x which is selected empirically on the basis of the present climate. Five solutions for $T(\varphi)$ were obtained, and these are shown in Fig. 21.3. Solution 1 is very close to the present climate, on the basis of which the model parameters were chosen. Solution 2, with a temperature 15°C lower, apparently corresponds to the climate of the glacial periods (the limit of polar ice $T = -10$°C here lies at a latitude $\varphi \approx 50°$). Solution 3, with much lower temperatures, corresponds to complete glaciation of the Earth. Solutions 4 and 5, which are asymmetrical relative to the equator, the Southern Hemisphere being warmer, are probably unrealistic; if empirical function $K(x)$ is replaced by a constant, then these solutions disappear. For variations in Q of ± 1, ± 2, and ± 4%, solution 3 turns out to be stable. Finally, if Q is reduced by 2%, both solutions 1 and 2 are unstable.

Now let us consider a model with diffusive meridional exchange (21.1), (21.2), (21.3′), (21.4), analyzed by North. Expressing the distribution of insolation as $S(x) = 1 + S_2P_2(x)$, he formulated an exact analytical solution of the model equations in hypergeometric functions:

$$I(x) = \begin{cases} A_1P_\nu(x) + QI_1(x), & x > x_c, \\ A_0f_\nu(x) + QI_0(x), & x < x_c: \end{cases}$$

$$P_\nu(x) = F\left(\frac{\nu+1}{2}, -\frac{\nu}{2}, 1, 1 - x^2\right);$$

$$f_\nu(x) = F\left(-\frac{\nu}{2}, \frac{\nu+1}{2}, \frac{1}{2}, x^2\right);$$

$$\nu = -\frac{1}{2} + \frac{1}{2}\left(1 - 4\frac{B}{D}\right)^{1/2};$$

$$I_k(x) = (1 - \alpha_k)\left[1 + \frac{S_2P_2(x)}{6(D/B) + 1}\right], \qquad (21.10)$$

where if ν is complex only the real parts of functions P_ν and f_ν are retained. Coefficients A_0 and A_1 and the value of x_c are found from the conditions $I(x_c + 0) = I(x_c - 0) = I_c$ and $I'(x_c + 0) = I'(x_c - 0)$, from which we get

$$Q = I_c(P_\nu f'_\nu - f_\nu P'_\nu)[I_0P_\nu f'_\nu - I_1P'_\nu f_\nu + (I'_1 - I'_0)P_\nu f_\nu]^{-1};$$

$$A_0 = (I_c - QI_0)f_\nu^{-1}, \qquad A_1 = (I_c - QI_1)P_\nu^{-1}, \qquad (21.11)$$

where all the functions pertain to the point $x = x_c$. For $x_c = 0$ (a completely glaciated Earth, $I_1 \leqslant I_c$), we have $A_1 = 0$ and $Q \leqslant I_c/I_1(0)$. For $x_c = 1$ (ice-free Earth), we have $Q \geqslant I_c/I_0(1)$. Figure 21.4 shows a plot of the

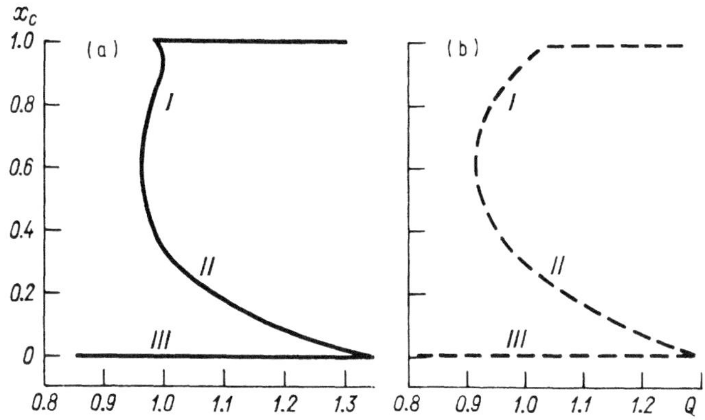

Fig. 21.4. Plots of functions $x_c(Q)$ corresponding to mean annual (a) and equinoctial (b) distributions of insolation, according to North (1975).

function $x_c(Q)$ corresponding to the mean annual distribution of insolation $S(x)$, where Q is measured in units of its present value. It is seen that, in some small interval of Q values around the present valuc, function $x_c(Q)$ has several branches (up to five), reflecting different climatic regimes. A value of $x_c(1) = 0.95$ corresponds to the present climate. Figure 21.4b shows the $x_c(Q)$ curve corresponding to the distribution of insolation at the time of the equinoxes $S(x) = 3/2(1 - x^2)$ (and to another value of D, selected for the best approximation of the temperature curve $T(x)$ of the present climate). Around the point $x_c = 1$ it behaves more naturally than does the curve in Fig. 21.4a: the ice retreats toward the pole with an increase in Q, and for a transition to complete glaciation of the Earth, judging by this curve, a considerable (7%) reduction of Q is necessary. It may be that the optimum solution lies somewhere between the two curves shown.

Stability of the solutions can be ascertained if we add to the right-hand side of equation (21.1) the term $R\partial T/\partial t$ with an arbitrary coefficient of thermal inertia R and then linearize this equation relative to a small perturbation of the steady-state regime (both $T(x)$ and x_c). In this way North established that the present climate and complete glaciation of the Earth (see Fig. 21.4b, branches I and III) are both stable, whereas the intermediate branch II (the climate of the glacial periods) is unstable, so that it cannot come into being via the evolution in time of fluctuations of the present climate. This inability of the model (at any rate in the version of Fig. 21.4b) to simulate the transition to a glacial period was attributed by North to the absence in the model of a number of important feedback relationships, such as $n[T]$, $D[T]$, the advection of salinity in seawater, the presence of sufficient moisture in the atmosphere to produce snowfall, and phase differences between the fluctuations in time of temperature and snowfall. A more detailed study of the stability of steady-state solutions was carried out by Drazin and Griffel (1977), who, for the mean annual distribution of insolation, came up with five solutions symmetrical with respect to the equator and three pairs of asymmetrical solutions; for present-day values of the parameters, one of the asymmetrical solutions turned out to be stable.

Golitsyn and Mokhov used the model being considered here (with values of $\alpha_w = 0.303 - 0.080\, P_2(x)$ and using a two-mode approximation with respect to an expansion in Legendre polynomials) to calculate the rate of entropy exchange with the surroundings. They found that this rate is a minimum for a D value 1/2.5 as large as the value selected by North to describe the present climate. For this value of D the climate has a temperature differential between equator and poles which is more contrasting than the actual differential, as was also the case for the model with meridional exchange according to Newton's law. Then, functional $\Phi\{T(x)\}$, for which the model equations (21.1), (21.2), and (21.3′) serve as Lagrangian equations, with the same temperature dependence as was used in formula (21.6), becomes

$$\Phi = \int_0^1 \left\{ - \left[QS(x) \left(1 - \alpha_0 + \frac{1}{2} \frac{\mathrm{d}\alpha}{\mathrm{d}T} T \right) - A + BT_0 - \frac{B}{2} T \right] \times \right.$$

$$\left. \times\ T + \frac{D(1 - x^2)}{2} \left(\frac{\mathrm{d}T}{\mathrm{d}x} \right)^2 \right\} \mathrm{d}x. \tag{21.12}$$

This gives the following condition for stability:

$$\delta^2\Phi = B \int_0^1 (\delta T)^2\, \mathrm{d}x - Q \int_{x_1}^{x_2} S\, \frac{\mathrm{d}\alpha}{\mathrm{d}T}\, (\delta T)^2\, \mathrm{d}x + \tag{21.13}$$

$$+ D \int_0^1 (1 - x^2) \left[\delta \left(\frac{\mathrm{d}T}{\mathrm{d}x} \right) \right]^2 \mathrm{d}x \geqslant 0,$$

which yields the same conclusions regarding the stability of steady-state solutions as condition (21.7) do.

CHAPTER 9

ZONAL MODELS

22. Zonal Models of the Atmosphere

As stated in the preceding section, because of the zonality of the daily insolation (the insolation being the energy base of the climate system), the zonal characteristics of the climate system constitute quite representative climatic characteristics. If just these zonal characteristics (the mean meridional profiles of the fields describing the climate system) are used, however, then we will inevitably neglect some important longitudinal differences, produced by the distribution of continents and oceans, together with the mighty seasonal (monsoon) effects caused by these differences. For the ocean and the land such longitudinal differences are determining factors: shores serve as insurmountable meridional barriers, so that zonal models are less representative, and this is also to a large degree true for the atmosphere-ocean-land system considered as a whole. For the atmosphere, on the other hand, nonzonal differences in the underlying surfaces are not an internal feature determining the state of the atmosphere; rather they are an external factor and do not create insurmountable barriers. This means that zonal models of the atmosphere are much more representative, and these will be the subject of the present section.

The most accurate means of constructing a zonal model of the atmosphere is zonal averaging of the equations of its dynamics (presented in §9). Let us denote the zonal mean values (which are independent of the longitude λ, and which thus may be functions only of the height z, the colatitude θ, and the time t) by bars over the symbols, and let us denote the deviations from these mean values (the fluctuations) by primes. If we average the equation of hydrostatics (9.1) and the continuity equation (9.2), then we obtain

$$\frac{\partial \bar{p}}{\partial z} = -g\bar{\varrho}; \tag{22.1}$$

$$\frac{\partial \bar{v}_z}{\partial z} + \frac{1}{a \sin\theta}\frac{\partial \bar{v}_\theta \sin\theta}{\partial \theta} = 0. \tag{22.2}$$

The second of these equations can be used to introduce the stream function ψ, describing the mean circulation in the plane of the meridian, provided we set

$$\bar{v}_z = -\frac{1}{a \sin\theta}\frac{\partial \psi}{\partial \theta}; \qquad \bar{v}_\theta = \frac{1}{\sin\theta}\frac{\partial \psi}{\partial z}. \tag{22.2'}$$

With the aid of these formulas, the vorticity of the mean meridional circulation Ω_λ can be expressed simply in terms of the stream function:

$$\Omega_\lambda = \frac{\partial \bar{v}_\theta}{\partial z} - \frac{1}{a}\frac{\partial \bar{v}_z}{\partial \theta} = \frac{1}{\sin\theta}\left(\Delta\psi - 2\,\frac{\text{ctg}\,\theta}{a^2}\,\frac{\partial \psi}{\partial \theta}\right);$$
$$\Delta\psi = \frac{\partial^2\psi}{\partial z^2} + \frac{1}{a^2 \sin\theta}\frac{\partial}{\partial\theta}\sin\theta\,\frac{\partial\psi}{\partial\theta}. \tag{22.3}$$

The zonal velocity component $\bar{v}_\lambda$ of the mean circulation is conveniently represented as

$$\bar{v}_\lambda = a\,(z, \theta, t)\, a \sin\theta, \tag{22.4}$$

where α is the angular velocity of the differential zonal rotation of the atmosphere relative to the Earth's surface. Next, by averaging the equations of horizontal motion (9.3) and (9.4), we get

$$\frac{\mathrm{d}\bar{v}_\theta}{\mathrm{d}t} - \frac{\text{ctg}\,\theta}{a}\,\bar{v}_\lambda^2 - 2\Omega\cos\theta\bar{v}_\lambda = -\frac{1}{a\varrho_0}\frac{\partial \bar{p}}{\partial\theta} + f_\theta + F_\theta, \tag{22.5}$$

$$\frac{\mathrm{d}\bar{v}_\lambda}{\mathrm{d}t} + \frac{\text{ctg}\,\theta}{a}\bar{v}_\theta\bar{v}_\lambda + 2\Omega\cos\theta\bar{v}_\theta = f_\lambda + F_\lambda, \tag{22.6}$$

where $\mathrm{d}/\mathrm{d}t = \partial/\partial t + \bar{v}_z(\partial/\partial z) + (\bar{v}_\theta/a)$ is the operator of the individual derivative with respect to the mean meridional circulation; for simplicity, we set $\overline{\varrho^{-1}\nabla_h p} \approx \varrho_0^{-1}\nabla_h\bar{p}$, where ϱ_0 is the main part of field ϱ, depending only on z. The quantities F_θ and F_λ are here defined by the formulas:

$$F_\theta = -\frac{\partial\overline{v_\theta' v_z'}}{\partial z} - \frac{1}{a\sin\theta}\frac{\partial \overline{v_\theta'^2}\sin\theta}{\partial\theta} + \frac{\text{ctg}\,\theta}{a}\,\overline{v_\lambda'^2}; \tag{22.5'}$$

$$F_\lambda = -\frac{\partial\overline{v_\lambda' v_z'}}{\partial z} - \frac{1}{a\sin\theta}\frac{\partial\overline{v_\theta' v_\lambda'}}{\partial\theta} - \frac{\text{ctg}\,\theta}{a}\,\overline{v_\theta' v_\lambda'} \tag{22.6'}$$

If f_θ and f_λ are the components of the divergence of the mean tensor describing the frictional stresses set up by small-scale processes, then $-\varrho_0 B_{ij} = -\varrho_0\overline{v_i' v_j'}$ is the tensor of the additional (so-called Reynolds) stresses, produced by large-scale nonzonal motions. The Reynolds stresses are, as a rule, much greater than the small-scale stresses, so that the latter can usually be neglected (at any rate, outside of the boundary layers). These Reynolds stresses play a very active role in the dynamics of the atmospheric circulation, especially in term F_λ of equation (22.6) (see §§10 and 11), since in some places they produce a negative viscosity, supplying the quasi-zonal jet streams with kinetic energy. If this phenomenon is not included in the model (as well as in the now obsolete description of the Reynolds stresses as dissipative forces), then it can hardly claim to give a qualitative reconstruction of the zonal circulation of the atmosphere.

Difficulties are encountered when attempting an accurate formulation of the boundary conditions for the velocity field of the zonal circulation at the Earth's surface. Let us write the equation of the surface relief as $z = h(\theta, \lambda) = \bar{h}(\theta) + h'(\theta, \lambda)$ let us allow linearization with respect to h', so that for any function $a(z)$ we will have $a(h) \approx a(\bar{h}) + (\partial a(\bar{h})/\partial \bar{h})h'$. Then the mean velocity field is

$$\overline{v_i(h)} \approx \bar{v}_i(\bar{h}) + \overline{\frac{\partial v_i'(\bar{h})}{\partial \bar{h}} h'}, \tag{22.7}$$

where the second terms here represent additional unknown quantities. In addition, whereas at the land surface all three velocity components must go to zero, at the surface of the ocean this will be true only of the vertical component. The horizontal components there have to be determined from the corresponding continuity conditions, and thus the ocean must be included in the model which left out using the zonal model. Accordingly, when constructing a zonal model of the atmosphere, it is advisable to limit the integration of the equations for the velocity field just to the region above the atmospheric boundary layer, the effects of this boundary layer being "parametrized" with the aid of formulas like (9.12) – (9.14) and coefficients c_T, c_H, and c_E and angle α being given as weighted means of the values for the continents and oceans.

The equation of state (9.5) and the formula for the entropy (9.8) are generally used in the zonal model, but with all the variables in them replaced approximately by their mean zonal values (although such a procedure is not equivalent to a precise zonal averaging of these quantities). Then, by averaging the equations describing the evolution of the entropy and the specific humidity, we get

$$\frac{d\bar{\eta}}{dt} = \left(\overline{\frac{\varepsilon + \mathcal{L}m}{T}}\right) - \left[\frac{\partial(\bar{Q}_z + \overline{v_z'\eta'})}{\partial z} + \frac{1}{a\sin\theta}\frac{\partial(\bar{Q}_\theta + \overline{v_\theta'\eta'})\sin\theta}{\partial\theta}\right]; \tag{22.8}$$

$$\frac{d\bar{q}}{dt} = -m - \left[\frac{\partial(\bar{I}_z + \overline{v_z'q'})}{\partial z} + \frac{1}{a\sin\theta}\frac{\partial(\bar{I}_\theta + \overline{v_q'\mathbf{q}'})\sin\theta}{\partial\theta}\right], \tag{22.9}$$

where in the entropy flux $\overline{v_k'\eta'}$ the contributions of the heat and moisture fluxes specified by formula (10.2) must be taken into account. Equation (9.9) for the specific humidity can be averaged similarly. Almost all the terms in the boundary condition for equation (22.8), namely equation (9.11) for the heat budget at the Earth's surface, can likewise be averaged directly. The equations for the moisture budget at the surfaces of the continents and oceans, on the other hand, behave so diversely as to preclude their simple zonal averaging. Consequently, for the moisture (and possibly for the heat as well) it is advisable to resort to parametrization of the effects of the atmospheric boundary layer in the manner of (9.12)–(9.14), as was suggested above when discussing the boundary conditions for the velocity field.

The right-hand sides of equations (22.5) and (22.6) and (22.8) and (22.9) include, first of all, the zonally averaged fluxes of momentum $\bar{\tau}_{ij}$, entropy $\bar{Q}_k$, and moisture $\bar{I}_k$ produced by small-scale processes (molecular processes and local turbulence) and, secondly, the fluxes $\overline{v_i'v_j'}$, $\overline{v_k'\eta'}$, and $\overline{v_k'q'}$ produced by large-scale nonzonal motions. The former and the latter fluxes have to be written differently. The former fluxes are considered traditionally as dissipative smoothing factors, and they are described with the aid of the hypotheses of the semiempirical theory of turbulence, which sees these fluxes as linear functions of the gradients of the corresponding averaged fields; the coefficients of these linear functions, which have the sense of coefficients of eddy viscosity, heat conductivity, and diffusion, are selected empirically. The main difficulty here is that in general quite a number of such coefficients are necessary. For example, if the flux $\bar{Q}_i$ of the entropy $\bar{\eta}$ (or of any other quantities described by scalar fields) is written as $\bar{Q}_i = - K_{i\alpha}(\partial\bar{\eta}/\partial x_\alpha)$, there is no basis for assuming either isotropy or symmetry of tensor K_{ij}, or even that its principal axes coincide with the spherical coordinate axes. Nevertheless, the latter is usually assumed and, what is more, this tensor is taken to be axisymmetric about the vertical axis, so that

$$\bar{Q}_\theta = - \frac{K_h}{a}\frac{\partial\bar{\eta}}{\partial\theta}\,; \quad \bar{Q}_\lambda = - \frac{K_h}{a \sin\theta}\frac{\partial\bar{\eta}}{\partial\lambda}\ (= 0);$$
$$\bar{Q}_z = - K\frac{\partial\bar{\eta}}{\partial z}\,, \tag{22.10}$$

where K_h and K_z are the horizontal and vertical coefficients of eddy diffusion (note that the values of K_z are in general different for heat, moisture, and passive substances). Things get even more complicated for the coefficients of eddy viscosity, which, as they relate the stress tensor $\bar{\tau}_{ij}$ to the tensor of the mean deformation rates $\bar{\varphi}_{ij}$, must form a fourth-rank tensor A_{ijkl}, introduced via the formula $\bar{\tau}_{ij} = - \frac{2}{3}\varrho_0 B^2\delta_{ij} + \varrho_0 A_{ij\alpha\beta}\bar{\varphi}^{\alpha\beta}$, where B^2 is the kinetic energy of turbulence per unit mass. Taking into account the symmetry of this tensor with respect to indices i, j and k, l and assuming (without any special basis) it to be axisymmetric relative to the vertical axis, we obtain (Kamenkovich, 1967):

$$\bar{\tau}_{\lambda z} = \varrho_0 A_z \frac{\partial \bar{v}_\lambda}{\partial z}; \qquad \bar{\tau}_{\theta z} = \varrho_0 A_z \left(\frac{\partial \bar{v}_\theta}{\partial z} + \frac{1}{a} \frac{\partial \bar{v}_z}{\partial \theta} \right); \tag{22.11}$$

$$\bar{\tau}_{\theta\lambda} = \varrho_0 A_h \sin \theta \frac{\partial \alpha}{\partial \theta}; \tag{22.12}$$

$$\bar{\tau}_{zz} = - \frac{2}{3} \varrho_0 B^2 + 2\varrho_0 A \frac{\partial \bar{v}_z}{\partial z};$$

$$\bar{\tau}_{\theta\theta} = - \frac{2}{3} \varrho_0 B^2 + 2\varrho_0 A_h \frac{1}{a} \frac{\partial \bar{v}_\theta}{\partial \theta} + \varrho_0 (A_h - A) \frac{\partial \bar{v}_z}{\partial z};$$

$$\bar{\tau}_{\lambda\lambda} = - \frac{2}{3} \varrho_0 B^2 + 2\varrho_0 A_h \frac{\text{ctg}\, \theta}{a} \bar{v}_\theta + \varrho_0 (A_h - A) \frac{\partial \bar{v}_z}{\partial z}, \tag{22.13}$$

where A_z, A_h, and A are the coefficients of eddy viscosity.

When constructing zonal models of the atmosphere, we encounter serious difficulties in describing the fluxes $\overline{v_i'v_j'}$, $\overline{v_k'\eta'}$, and $\overline{v_k'q'}$ produced by large-scale nonzonal motions, that is, mainly by synoptic eddies and Rossby waves. Especially important here is the quantity $\overline{v_\theta' v_\lambda'}$, representing the meridional flux of zonal momentum, which is the source of the "negative viscosity" supplying kinetic energy $\bar{K}$ to the mean zonal currents at a rate

$$\frac{\partial \bar{K}}{\partial t} = \varrho_0 \overline{v_\theta' v_\lambda'} \sin \theta \frac{\partial \alpha}{\partial \theta}. \tag{22.14}$$

If we used formula (22.12) for $\bar{\tau}_{\theta\lambda} = - \varrho_0 \overline{v_\theta' \; v_\lambda'}$, then we would get

$$\frac{\partial \bar{K}}{\partial t} = - \varrho_0 A_h \left(\sin \theta \frac{\partial \alpha}{\partial \theta} \right)^2, \tag{22.14'}$$

so that to ensure an increase in the kinetic energy of the zonal current ($\partial \bar{K}/\partial t > 0$) we would somehow have to specify a negative eddy viscosity $A_h < 0$. Therefore, formula (22.12) does not solve the problem.

A more justifiable parametrization of the momentum flux $-\varrho_0 \overline{v_\theta' v_\lambda'}$ produced by synoptic processes was proposed back in 1965 by Williams and Davis. They noted that synoptic processes arise due to a baroclinic instability of the zonal currents, so that their effects must be determined by the mean meridional temperature gradient $1/a(\partial \bar{T}/\partial \theta)$. Accordingly, Williams and Davis assumed that

$$\overline{v_\theta' v_\lambda'} = - k \frac{L}{a} \frac{\partial R\bar{T}}{\partial \theta}; \qquad L = \frac{c}{\Omega}, \tag{22.15}$$

where R is the gas constant, L is the typical horizontal scale of the synoptic processes (c is the velocity of sound, and Ω is the angular velocity of the Earth's rotation), and k is a dimensionless exchange coefficient (assumed to increase with height, according to the law $k = k_0 \, (z/H)^2$, where H is the thickness of the homogeneous atmosphere). Formula (22.15) already ensures the appearance of negative-viscosity effects in the latitude zones where the main jet streams form. The remaining fluxes $\overline{v_i'v_j'}$, $\overline{v_k'\eta'}$, and $\overline{v_k'q'}$ produced by synoptic motions were described by Williams and Davis with the aid of the usual formulas, similar to (22.10) and (22.11), with positive coefficients of virtual viscosity, heat conductivity, and diffusion.

Williams and Davis (1965) used the above parametrization of the flows produced by synoptic processes to calculate the first numerical model of the zonal circulation of the atmosphere with a negative horizontal viscosity. It should be noted that in their work, and in most later works, the equation for the entropy (22.8) was replaced by the usual equation of convective heat conduction with sources present; in some works, on the other hand, the potential temperature was used in this equation instead of the entropy. Later this same approach was applied for a "dry" atmosphere (zero humidity) by Dolzhanskii (1969, 1971). In the first of these works the influx of heat to the atmosphere was described by Newton's law as

$$\bar{\varepsilon} = c_p \frac{\bar{T}_r - \bar{T}}{\tau_r}, \tag{22.16}$$

where $\bar{T}_r$ is the temperature of radiation equilibrium, and τ_r is the relaxation time, of the order of 10^6 s (more detailed calculations using this method were carried out later by Kirichkov, 1978). In the second work Dolzhanskii described the heat influx in more detail with the aid of transmission functions (5.7) and (5.8). These models gave a quite good picture of the mean annual zonal temperature field in the atmosphere (Fig. 22.1). Inspection of the figure shows that the calculated temperature essentially coincides with the observed temperature everywhere, except for the upper troposphere of the polar regions, which appears to be a little warmer than in reality. The models also reproduced successfully the observed tricellular structure of the meridional circulation (with direct cells in the low and high latitudes and a backward cell in the middle latitudes from 35 to 65°), although the intensities obtained for both the meridional circulation and the zonal flow were somewhat low.

The same parametrization of the fluxes produced by synoptic processes was used in the model of the zonal circulation devised by Dymnikov, Perov, and Lykosov (1979). They described the interaction of the atmosphere with the underlying surface using formulas of the type of (9.12), with special semi-empirical expressions for coefficients c_τ and $c_H = c_E$ over the continents and oceans. For both stable and unstable stratifications they specified the temperature of the ocean surface, and for the land the temperature was found from the heat-budget equation (9.10) with a special choice of the albedo values A and with a calculation, where necessary, of the melting of snow and ice and of variations in the thickness of the snow cover. The relative air humidity over an ocean or snow surface was taken to be unity, and elsewhere it was found together with the soil humidity from additional relations. The number of nonconvective clouds was specified as a linear function of the relative humidity. For regions with dry or moist convection certain

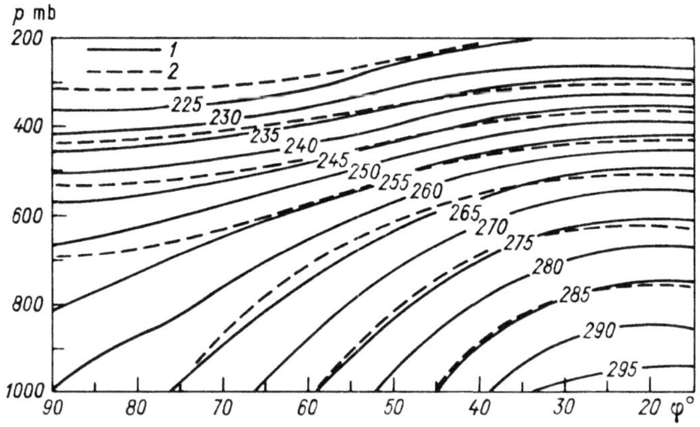

Fig. 22.1. Mean annual zonal temperature field in troposphere: (1) actual, (2) calculated, according to Kirichkov (1978).

conditions of correspondence of the vertical temperature and humidity profiles were assumed, including the amount of condensed moisture. The large-scale condensation was calculated with the aid of the simplified equations (22.8) and (22.9), written in the form

$$\frac{\partial \bar{T}}{\partial t} = \frac{\mathscr{L}}{c_p}(\bar{C} - \bar{E}); \qquad \frac{\partial \bar{q}}{\partial t} = -(\bar{C} - \bar{E}), \tag{22.17}$$

where C and E are the condensation and evaporation. The horizontal small-scale diffusion was described in the manner of (22.10)–(22.13). The long-wave cooling of the atmosphere was described with the aid of transmission functions; the fluxes of short-wave radiation in the regions were calculated without Rayleigh scattering, and outside the regions taking Rayleigh scattering into account (absorption by ozone was also taken into account). The equations were solved numerically in $\varphi\sigma$ coordinates (where $\sigma = p/\mathcal{P}$, and $\mathcal{P}$ is the surface pressure), after first being "symmetrized" by a conversion from v_θ, v_λ, T, q to $\mathcal{P}^{1/2}v_\theta$, $\mathcal{P}^{1/2}v_\lambda$, $\mathcal{P}^{1/2}T$, $\mathcal{P}^{1/2}q$, followed by a splitting up of the processes of transfer, diffusion, and adaptation, that is, these processes were described individually during three successive time steps.

Numerical calculations were carried out for a five-level model (heights of 950, 800, 500, 200, and 50 mb) with a latitude interval of 5° over the entire globe and with a time interval of 40 min. The mean annual zonal temperature of the underlying surface was specified, the initial state of the atmosphere was assumed to be at rest, and the initial distributions of $\bar{T}$ and $\bar{q}$ were taken to be functions of height alone. A steady-state zonal circulation close to that observed in nature was established for 70 days in the troposphere, except that the temperature in the polar regions was lower (because the temperature of the underlying surface there was assumed to be too low) and the intensity of the zonal flow was somewhat high (because the meridional temperature differential was assumed to be too high), as was the intensity of the tricellular (in the hemisphere) meridional circulation. When the negative viscosity (22.15) was dropped from the model, the backward cell in middle latitudes of the meridional circulation was broken down into three cells and the maximum of the zonal flow became attenuated and moved toward the equator. Somewhat later Dymnikov, Galin, and Perov (1980) used this model to calculate the effect of a doubling of the CO_2 concentration in the atmosphere.

Other formulas have also been proposed to parametrize the fluxes produced by synoptic processes. For instance, Saltzman and Vernekar (1968, 1971–1972, 1975) employed approximate formulas for baroclinically unstable perturbations of the zonal circulation to obtain the expression

$$\overline{v'_\theta v'_\lambda} = -\left(\sin\theta\,\frac{\partial\alpha}{\partial\theta}\right)^{-1} \times \\ \times \left[C\,\langle\bar{v}_\lambda^2\rangle - \left(f + \frac{\operatorname{ctg}\theta}{a}\bar{v}_\lambda\right)\bar{v}_\theta\bar{v}_\lambda + \Phi\right], \qquad (22.15')$$

where Φ is the contribution of standing waves. Formula (22.15′), which does not contain explicitly the meridional temperature gradient (it enters in via $\bar{v}_\lambda$), was used to construct numerical models of the zonal circulation of the atmosphere in the present epoch, as well as during the glacial periods of the Pleistocene.

Stone (1972, 1973), who also began by calculating baroclinically unstable perturbations, made use of the thermal wind U and the Richardson number, defined as

$$U = -\frac{gH}{fT_0}\frac{\partial\Theta_0}{\partial y}; \qquad \mathrm{Ri} = \frac{gH^2}{T_0U^2}\frac{\partial\Theta_0}{\partial z}, \qquad (22.18)$$

to derive the expressions:

$$\begin{aligned}
\overline{u'v'} &= -k_1\,U^2(1+\mathrm{Ri})^{3/2}\left[\left(\frac{z}{H}-\frac{1}{2}\right)^2+\frac{1}{12}\right];\\
\overline{u'w'} &= -k_2 f\,HU(1+\mathrm{Ri})^{1/2}\left[\left(\frac{z}{H}-\frac{1}{2}\right)^2+\frac{1}{4}\right];\\
\overline{v'w'} &= -k_3 f\,HU(1+\mathrm{Ri})\left[\left(\frac{z}{H}-\frac{1}{2}\right)^2-\frac{1}{16}\right];\\
\overline{v'\Theta'} &= k_4\,HU\,\frac{\partial\Theta_0}{\partial z}\,\frac{(1+\mathrm{Ri})^{1/2}}{\mathrm{Ri}}\,\frac{y}{L}\left(1-\frac{y}{L}\right);\\
\overline{w'\Theta'} &= k_5 f\,H^2\,\frac{\partial\Theta_0}{\partial z}\,\frac{(1+\mathrm{Ri})^{1/2}}{\mathrm{Ri}}\,\frac{z}{H}\left(1-\frac{z}{H}\right),
\end{aligned} \qquad (22.19)$$

where $k_1, \ldots, k_5$ are numerical constants. Stone used these formulas to calculate a model of the steady-state zonal circulation with heat influxes, in middle latitudes according to Newton's law (22.16).

We should also mention the result of Sela and Wiin-Nielsen (1971), which ascribes a positive virtual viscosity to the meridional transfer of potential vorticity by synoptic processes. Thus the averaged equation (10.7) for the potential vorticity can be written as

$$\frac{\partial \bar{\omega}_*}{\partial t} + \frac{\bar{v}_\theta}{a} \frac{\partial \bar{\omega}_*}{\partial \theta} = \frac{1}{a^2 \sin \theta} \frac{\partial}{\partial \theta} K_h \sin \theta \frac{\partial \bar{\omega}_*}{\partial \theta} + \tag{22.20}$$

$$+ \operatorname{curl}_z \bar{\mathbf{f}} + \frac{\partial}{\partial z} \frac{f \bar{\varepsilon}_\eta}{\partial \eta_0 / \partial z},$$

where, as in (10.4), **f** is the viscous force produced by small-scale processes, and E_η is the rate of heat influx to a unit mass, divided by the temperature (right-hand side of equation (9.6)). With the aid of this equation Ohring and Adler (1978) were able to construct a simplified (two-level) model of the zonal circulation of the atmosphere, with a quite detailed description of the various kinds of heat influxes. They also used this model to perform several experiments on the sensitivity of the climate to variations in the external parameters.

It should be stressed at this point, however, that the parametrization of the single-point second moments or of the fluxes $\overline{v_i'v_j'}$, $\overline{v_k'\eta'}$, and $\overline{v_k'q'}$ produced by synoptic processes, which constitute additional unknowns in the averaged equations of hydrodynamics (22.5) and (22.6), (22.8) and (22.9), involves a number of arbitrary hypotheses. This arbitrariness can be reduced substantially if the averaged equations are supplemented by the dynamic equations for the second moments, derived from the equations of hydrodynamics (the so-called Friedman–Keller equations). It is true that such a system of equations is still not a closed system, since new additional unknowns appear in it: the single-point third moments of the fluctuations, as well as the second moments relating the velocity fluctuations to the fluctuations of the stress gradients. These new unknowns require us to make certain hypotheses involving some degree of arbitrariness. Now, however, the arbitrariness will have much less effect on the principal unknowns ψ, α, η, and $\bar{q}$. One possible simplification of the Friedman–Keller equations for the single-point second moments is to neglect the third moments in these equations. Such simplified equations for the second moments can be derived from the equations of hydrodynamics, linearized with respect to the fluctuations. Let us write these in tensor form, without applying the equation of hydrostatics to the fluctuations, and let us use instead the more accurate Boussinesq approximation:

$$\frac{\partial v_i'}{\partial t} + \bar{v}_\alpha \nabla^\alpha v_i' + v_\alpha' \nabla^\alpha v_i + \psi_i' - g\zeta' e_i = -\frac{1}{\varrho_*}\nabla_i P' + \frac{1}{\varrho_0}\nabla^\alpha \tau_{i\alpha}'; \tag{22.21}$$

$$\frac{\partial \zeta'}{\partial t} + \bar{v}_\alpha \nabla^\alpha \zeta' + v_\alpha' \nabla^\alpha \zeta_1 = \varepsilon_\zeta'; \tag{22.22}$$

$$\frac{\partial q'}{\partial t} + \bar{v}_\alpha \nabla^\alpha q' + v_\alpha' \nabla^\alpha \bar{q} = \varepsilon_q'; \tag{22.23}$$

where ψ_i are the components of the Coriolis acceleration, e_i are the components of the unit vertical vector,

$$\zeta = \frac{1}{\varrho_0 c_0^2}\left[(p - p_0) - c_0^2(\varrho - \varrho_0)\right]$$

is the normalized and linearized deviation of the entropy η from its static value η_0, ϱ_* is the potential density, defined by the relation $(d \ln \varrho_*/dz) = (d \ln \varrho_0/dz) + (g/c_0^2)$; $P = (\varrho_*/\varrho_0)(p - p_0)$ is the normalized nonhydrostatic part of the pressure, $\zeta = \bar{\zeta} - \zeta_0$, where ζ_0 is found from the Väisälä-Brunt frequency N using the relation $d\zeta_0/dz = N^2/g$, and ε_ζ and ε_q are the normalized right-hand sides of equations (22.8) and (22.9).

The second moments $\overline{a'b'}$ will be denoted by the symbols B_{ab}. From equations (22.21) it is easy to obtain the following equations for the moments of the velocity field B_{ij}:

$$\begin{aligned}
&\frac{\partial B_{ij}}{\partial t} + \bar{v}_\alpha \nabla^\alpha B_{ij} + (B_{i\alpha}\nabla^\alpha \bar{v}_j + B_{j\alpha}\nabla^\alpha \bar{v}_i) + \\
&\quad + \left(\overline{\psi_i' v_j'} + \overline{\psi_j' v_i'}\right) - g(B_{i\zeta} e_j + B_{i\zeta} e_i) \\
&= -\frac{1}{\varrho_*}\left(\overline{v_i'\nabla_j P'} + \overline{v_j'\nabla_i P'}\right) + \frac{1}{\varrho_0}\left(\overline{v_i'\nabla^\alpha \tau_{i\alpha}'} + \overline{v_i'\nabla^\alpha \tau_{i\alpha}'}\right) \\
&\approx -c_1\frac{B}{L}\left(B_{ij} - \frac{1}{3}B^2 g_{ij}\right) - c_2\frac{B^3}{L}\left(e_i e_j - \frac{1}{3}g_{ij}\right) - \\
&\quad - 2\frac{B^3}{L}\left[(c_3 - 3c_4)e_i e_j + c_4 g_{ij}\right],
\end{aligned} \tag{22.24}$$

where the middle part contains the single-point second moments relating the velocity fluctuations to the fluctuations of the stress gradients. On the right-hand side of (22.24) these moments are expressed in terms of B_{ij} by semi-empirical formulas applied by Monin (1965) in the theory of the turbulent boundary layer of the atmosphere (he also determined the numerical constants c_1, c_2, c_3, c_4). Here $B^2 = B_{zz} + B_{\theta\theta} + B_{\lambda\lambda}$ is the mean-square fluctuating velocity, L is the vertical scale of the velocity fluctuations, and g_{ij} is the metric tensor. The first term on the right-hand side describes the isotropizing effect of the fluctuations in pressure; the second term, on the other hand, reflects the tendency toward anisotropy at the underlying surface; the third term describes the anisotropic viscous dissipation of the fluctuation energy. Moreover, equations for moments $B_{1\zeta}$ can be obtained from (22.21) and (22.22) in the form

$$\begin{aligned} &\frac{\partial B_{i\zeta}}{\partial t} + \bar{v}_\alpha \nabla^\alpha B_{i\zeta} + B_{\alpha\zeta}\nabla^\alpha \bar{v}_i + B_{i\alpha}\nabla^\alpha \zeta_1 + \overline{\psi'\zeta'} - gB_{\zeta\zeta}e_i \\ &= -\frac{1}{\varrho_*}\overline{\zeta' \nabla_i P'} + \frac{1}{\varrho_0}\overline{\zeta' \nabla^\alpha \tau'_{i\alpha}} + \\ &+ \overline{v'_i \varepsilon'_\zeta} \approx -c_5 \frac{B}{L} B_{i\zeta} - c_6 \frac{B^2}{L} B^{1/2}_{\zeta\zeta} e_i . \end{aligned} \tag{22.25}$$

The equation for the entropy dispersion $B_{\zeta\zeta}$ is obtained from (22.22) in the form

$$\frac{\partial B_{\zeta\zeta}}{\partial t} + \bar{v}_\alpha \nabla^\alpha B_{\zeta\zeta} + 2B_{\alpha\zeta}\nabla^\alpha \zeta_1 = \overline{2\zeta' \varepsilon'_\zeta} \approx -2c_7 \frac{B}{L} B_{\zeta\zeta}, \tag{22.26}$$

where c_5, c_6, c_7 are new numerical constants. Equations (22.25′) and (22.26′) for moments B_{iq} and B_{qq} can be derived analogously to (22.25) and (22.26). Once the vertical scale of the fluctuations L has been specified, equations (22.5) and (22.6), (22.8) and (22.9), (22.24–22.26), (22.25′) and (22.26′) form a closed system of equations in unknowns ψ, α, $\bar{\zeta}$, $\bar{q}$, B_{ij}, $B_{i\zeta}$, $B_{\zeta\zeta}$, B_{iq}, and B_{qq}, which also constitutes a model of the zonal circulation of the atmosphere. This approach was applied by Monin as early as 1958 to describe the fluctuations of the zonal circulation in the simplest case of a barotropic atmosphere, idealized as an incompressible two-dimensional spherical shell without small-scale dissipation. In this case equation (22.6) and equations (22.24) for $B_{\theta\lambda}$ and $E = \frac{1}{2}(B_{\theta\theta} + B_{\lambda\lambda})$ become

$$\begin{aligned} &\frac{\partial \alpha}{\partial t} = -\frac{1}{a^2 \sin^3\theta}\frac{\partial B_{\theta\lambda}\sin^2\theta}{\partial\theta}; \\ &\frac{\partial B_{\theta\lambda}}{\partial t} = -E\sin\theta\frac{\partial\alpha}{\partial\theta}; \qquad \frac{\partial E}{\partial t} = -B_{\theta\lambda}\sin\theta\frac{\partial\alpha}{\partial\theta}, \end{aligned} \tag{22.27}$$

where we can assume approximately that $B_{\theta\theta} \approx B_{\lambda\lambda}$ and also that in the geostrophic approximation

$$\overline{\frac{v'_\theta}{a\sin\theta}\frac{\partial P'}{\partial\lambda}} + \overline{\frac{v'_\lambda}{a}\frac{\partial P'}{\partial\theta}}$$

$$\approx \frac{1}{2\Omega \cos \theta}\left[\overline{\left(\frac{1}{a}\frac{\partial P'}{\partial \theta}\right)^2} - \overline{\left(\frac{1}{a \sin \theta}\frac{\partial p'}{\partial \lambda}\right)^2}\right] \approx 0;$$

$$\overline{\frac{v'_\theta}{a}\frac{\partial P'}{\partial \theta}} + \overline{\frac{v'_\lambda}{a \sin \theta}\frac{\partial P'}{\partial \lambda}} \approx 0.$$

According to equations (22.27), the nonnegative quantity $U^2(\theta) = E^2 - B_{\theta\lambda}^2$ does not vary with time. If, on the basis of the first equation (22.27), we now set

$$\alpha = -\frac{\Omega}{\sin^3 \theta}\frac{\partial \psi}{\partial \theta}; \qquad B_{\theta\lambda} = -\frac{\Omega a^2}{\sin^2 \theta}\frac{\partial \Psi}{\partial t}, \tag{22.28}$$

then the dimensionless function Ψ will satisfy the nonlinear hyperbolic equation

$$\frac{\partial^2 \Psi}{\partial t^2} = \left[\frac{U^2(\theta)}{a^4} + \left(\frac{\Omega}{\sin^2 \theta}\frac{\partial \Psi}{\partial t}\right)^2\right]^{1/2} \sin^3 \theta \frac{\partial}{\partial \theta}\frac{1}{\sin^3 \theta}\frac{\partial \Psi}{\partial \theta}, \tag{22.29}$$

describing the nonlinear propagation of waves along the meridian with a velocity $[E(\theta, t)]^{1/2}$. This constitutes the mechanism of the barotropic oscillations of the zonal circulation of the atmosphere. Equations (22.24)–(22.26) were applied more comprehensively by Monin and Simuni (1980) to calculate the steady-state zonal circulation in the convective layer of the Sun, with a view toward explaining its differential rotation.

To conclude this section, let us just mention some models which are intermediate between zonal and three-dimensional models. In these the zonal circulation and the nonzonal perturbations are described individually, but all just with a very coarse spatial resolution (only a small number of levels along the vertical are used and only a small number of spectral components of the hydrodynamic fields are taken into account).

In contrast to the integral-parameter models of Chapter 7, these models could be called small-component models. One of the most detailed such models was constructed by Galin and Kirichkov (1979). In it the field of the stream function ψ of the horizontal motions is described by 14 components: this model is a two-level one, and at each level the zonal part of ψ is represented at a linear combination of three Legendre polynomials $\psi_1(t) P_1(\cos \theta) + \psi_3(t)P_3(\cos \theta) + \psi_5(t)P_5(\cos \theta)$, while the nonzonal part is expressed as a single sinusoidal wave with a longitudinal wave number m:

$$\psi' = [A_1(t)P^m_{m+1}(\cos \theta) + A_2(t)P^m_{m+3}(\cos \theta)] \sin m\lambda + $$
$$+ [B_1(t)P^m_{m+1}(\cos \theta) + B_2(t)P^m_{m+3}(\cos \theta)] \cos m\lambda. \tag{22.30}$$

For the 14 coefficients (ψ_1, ψ_3, ψ_5, A_1, A_2, B_1, B_2, at the two levels), from the equations of hydrodynamics with an influx of heat according to Newton's law we derive a system of ordinary nonlinear differential equations. Depending on the value assigned to $\delta\bar{T}_r$, the difference between the temperatures of radiation equilibrium at the equator and the pole, the following solutions of this system are possible: (1) a stable steady-state zonal solution (Hadley circulation); (2) a solution in the form of a traveling wave of constant amplitude; (3) a solution in the form of a traveling wave with a periodically varying amplitude. According to calculations, for a wave number $m = 6$ corresponding to maximum baroclinic instability of the Hadley circulation, this wave has a fundamental period of 10–15 days, forming the so-called "index cycle", but in some narrow interval of $\delta\bar{T}_r$ values a period of 4–5 months is also superimposed onto it. Finally, we note that the solution in the form of a single traveling wave of constant amplitude, despite all its over-simplification, reproduces quite satisfactorily the observed zonal temperature fields, the zonal velocity, and the tricellular meridional circulation, as well as the atmospheric energy cycle.

CHAPTER 10

THREE-DIMENSIONAL MODELS

23. The Models of the Geophysical Fluid Dynamics Laboratory

After the first numerical experiment on the general circulation of the atmosphere by Phillips (1956), numerical models of the atmosphere rapidly achieved great popularity. At present numerous research and operational meteorological centers in many countries have at their disposal such models of their own construction. Especially important with regard to the development of this method were the models worked out at the Geophysical Fluid Dynamics Laboratory (Princeton, U.S.A.) under the direction of J. Smagorinsky. These included a nine-level model of a dry atmosphere by Smagorinsky, Manabe, and Holloway (1965) and a model of the general circulation with a hydrologic cycle by Manabe, Smagorinsky, and Strickler (1965).

In these models turbulent mixing was described with the aid of exchange coefficients, assumed to be proportional to the velocity gradients as per formulas of the Prandtl type. The coefficient of horizontal exchange was defined as $K_h = 2k_0^2(\Delta/2m)^2\,|D|$, where k_0 is a number of the order of unity analogous to the Kármán constant, Δ is the interval of the horizontal grid, m is a scale factor for the particular stereographic-projection map used, and D is the total deformation rate of the horizontal motion. The coefficient of vertical exchange is defined as $K_z = l^2\,|\partial \mathbf{v}_h/\partial z|$, where l is assumed proportional to $z + z_0$ for $z \leqslant h = 75$ m, with $z_0 = 1$ cm, proportional to $H - z$ for $h \leqslant z \leqslant H = 2.5$ km, and equal to zero for $z > H$. The vertical coordinate was taken to be the normalized pressure $\sigma = \varphi/\mathcal{P}$, with $\dot{\sigma} = 0$ at $\sigma = 0$ and $\sigma = 1$; at the lower limit of the atmosphere a simplified equation of the mean diurnal heat budget without a heat flux into the soil was selected.

In the "dry" model "convective adjustment", which was then a popular concept, was introduced: when $[\partial T/\partial z]$ exceeded the moist-adiabatic gradient $[\partial T/\partial z]_a$, then $\partial T/\partial z$ was replaced by $(\partial T/\partial z)_a$, on condition that the total potential energy remained unchanged. The flux of long-wave radiation was taken to be a function of temperature and of the climatological vertical distributions of water vapor, carbon dioxide, ozone, and clouds, which in turn depend on the latitude (like the value specified for the albedo of the Earth's surface). The flux of short-wave radiation was calculated for the mean diurnal cosine of the Sun's zenith angle. The number of nodes between the equator and the pole in the rectangular network of points used on the stereographic map was $N = 20$. Levels along the vertical were taken for values of $\sigma = 0.009, 0.074, 0.189, 0.336, 0.500, 0.664, 0.811, 0.926$, and 0.991. At time $t = 0$ the atmosphere was assumed to be at rest and isothermal with $T = 289$

K. Integration of the equations was performed with a time interval of 10 min for a period of 300 days.

The results of the integration over a 70-day period (from the 231st day to the 300th day) were processed statistically and compared with actual data. Calculations were made of the mean hemispheric values of the kinetic and potential energies and their rates of generation, transformation, and dissipation, as well as of the angular momentum and its rate of variation. In addition, the mean values over the hemisphere of the following were calculated: the zonal temperature and velocity, the meridional circulation, the distributions of kinetic and potential energy, the longitudinal spectra of the horizontal velocity components, the zonal distributions of the heat-budget components, the momentum, and the angular momentum, and all the elements of the energy cycle (the zonal and nonzonal kinetic and potential energies and the rates of their conversion, generation, and dissipation); empirical data on the energy cycle were given before in Fig. 11.1. This calculation program forms a methodological basis for evaluating the qualities of numerical models describing the general circulation of a dry atmosphere.

In the "wet" model the radiation fluxes were, as previously, determined from the climatological distributions of absorbing substances, rather than from the calculated distributions. However, the humidity field, the thermal effects of its transformations, and the precipitation had to be calculated. The Earth's surface was assumed to be absolutely moist and without any heat capacity. The nonconvective condensation of water vapor was assumed to take place when, and only when, the relative humidity q_{rel} reached a value of unity; moreover, it was assumed that q_{rel} never exceeds unity and that all the condensed vapor instantly falls as precipitation. "Convective adjustment" was carried out (for $q_{rel} \geqslant 1$ with condensation). The condensation δr (in terms of the mixing ratio r), the temperature correction δT for convective adjustment, and the precipitation rate P were found using the equations:

$$\frac{\partial}{\partial p}\,\Theta_e(T+\delta T, r+\delta r, p) = 0; \qquad r+\delta r = r_m(T+\delta T, p);$$

$$\frac{1}{g}\int_{p_l}^{p_u}(c_p\delta T + \mathscr{L}\delta r)\,\mathrm{dp} = 0; \tag{23.1}$$

$$P = -\frac{1}{g}\int_0^{\mathcal{P}}\delta r\,\mathrm{d}p, \tag{23.2}$$

where θ_e is the equivalent potential temperature, r_m is the mixing ratio at saturation, and p_u and p_l are the pressures at the upper and lower boundaries of the unstable layer. The wet model reproduced the atmospheric circulation somewhat better than the dry model did, and, what is the main thing, it described fairly well a number of characteristics of the hydrologic cycle in the atmosphere. For example, the mean annual precipitation for the Earth according to the model was 106 cm/yr, which agrees well with the empirical estimate of 104 cm/yr.

Later some even more detailed numerical models of the general circulation of the atmosphere were constructed at the Geophysical Fluid Dynamics Laboratory. These included an 11-level model with a seasonal variation devised by Manabe and Holloway (1975), and the first numerical model of the entire climate system (atmosphere-ocean-land system), devised by Manabe and Bryan (1969). The latter model included both an interacting atmosphere and ocean and the elements of the land hydrology. The atmospheric block in this model was basically the same as in the above-described wet model. Hydrological processes are described with the aid of the simplified empirical relations mentioned at the end of §16. For instance, the evaporation from the ocean, from a highly moistened land surface (including dense vegetation), and from ice or snow is assumed to be proportional to $r_m - r(h)$, as indicated by the third formula of (9.12). For a soil humidity W less than the critical value $W_{cr} = \frac{3}{4}W_s$ (where W_s is the moisture capacity of the soil), the evaporation will be less by a factor of W_{cr}/W. For $W = W_s$ and a seepage rate $V_z = P - E > 0$, it is assumed that $\partial W/\partial t = 0$; for $W < W_s$ the following equation is used:

$$\partial W/\partial t = P - E. \tag{23.3}$$

The dynamics of a snow cover is described by equation (17.1), in which E_{wi} is found from the heat-budget equation (9.10), for $T_s = T_{wi}$ and $H_s = 0$. The moisture content of snow is assumed to be zero, so that the soil humidity under the snow cover is determined from the relations $\partial W/\partial t = 0$ and $V_z = P - E_{wi}$ for $W = W_s$ and $\partial W/\partial t = P - E_{wi}$ for $W < W_s$.

Calculations for such an atmospheric block [cf. block diagram in Fig 24.1] were carried out on an arbitrary spherical region (in Mercator projection) 120° longitude in extent (with cyclic repetition), in which there are equal amounts of land and sea between latitudes ±66.5°, but only land at latitudes higher than these (up to ±81.7°). The initial state of the atmosphere was assumed to be a quasi-equilibrium state, obtained from the wet model, and the results of integrating the equations over a 100-day period (243rd day to 343rd day) were statistically processed. This model gave a faithful reproduction of the qualitative characteristics of the distribution of the components of the heat and water budget at the Earth's surface, including the distribution of precipitation with an equatorial rain belt and subtropical deserts, bordered by a belt with low precipitation that terminates in a relatively rainy region along the east coast of the continent. At the same time, the absence of seasonal variations and of a meridional transfer of heat by ocean currents led to the development at high latitudes of a surplus snow cover and extremely low temperatures.

The oceanic block in the model of Manabe and Bryan comprised the following: (1) quasi-static equations of hydrodynamics in the Boussinesq approximation, with a vertical eddy viscosity $K_z = 1.5$ cm²/s and a horizontal eddy viscosity $A_h = 10^8$ cm²/s (let us note at the outset that this value for A_h seems somewhat high); (2) a realistic equation of state $\varrho(T, p, S)$, which differs in form slightly from (12.3); (3) as simplified thermodynamic equations (12.4) and (12.5), equations of convective diffusion for T and S with diffusion coefficients $K_z = 1.5$ cm²/s and $K_h = 2.5 \times 10^7$ cm²/s, using the convective adjustment procedure for T and S; (4) zero boundary conditions for V_h, $\partial T/\partial n$, and $\partial S/\partial n$ at the ocean shores and for v_z, $\partial T/\partial z$, and $\partial S/\partial z$ at the ocean bottom $z = -4$ km, $v_z = 0$ and the value of $\partial V_h/\partial z$ determined from the coupling [synchronization] with the atmosphere at the ocean surface; (5) an equation for the thickness h_i of the pack ice:

$$\frac{\partial h_i}{\partial t} + \operatorname{div}(\gamma \bar{V}_h h_i) = A_h \Delta h_i + P_s - E - (Q_1 + Q_2)\varrho_i^{-1}\mathscr{L}_{wi}^{-1} \quad (23.4)$$

where $\gamma = 1$ for $h_i < 3$ m and $\gamma = 0$ for $h_i > 3$ m; $\bar{V}_h$ is the mean velocity of the current in the upper 67-meter layer of the ocean; and Q_1 and Q_2 are the heat fluxes into the ice at its upper and lower surfaces. In the first stage of the calculations the frictional stress of the wind, E, P, and the temperature of the ocean surface were specified from the data of the atmospheric block as an average over the last 100 days of its integration, for an inert ocean without horizontal transport of heat, and it was assumed that

$$Q_1 = \frac{k(T_1 - 271.2)}{h_i + k/\varepsilon}; \qquad Q_2 = \frac{k(T_2 - 271.2)}{z_0 - z_1}, \quad (23.5)$$

where T_1 is the temperature of the upper surface of the ice, $k = 5 \times 10^{-3}$ cal/(cm s) $= 20.95 \times 10^{-3}$ W/(cm s), and $\varepsilon = 2.93 \times 10^{-3}$ cal/(cm² s) $=$ 122.77 W/m². In the second stage of the calculations, for an ice-free surface it was assumed that $K_z \partial T/\partial z = Q_1$, where Q_1 was found from the equation of the heat budget, with T_1 taken to be the mean temperature of the upper mixed layer; in the presence of ice the procedure was more complicated, in order to satisfy the conditions $T_1 \leqslant 273.1$ K and $T_2 = 271.2$ K (the freezing points of fresh water and seawater). For $\partial S/\partial z$ at $z = 0$ in regions with ice the value $-S[P_w + (Q_1 + Q_2)\varrho_i^{-1}\mathscr{L}_{wi}^{-1}]$, was taken, and in regions without ice the value $-S(P_w + P_s - E)$ was assumed.

In the model of the atmosphere-ocean-land system the equations of the atmospheric block were integrated according to a nine-level scheme with time intervals of 10 min over a period of a year; the equations of the oceanic block were integrated according to a five-level scheme with 166.7 min intervals over a period of 100 years. An artificial "synchronization" was carried out, that is, a coupling of the state of the ocean in each interval with the time-smoothed state of the atmosphere in each 18th interval (smoothing over the time was done with a procedure equivalent to an exponential filter with a weighting function equal to $\lambda^{-1} e^{t/\lambda}$ for $t \leqslant 0$ and equal to 0 for $t > 0$ for a time constant λ in one week). Such a synchronization was necessary because of the

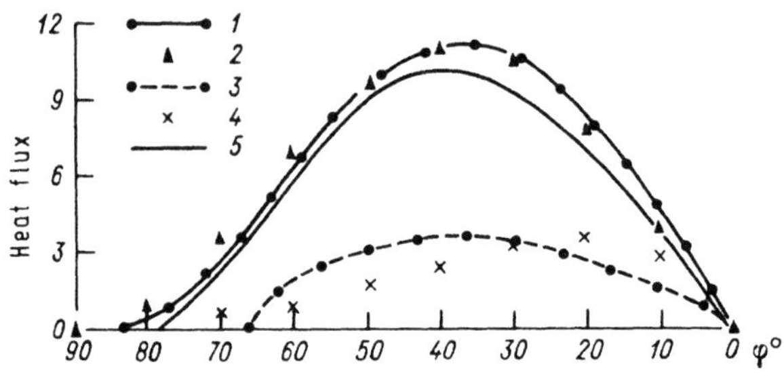

Fig. 23.1. Meridional heat flux in 10^{19} cal/day (or 4.19×10^{13} MJ/day) at different latitudes: (1) according to model of atmosphere–ocean–land system; (2) actual (estimated by Houghton, 1954); (3) according to model in ocean; (4) actual oceanic (estimated by Budyko, 1963); (5) according to isolated atmospheric block of model.

very great inertia of the ocean, the initial state of which was obtained as a consequence of the above-mentioned first stage of the calculations with a constant temperature of the ocean surface.

The results of integrating the model equations over the last 100 atmosphere days were statistically processed. The calculations showed that the state of the atmosphere obtained taking the oceanic circulation into account was somewhat closer to reality than the result obtained with an isolated atmospheric block. The meridional heat transport by ocean currents (which comprised about half of the atmospheric transport, see Fig. 23.1) reduced the meridional temperature difference, the intensity of the zonal circulation, the vertical gradient of the wind velocity in the troposphere, the baroclinic instability of the atmosphere, the kinetic-energy level of the large-scale turbulence in the middle latitudes, the intensity of the backward cell of the meridional circulation existing there, the intensities of the subtropical and polar baric maxima and minimum in the middle latitudes, the exchange of angular momentum between the atmosphere and the solid Earth, and the meridional flux of momentum.

Heating of cold continental air by the ocean tended to enhance the cyclogenesis at the east coast of a continent in the high latitudes. Cool equatorial upwelling in the ocean tended to lessen the tropical precipitation over it and to enhance that over the continent. In the subtropical and middle latitudes the transport of warm water by the subtropical gyre increased the fluxes of sensible and latent heat from ocean to atmosphere as well as the amount of precipitation along the east coast of the continent (which confined the subtropical desert to the western half of the continent). In the northern latitudes the transport of warm water by the subarctic gyre had a similar effect along the west coast of the continent. The rain belts of the middle latitudes and the boundaries of the snow were shifted poleward somewhat.

This numerical experiment was, however, unsuccessful, in that equilibrium of the thermal state of the ocean was not achieved: at the end of the experiment heat was still being supplied to the ocean from the atmosphere at a rate of 1.5×10^{-2} cal (cm^2 min), or 10 W/m^2, that is, a little less than 1% of the solar constant, which corresponds to a mean rate of heating of the ocean water of about 2°C per century. At the same time, the fact that, unlike the previous models, the values of T and S at the ocean surface did not have to be specified constituted an important step forward; let us note, in particular, the very good fit of the salinity field in the ocean with the halocline in the arctic zone, with the surface maximum of S in the subtropics, and with the weak minimum of S at a depth of about 1 km. Since the meridional transport of heat by ocean currents associated with the vertical heat exchange in the ocean proved to be close to the actual transport, it appears that the value of $K_z = 1.5$ cm^2/s chosen was a good one.

On the whole, the model of the atmosphere–ocean–land system of Manabe and Bryan can be said to be very promising, with great possibilities for the improvement of many of its features: synchronization of the ocean and atmosphere, establishment in the ocean of a quasi-steady thermal regime, geometry of the continents and oceans, description of the processes of land hydrology, choice of the values of A_h and K_h, choice of the distributions of radiation-absorbing substances, and allowing for seasonal fluctuations.

A number of improvements were introduced in the following version of the model of the mean annual state of the atmosphere–ocean–land system (Manabe, Bryan, and Spelman, 1975; Bryan, Manabe, and Pacanowski, 1975). This model was calculated in the real geometry of the continents and oceans on a sphere with a 500 km interval of the horizontal grid, 9 levels in the atmosphere and 12 in the ocean, with synchronization of 1.3 atmospheric years with 430 oceanic years, with calculation of the radiative heat fluxes for a constant CO_2 concentration and specified climatic zonal distributions of ozone and clouds, but for a water-vapor distribution to be calculated in the model. Later Manabe, Bryan, and Spelman (1979) introduced into this model the seasonal fluctuations (via the insolation and the ozone distribution and taking into account the seasonal variations of the specified latitudinal distributions of the albedos of the continents and oceans, due to the appearance and melting of snow and ice).

The model with seasonal fluctuations was integrated over 4.2 atmospheric years (with a horizontal grid interval in the atmosphere of 500 km during the first 3 years and 250 km during the last 1.2 years), synchronized with 1200 oceanic years (with a horizontal grid interval in the ocean of 500 km). The relaxation time in the oceanic block was shortened by means of an artificial enhancement of the temperature and salinity reactions of the deep layers of the ocean (a fivefold enhancement at the greatest depths). Using a simpler model, it was verified that this measure does not distort the limiting, statistically equilibrium, state of the ocean. The initial state of the atmosphere–ocean–land system was selected on the basis of a long-term integration of the model with a mean annual insolation (starting from a dry isothermal atmosphere with a temperature of 6.8°C and a homogeneous ocean with a temperature of 2.5°C and a salinity of 34.8‰). The boundary conditions for the atmospheric block at the ocean surface were specified on the basis of the data on the temperature of the mixed layer and the thickness of the sea ice, the mean monthly values of which were stored in the atmospheric bank. The boundary conditions for the oceanic block were found from the mean monthly values of the momentum and moisture fluxes, stored in the atmospheric bank, together with the temperature, the water-vapor mixing ratio, the wind velocity, and the radiative heat flux at the lower level in the atmosphere (necessary to calculate the total heat flux; its direct specification led to calculational instability).

During the 1200 years of the integration the temperature in the upper layers of the ocean attained a statistically equilibrium state, but in the deep layers it continued to increase (albeit slowly, about 0.1°C per century, which corresponds to a heat flux from the atmosphere of less than 0.25% of the solar constant). The salinity, especially in the upper layers of the ocean, approached the statistically equilibrium values even more gradually (since there is no direct feedback between it and the hydrologic cycle, which determines the values of $P - E$). In the upper layers the values obtained for the salinity were too low (less than 34‰).

The model reproduces satisfactorily a number of the features of the seasonal fluctuations of the state of the atmosphere–ocean–land system (see example in Fig. 23.2). It improves certain characteristics of the reproduced climate, in comparison with the model with the mean annual insolation; in particular, it gives a realistic thickness of the pack ice in the Arctic (a little less than 3 m) and it eliminates the continual accumulation of snow in the high latitudes on the continents of the Northern Hemisphere. At the same time, for instance, the meridional heat fluxes in the model ocean, as well as the amplitudes of the seasonal fluctuations of the ocean-surface temperature T_w in the middle latitudes of the Northern Hemisphere, were found to be in general somewhat too low. The values obtained for T_w around Antarctica, on the other hand, were about 6 to 10°C too high, so that further refinements of the model would be desirable.

Models of the atmosphere–ocean–land system similar to the above models of the Geophysical Fluid Dynamics Laboratory have been constructed at the Rand Corporation by Alexander and Schlesinger (1975) and at the National Center for Atmospheric Research (Boulder, Colorado) by Kasahara and Washington (1971) (six-level model of the general circulation of the atmosphere) and by Washington, Semtner, *et al.* (1980) (model of seasonal variations of atmosphere–ocean–land system). In the latter model, by the way, the method used to synchronize the atmosphere and the ocean was different from that of Manabe and Bryan (and, in the opinion of the authors, ten times as economical). Washington and his colleagues integrated the atmospheric block (according to an eight-level scheme) for January, April, July, and October.

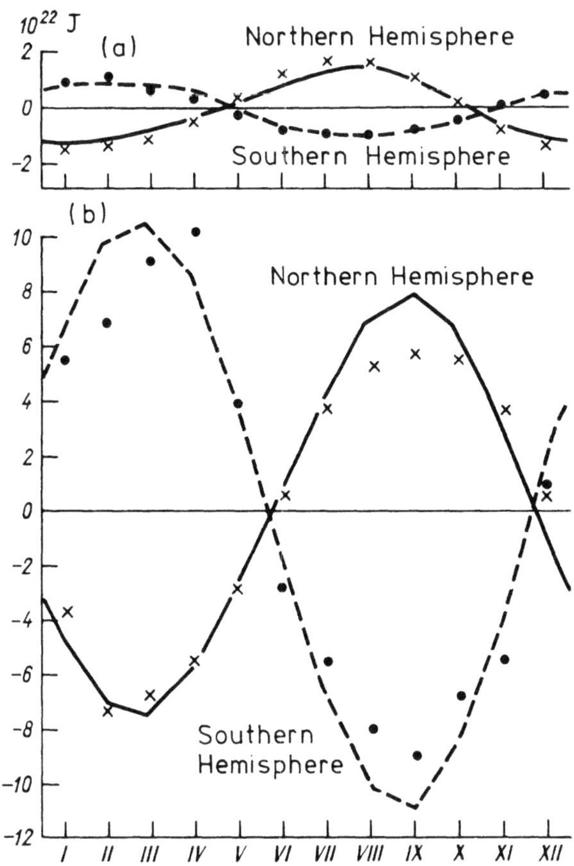

Fig. 23.2. Seasonal fluctuations of heat content, according to model (dashed curves) and according to actual data of Levitus *et al.* (1978) (crosses): (a) atmosphere (internal and potential energy) plus heat below 50 mb level; (b) ocean.

Then they determined the annual and semiannual harmonics of the effect of the atmosphere on the ocean and used these to integrate (according to a four-level scheme) the oceanic block over a period of 5 years. This was followed by a new integration of the atmospheric block. The cycle of computations was repeated over and over until the entire atmosphere–ocean–land system reached a state of statistical equilibrium.

24. The Model of the Oceanology Institute of the U.S.S.R. Academy of Sciences

The next model of the atmosphere–ocean–land system (after those of Manabe and Bryan, 1969, 1975) was apparently the one constructed in 1974–1975 at the Oceanology Institute of the U.S.S.R. Academy of Sciences. The results of calculations according to this model were published from 1976 on. The main distinguishing feature of the model was that, instead of an artificial "synchronization" of the ocean and atmosphere, the ocean was divided into an upper active layer (of thickness $h_a = 350$ m), naturally (synoptically) synchronized with the atmosphere, and a deep part, the climatic state of which is to be calculated separately. This approach was intended to describe, first of all, comparatively short-period processes in the atmosphere–ocean–land system, including the seasonal fluctuations and the interannual variability, with regard to which the deep ocean can, as a first approximation, be assumed invariable.

A block diagram of the model of the Oceanology Institute is given in Fig. 24.1. Four blocks are shown: free atmosphere, atmospheric boundary layer, active layer of ocean, and deep ocean. Arrows leading from small circles with variables in them indicate the content and direction of the information exchange between blocks. In each block are shown the variables calculated using the prognostic equations as well as the processes described by the parametrization. Below, as an example, we will describe this model in somewhat more detail than the preceding models.

One of the important differences between the models of the Oceanology Institute (IOAS) and the Geophysical Fluid Dynamics Laboratory (GFDL) is a systematic use in the former of parametrization of the interaction between the atmosphere and the underlying surface, on the basis of similarity theory for the Ekman boundary layer. Here the atmospheric boundary layer was assumed to be submerged in the lower working layer of the atmosphere, which comprises 15% of its mass. For this parametrization formulas (9.12) – (9.14) and Table 9.1 were used. The heat-budget equation (9.10) for the oceans was used to determine the heat flux into the depths H_s, a quantity needed in order to integrate the equations of the active-ocean-layer block. On land or at the surface of sea ice it was assumed that $H_s = 0$. The surface temperature T_s was also found from (9.10). In regions of perennial ice for $T_s > 0$°C this temperature was replaced by zero, on the assumption that H_s is used to melt the ice; interestingly enough, in experiments where this was not done, the summer surface temperature T_s in the Antarctic could jump as high as 30°C! The air humidity at the ocean surface, and also when it was raining on the land, was assumed to be saturating, but 2 hr after a rain on land it was assumed to be only half of the saturation value. The albedo of ice and snow was taken to be 0.7, of a vegetation cover 0.2, and of a water surface 0.1.

In the atmospheric block of the IOAS model $\sigma = p/\mathcal{P}$ was used as the vertical coordinate, where $\mathcal{P}$, the surface pressure, is one of the fields to be calculated. The surface pressure was computed with the aid of the so-called *tendency equation*, obtained by integrating the continuity equation over the entire depth of the atmosphere:

$$\frac{\partial \mathcal{P}}{\partial t} + \frac{1}{a \sin \theta} \int_0^1 \left(\frac{\partial \mathcal{P} v_\theta \sin \theta}{\partial \theta} + \frac{\partial \mathcal{P} v_\lambda}{\partial \lambda} \right) d\sigma = 0. \qquad (24.1)$$

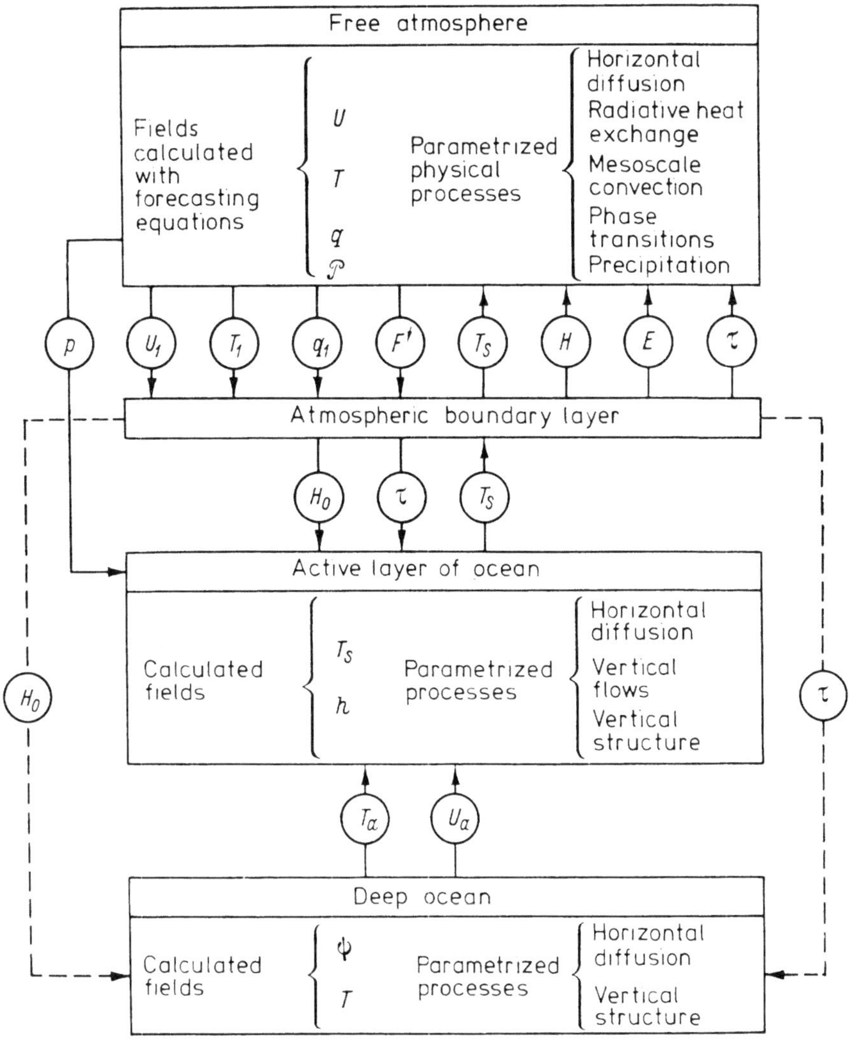

Fig. 24.1. Block diagram of combined circulation of atmosphere and ocean.

In the numerical experiments only a very rough vertical division of the atmosphere was carried out: into four layers with boundaries at heights of 1.5, 4.5, and 11 km. Then, in contrast to the GFDL model, the cloud amount was not specified climatically, but rather was determined at each time stage: the three indicated boundaries between layers were assigned cloud amounts n_1, n_2, n_3 according to the empirical formulas of Smagorinsky:

$$n_1 = -2 + \frac{10}{3}\frac{q_1}{q_{m1}} \quad ; \qquad n_2 = -0.7 + 2\,\frac{q_2}{q_{m2}} \quad ;$$

$$n_3 = -0.43 + 1.73 \frac{q_3}{q_{m3}} . \tag{24.2}$$

The total cloud amount will then be $n = 1 - (1 - n_1)(1 - n_2)(1 - n_3)$. The radiative heat fluxes were calculated using a comparatively simplified scheme, but allowing for the calculated cloud amount (24.2). The fluxes of short-wave radiation $F^{\downarrow}_{s,i}$ at the lower boundary of the ith atmospheric layer were found with the aid of the formula

$$\begin{aligned} F^{\downarrow}_{s,i} &= F^{\downarrow}_{s,i+1} [n_i P^c_i + (1 - n_i) P^b_i]; \\ F^{\downarrow}_{s,i+1} - F^{\downarrow}_{s,i} &= F^{\downarrow}_{s,i+1} [n_i \Pi^c_i + (1 - n_i) \Pi^b_i], \end{aligned} \tag{24.3}$$

where P^c_i, P^b_i and Π^c_i, Π^b_i are the coefficients of transmission and absorption in cloudy and cloudless cells of the horizontal grid (these are specially selected for each atmospheric layer of the model); $F_{s,1}$ is the flux of short-wave radiation incident on the Earth's surface; and $F^{\downarrow}_{s,5}$ is the flux at the upper limit of the atmosphere (that is, $F_{s,5}$ is the mean daily insolation; it was computed for a solar constant of 1381.6 W/m^2, taking into account the latitude of the place and the time of year, reckoned from the summer solstice in the Northern Hemisphere 21 June).

When calculating the fluxes of long-wave radiation $F^{\downarrow}_{L,i}$ and $F^{\uparrow}_{L,i}$, it was taken into account that the effective absorbing masses of water vapor $m_i = (P/g) \int_{\sigma_i}^{\sigma_{i+1}} q\sigma^{1/2}\, d\sigma$ (in g/cm^2; cf. formula (5.6)) in the three lower atmospheric layers are large, and the transmission function for water vapor was approximated using the formula

$$\begin{aligned} P = {} & \exp[0.15(m - m_i)](0.471 \exp[-0.626 m_i^{1/2}] + \\ & + 0.529 \exp[-8.94 m_i^{1/2}]) \end{aligned} \tag{24.4}$$

(cf. formula (5.9)). In the upper atmospheric layer, on the other hand, m_i is small, and the transmission coefficient for this layer was taken to be considerably lower than according to formula (24.4). Accordingly, the fluxes of long-wave radiation were computed with the aid of formulas of the following type:

$$\begin{aligned} F^{\uparrow}_{L,i+1,c} &= n_i B(T_i); \\ F^{\uparrow}_{L,i+1,b} &= (1 - n_i)[F^{\uparrow}_{L,i,c} P(m_i) + B(T_i) P(m_i) + \\ & \quad + F^{\uparrow}_{L,i,b} \exp(-0.15 (m_1 + \cdots + m_{i-1}))], \end{aligned} \tag{24.5}$$

where in the second formula the first term inside the square brackets describes the flux transmitted through the ith layer from the shielded part of its lower boundary, the second term describes the radiation of the layer itself, and the third term describes the radiation of lower-lying layers transmitted by it. The total heat influx due to long-wave radiation will then be equal to the difference

$$(F^{\uparrow}_{L,i+1,c} + F^{\uparrow}_{L,i+1,h}) - (F^{\uparrow}_{L,i,c} + F^{\uparrow}_{L,i,h})$$

plus a similar difference of the downward fluxes.

The heat influx due to phase changes of moisture are calculated for each pair of adjacent atmospheric layers. At values higher than the adiabatic temperature gradient γ (dry-adiabatic g/c_p for $\bar{q} < 0.95\bar{q}_m$ or wet-adiabatic

$$\frac{g}{c_p}\left(1 + \frac{\mathscr{L}\bar{q}}{R\bar{T}}\right)\left(1 + \frac{\mathscr{L}^2\bar{q}}{c_p R_v \bar{T}^2}\right)^{-1}$$

for $\bar{q} \geqslant \bar{q}_m$, where the bar indicates a weighted-mean value over the two layers), that is, for $T_i\sigma_i^{-R\gamma/g} > T_{i+1}\sigma_{i+1}^{R\gamma/g}$, the layers were assigned new values of temperature and moisture content, $\hat{T}$ and $\hat{q}$, found from the equations

$$\begin{aligned}
&\delta\sigma_i\sigma_i^{-R\gamma/g}(\hat{T}_i - T_i) + \delta\sigma_{i+1}\sigma_{i+1}^{-R\gamma/g}(\hat{T}_{i+1} - T_{i+1}) = 0;\\
&\delta\sigma_i(\hat{q}_i - q_i) + \delta\sigma_{i+1}(\hat{q}_{i+1} - q_{i+1}) = 0;\\
&\hat{T}_i\sigma_i^{-R\gamma/g} = \hat{T}_{i+1}\sigma_{i+1}^{-R\gamma/g};\\
&(\hat{T}_i - T_i)(T_i - T_{i+1})^{-1} = (\hat{q}_i - q_i)(q_i - q_{i+1})^{-1};
\end{aligned} \tag{24.6}$$

the first two equations reflect the conservation of total entropy and total moisture content, the third describes the establishment of adiabatic equilibrium, and the fourth represents the similarity of the increments in temperature and moisture content.

Next the drop in moisture content due to the fall of precipitation was calculated: $\delta q_i = 0$ for $\hat{q}_i < c_0\hat{q}_{mi}$ and $\delta\hat{q}_i = \hat{q}_i - c_0\hat{q}_{mi}$ for $\hat{q}_i \geqslant c_0\hat{q}_{mi}$, where $c_0 = 0.7$ for convective precipitation and $c_0 = 1$ for nonconvective precipitation, together with the corresponding temperature rise $\delta T_i = (\mathscr{L}/c_p)\delta q_i$ (this entire procedure should in principle be an iteration procedure). The heat influx $Q_c + Q_p$ and the moisture influx $P_c - P_p$ were then found by dividing δT_i and δq_i by the time of convective adjustment (assumed in the computations to be 2 hr). The precipitation P was obtained by summing the P_p values over all the layers. Finally, the coefficient of horizontal large-scale turbulent exchange was assumed to be

$$K_h = 0.1\varepsilon^{-1/3}l^{4/3}U/U_0, \tag{24.7}$$

where $\varepsilon = 5 \times 10^{-4}$ W/kg is the mean rate of dissipation of kinetic energy in the atmosphere, l is the interval of the horizontal grid, U is the total wind velocity, and $U_0 = 17$ m/s is its mean value.

In the block of the active layer of the ocean (see Fig. 24.1), because of the restricted possibilities of the computer, the salinity of the seawater was not taken into account in the numerical experiments (although it would be easy, in principle, to add the salinity field to the number of calculated fields). The effect of vertical motions was not taken into account either, while the mean (over the depth of the active layer) horizontal currents were determined as the sum of the climatic components, borrowed from the block of the deep ocean, plus the Ekman drift, which is easily calculated from the frictional stress of the wind, according to the rules presented in §14 in connection with formulas (14.5). The vertical profiles of the water temperature were assumed to possess the similarity described by formula (12.8), the temperature T_1 at the lower boundary of the active layer being selected from the block of the deep ocean. The temperature T_s in the upper mixed layer and the thickness h of this layer were determined using the equation of convective heat conduction and the equation of the turbulent-energy budget, integrated over the thickness of the active layer and essentially differing from (14.20), (14.21) only by the presence of the current. These equations have the form

$$\frac{\bar{d} T_s}{dt} = K\Delta_h T_s + \frac{H_s}{c_w \varrho_w h_1 F_1}\left(1 + \frac{c_1 F_1}{F_3}\right) - - \frac{\mathcal{E}_s}{g\alpha\varrho_w h_1^2} \frac{c_1 c_3}{F_3}; \tag{24.8}$$

$$\frac{\bar{d} h}{dt} = K\,\Delta_h h - \frac{H_s}{c_w \varrho_w (T_s - T_1)} \frac{F_2}{F_3} + \frac{\mathcal{E}_s}{g\varrho_w h_1} \frac{c_3 F_1}{F_3}, \tag{24.9}$$

where $K = 2.5 \times 10^7$ cm^2/s; $\mathcal{E}_s$ is the kinetic-energy flux from atmosphere to ocean; F_1, F_2, and F_3 are standard functions of $\hat{h} = h/h_1$, having the form

$$F_1 = 1 - c_1(1 - \hat{h}); \quad F_2 = (1 - \hat{h})F_1 - \frac{1}{2} + \frac{c_4}{2}\hat{h}^2 + c_2(1 - \hat{h})^2;$$

$$F_3 = (c_1 - 2c_2)(1 - \hat{h})F_1 - c_1 F_2,$$

with $c_1 \approx 0.73$; $c_2 \approx 0.29$; $c_3 \approx 0.13$; $c_4 \approx 0.9$. Moreover, it was assumed that, if the heat-budget equation (9.10) yields $T_s < -1.8$°C, then ice must be present on the surface of the ocean. In such a case $H_s = 0$, $\mathcal{E}_s = 0$, and the temperature of the mixed layer beneath the ice is -1.8°C (while the temperature T_s of the upper surface of the ice was found from (9.10) for $H_s = 0$).

The two-dimensional model of the general circulation of the World Ocean devised by Kagan *et al.* (1974, 1975) was used as the block of the deep ocean in the numerical experiments according to the IOAS model. In this model the dependence of ϱ_w only on the temperature was taken into account, and the water temperature was assumed to vary only in the upper layer of the ocean (of thickness $h_b = 2$ km). The temperature there was described by the formula $T = T_b + A(z)\Theta(\theta, \lambda, t)$, where T_b is the fixed abyssal temperature, $A(z)$ is a standard function of the depth (of type (12.8), but with a fixed thickness of the mixed layer), and Θ is determined from the equation of horizontal heat transfer. The latter was taken to be

$$\frac{\partial\Theta}{\partial t} + \frac{1}{a\sin\theta}\left[\frac{\partial}{\partial\theta}(v_{\theta_0} + v_\theta)\Theta\sin\theta + \frac{\beta_1\theta}{f}\frac{\partial\Theta}{\partial\lambda}\right] +$$

$$+ \frac{1}{a \sin\theta} \frac{\partial}{\partial\lambda} \left[(v_\lambda + v_{\lambda_0})\Theta - \frac{\beta_1\Theta}{f} \frac{\partial\Theta}{\partial\theta} \right]$$

$$= K_1 \, \Delta_h\Theta + \frac{H_s}{c_w\varrho_w\beta_2} , \tag{24.10}$$

where $K_1 = 10^8 \text{ cm}^2/\text{s}$;

$$\beta_1 = g\alpha\beta_3/2a\varrho_w; \qquad \beta_2 = \int_0^{h_w} A \, dz; \qquad \beta_3 = \beta_2 - \frac{1}{h_w} \int_0^{h_w} Azdz;$$

$h_w = h_w(\theta, \lambda)$ is the depth of the ocean; $v_\theta = -(1/h_w \sin\theta)(\partial\psi/\partial\lambda)$ and $v_\lambda = (1/h_w)(\partial\psi/\partial\theta)$ are the mean (over the depth of the ocean) barotropic components of the current velocity; and v_{θ_0} and v_{λ_0} are the mean (over the depth) components of the drift-current velocity. The latter are found from the frictional stress of the wind $(\tau_{\theta_0}, \tau_{\lambda_0})$ with the aid of the formulas

$$v_{\theta_0} = \frac{\tau_{\lambda_0}}{\varrho_w h_w f} \left[1 - \frac{h_w A\,(0)}{\beta_2} \right];$$

$$v_{\lambda_0} = -\frac{\tau_{\theta_0}}{\varrho_w h_w f} \left[1 - \frac{h_w A\,(0)}{\beta_2} \right]. \tag{24.11}$$

The following equation was used for the stream function ψ of a barotropic current:

$$\frac{\partial\Delta_2\psi}{\partial t} + \frac{1}{a} \frac{\partial v_\theta\Delta_2\psi}{\partial\theta} + \frac{1}{a\sin\theta} \frac{\partial v_\lambda\Delta_2\psi}{\partial\lambda} + \frac{\partial f v_\theta \sin\theta}{\partial\theta} + \frac{\partial f v_\lambda}{\partial\lambda}$$

$$= -\frac{\partial}{\partial\theta} \sin\theta \, \frac{\tau_{\lambda_0} - \tau_{\lambda_1}}{\varrho_w h_w} + \frac{\partial}{\partial\lambda} \frac{\tau_{\theta 0} - \tau_{\theta 1}}{\varrho_w h_w} +$$

$$+ \frac{g\alpha}{a\varrho_w} \left(\frac{\partial}{\partial\lambda} \beta_3 \frac{\partial\Theta}{\partial\theta} - \frac{\partial}{\partial\theta} \beta_3 \frac{\partial\Theta}{\partial\lambda} \right);$$

$$\Delta_2\psi = \frac{\partial}{\partial\theta} \frac{\sin\theta}{h_w} \frac{\partial\psi}{\partial\theta} + \frac{1}{\sin\theta} \frac{\partial}{\partial\lambda} \frac{1}{h_w} \frac{\partial\psi}{\partial\lambda} , \tag{24.12}$$

where $(\tau_{\theta 1}, \tau_{\lambda 1})$ is the near-bottom frictional stress. The boundary conditions in this case were the requirements that the mean (over the depth) fluxes of heat and momentum on the contours Γ_i of the ocean shores go to zero:

$$\left[\left(a v_{\lambda 0}\Theta - \frac{K_1}{\sin\theta} \frac{\partial\Theta}{\partial\lambda} - \frac{a\beta_1\Theta}{f} \frac{\partial\Theta}{\partial\theta} \right) \cos(n, \lambda) + \right.$$

$$\left. + \left(a v_{\theta 0}\Theta - K_1 \frac{\partial\Theta}{\partial\theta} + \frac{a\beta_1\Theta}{f\sin\theta} \frac{\partial\Theta}{\partial\lambda} \right) \sin(n, \lambda) \right]_{\Gamma_i} = 0; \tag{24.13}$$

$$[\psi]_{\Gamma_1} = 0; \int_{\Gamma_2} \left(\frac{\partial v_\lambda}{\partial t} + \frac{g\alpha\beta_3}{a\varrho_w \sin\theta} \frac{\partial \Theta}{\partial \lambda} - \frac{\tau_{\lambda 0} - \tau_{\lambda 1}}{\varrho_w h_w} \right) d\lambda = 0, \tag{24.14}$$

where Γ_2 is the contour of Antarctica, and Γ_1 represents the other shores, the ocean being approximated as a doubly connected region.

The equations of the deep-ocean block (see Fig. 24.1) were integrated on a latitude-longitude grid with a 5° interval, using the stationing method for zero initial conditions $\psi = \Theta = 0$. Here the values of $(\tau_{\lambda 0}, \tau_{\theta 0})$ and H_s were determined on the basis of present values of the mean annual fields of the atmospheric pressure, air temperature, and radiation budget at the ocean surface (for a Bowen ratio of 0.5). We recall that from the results of this calculation for the combined model of the atmosphere–ocean–land system, only three two-dimensional fields v_θ, v_λ, and T_1 were necessary.

In the numerical experiments for a combined atmosphere–ocean–land model the equations of the four-layer atmospheric model and the block of the active layer of the ocean were integrated on the spherical grid of Kurihara (1968) with a horizontal interval of about 1000 km (there were 360 points in each spherical layer, giving a total of 7560 degrees of freedom; experiments with such a low spatial resolution were, of course, only tentative in nature). The initial conditions were assumed to be: zonal distributions of the air temperature and mixed-layer temperature which were close to the mean annual values, an adiabatic vertical distribution of the air temperature, no wind, and uniform distributions of the sea-level atmospheric pressure (1013 mb), the relative humidity of the air (0.5), and the thickness of the upper mixed layer (50 m). The integration was carried out at time intervals of 20 min in the atmosphere and 2 hr in the active layer of the ocean, with a natural synchronization of these. The calculation was made for a time period of 1000 days, taking into account the annual variation of the insolation.

A statistically equilibrium quasi-periodic regime in the atmosphere–ocean–land system was reached in about a year. By this time the mean (over the mass of the troposphere) wind velocity was 17 m/s (but in the upper layer of the atmosphere the mean velocity reached 40 m/s); the temperature was 244 K (but T_s was 282 K, the minimum monthly means being T_s = 234 K in Antarctica and the maximum monthly means being T_s = 308 K in North Africa); the mean specific humidity was 1.6 g/kg; the overall cloud amount was 0.47; the radiative cooling of the atmosphere and the compensating phase-convective heat influx were about 0.7°C/day; evaporation and precipitation were about 3.1 mm/day; the total short-wave and long-wave radiative heat flux taken up by the Earth's surface was 470 W/m^2 (about 80% of this is carried off by the long-wave emission of the Earth's surface, the rest going to evaporation and turbulent heat exchange); and the thickness of the upper

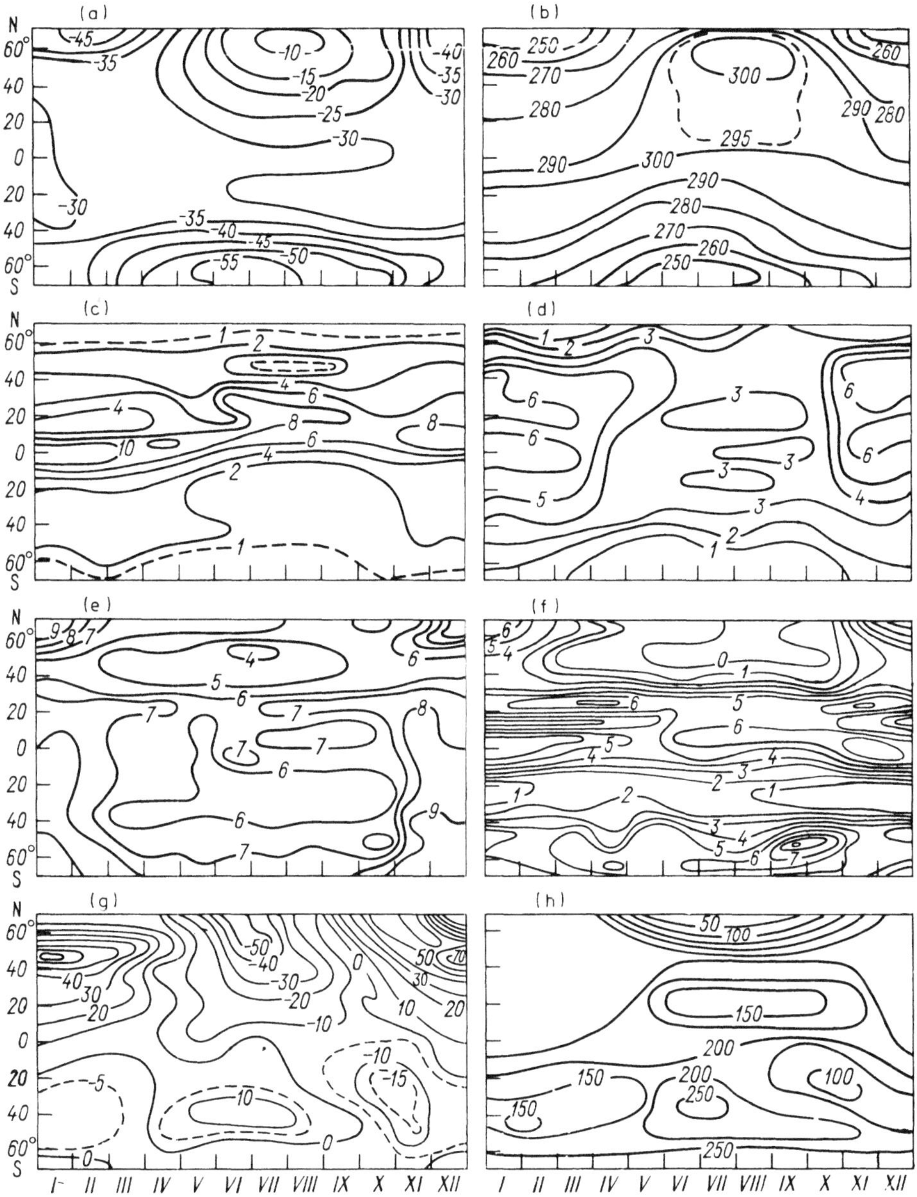

Fig. 24.2. Seasonal fluctuations (annual isopleths) of zonal characteristics of atmosphere–ocean–land system, according to IOAS model: (a) mean (over the mass) temperature of atmosphere (°C); (b) temperature of Earth's surface (Kelvins); (c) rate of precipitation (mm/day); (d) evaporation rate (mm/day); (e) total cloud amount over oceans; (f) total cloud amount over continents; (g) vertical mass flux at ocean surface; (h) thickness of mixed layer (m).

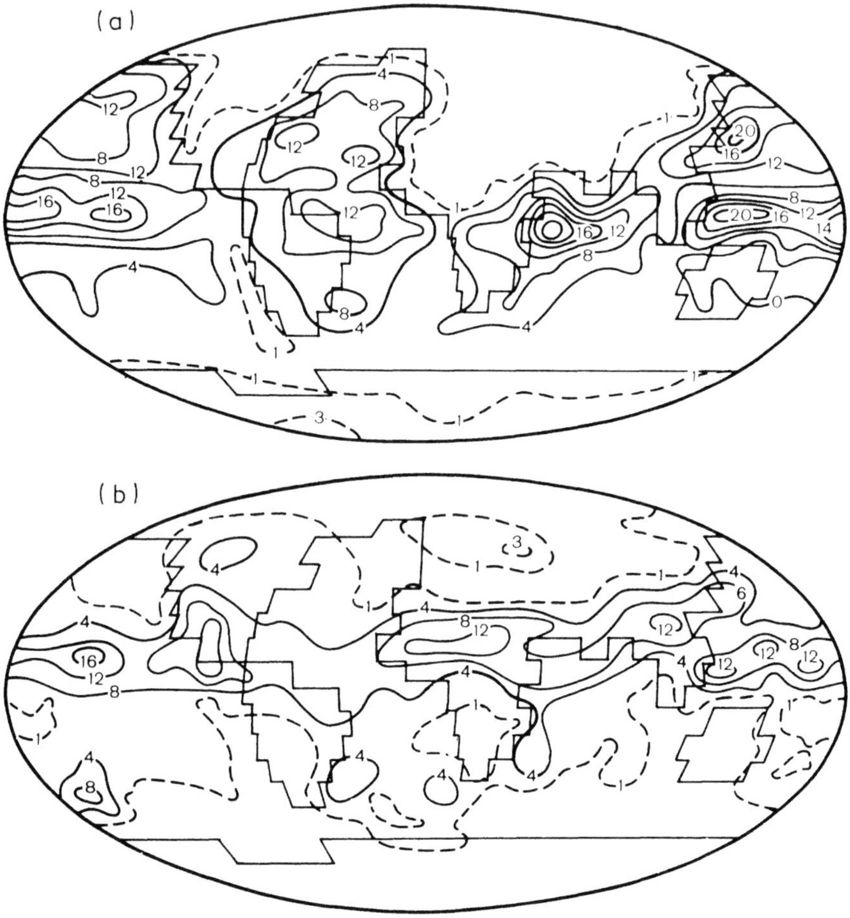

Fig. 24.3. Global precipitation distributions P (mm/day) in January (a) and July (b), according to IOAS model.

mixed layer was 82 m. Over the ocean the pressure was found to be lower than overland (994 mb as compared with 1040 mb), the air temperature was 5°C higher, and the specific humidity (2.1 g/kg as compared with 1.2 g/kg), cloud amount (0.8 as compared with 0.4), and evaporation and precipitation (6 mm/day as compared with 3.1 mm/day) were all about twice as great.

Let us now consider some nonstandard results of a numerical experiment with the IOAS model. Figure 24.2 shows annual isopleths of several zonal characteristics of the atmosphere–ocean–land system, illustrating their seasonal fluctuations. The isopleths of the mean (over the mass) atmospheric temperature (Fig. 24.2a) reveal that maximum amplitudes of the annual variation were reached in both hemispheres at latitudes of 60°. The asymmetry about the equator is due to the distribution of the continents (in particular, the maximum temperatures were obtained in summer over the

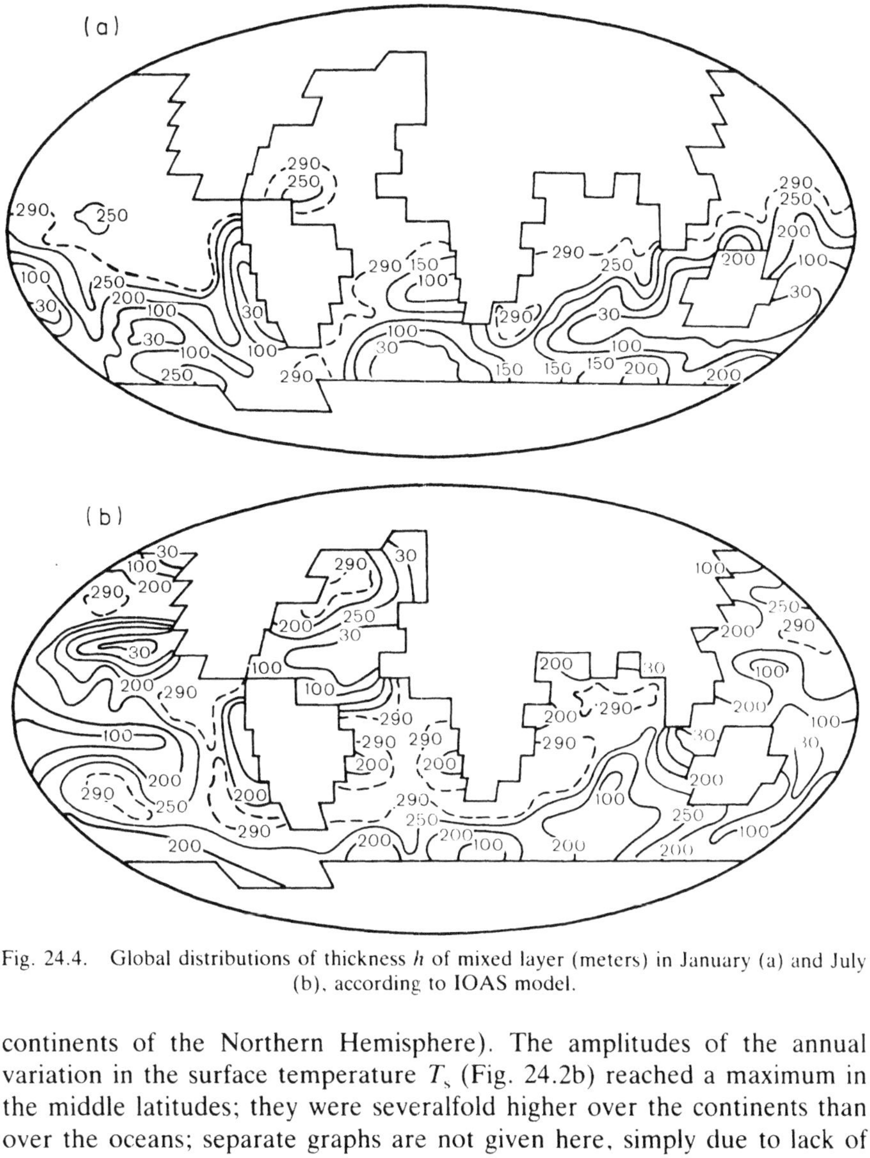

Fig. 24.4. Global distributions of thickness h of mixed layer (meters) in January (a) and July (b), according to IOAS model.

continents of the Northern Hemisphere). The amplitudes of the annual variation in the surface temperature T_s (Fig. 24.2b) reached a maximum in the middle latitudes; they were severalfold higher over the continents than over the oceans; separate graphs are not given here, simply due to lack of space.

The amount of precipitation P (Fig. 24.2c) proved to be highest in the equatorial zone (especially over the oceans of the Northern Hemisphere in winter, whereas over the continents P reached a maximum in the tropics in summer; in the middle latitudes over the continents, and in summer over the oceans, the precipitation was found to be comparatively low). These features of the precipitation distribution are also illustrated by the calculated global charts for January and July (Fig. 24.3). The evaporation E (Fig. 24.2d) was found to be highest over the oceans of the middle latitudes in winter, when the temperature difference $T_w - T_a$ is great (in summer E increased over the continents). The amount of cloudiness was, on the average, appreciably greater over the oceans (Fig. 24.2e) than over the continents (Fig. 24.2f); it was high throughout the year at the equator and had winter maxima in the middle and high latitudes. The vertical mass flux at the ocean surface, described back in §13, was found to have the greatest seasonal variability in the middle latitudes, to have its minimum for the year at the equator (due to the smallness of $T_w - T_a$ and the radiation-shielding effect of the thick clouds), and to be reduced in winter in the high latitudes, because of the presence there of ice (Fig. 24.2g). Finally, Fig. 24.2h and the charts in Fig. 24.4 show isolines of the thickness h of the mixed layer. In winter the upper mixed layer of the ocean spreads practically throughout the active layer, and with the advent of spring warming h decreases to tens of meters in the middle latitudes and to 120–150 m in the subtropics; in summer h is still reduced, with a very high variability in space.

Because it reproduces satisfactorily the present-day climate, has a comparatively low inertia, and involves a natural synoptic synchronization of the atmosphere and the active layer of the ocean, the model of the Oceanology Institute, and other models of the same type, are very suitable for such things as the utilization of data obtained in nature (for instance, the data of the first GARP experiment).

25. The Model of the Siberian Branch of the U.S.S.R. Academy

A group of specialists at the Siberian Branch of the U.S.S.R. Academy of Sciences, under the direction of G. I. Marchuk, constructed a model of the present-day circulation of the atmosphere and ocean, on the basis of a program published in 1975. In 1980 the first results of numerical experiments according to this model appeared. A hydrodynamic model of the general circulation of the atmosphere, published as a pair of preprints in 1977 and as a journal article in 1979, served as the atmospheric block of this model.

In this model, as in that of Manabe and Bryan (1969), the ocean and atmosphere were also artificially "synchronized": one atmospheric year was equated to approximately 100 oceanic years. More precisely, for time intervals in the atmosphere of 40 min and in the ocean of 2 days, the "coupling" of the atmospheric and oceanic fields was carried out over every 6 atmospheric intervals and every 7 oceanic intervals, the atmospheric data being transferred to the ocean with time smoothing via an exponential filter equivalent to the one described in §23.

For this coupling, the effect of the ocean on the atmosphere was determined by the temperature of the ocean surface $T_s = T_w$, found from the thermodynamic equation for the ocean, taking into account heat transport by ocean currents and with the simplifying assumptions that the effect of the salinity is not great and that the mixed layer has a constant thickness $h \sim 100$ m. The surface temperature was used to find the radiative heat flux from ocean to atmosphere and the turbulent fluxes of sensible and latent heat. The heat flux from atmosphere to ocean H_s was determined with the aid of the heat-budget equation (9.10). The exchange of moisture was reduced to evaporation, precipitation, and runoff from the land, and the exchange of momentum was reduced to its transfer from ocean to atmosphere by the frictional stress of the wind (calculated in the atmospheric block, neglecting water currents at the ocean surface).

The atmospheric block of the model of the Siberian Branch (with a vertical coordinate σ) contains the same fundamental equations as both the "wet" model of the Geophysical Fluid Dynamics Laboratory and the model of the Oceanology Institute. These equations include the evolution equations for v_θ, v_λ, T, q, and $\mathcal{P}$, where, for convenience in devising absolutely reliable finite-difference methods of solving them, at the same time obeying the integral conservation laws, the first four equations were "symmetrized" via a conversion to the variables $\mathcal{P}^{1/2}v_\theta$, $\mathcal{P}^{1/2}v_\lambda$, $\mathcal{P}^{1/2}T$, and $\mathcal{P}^{1/2}q$.

The parametrization of the small-scale processes in the atmospheric block was carried out just as in §22 for the zonal model of the atmospheric circulation devised by Dymnikov *et al.* (1979). Here we will just add a few details. In formulas of the type of (9.12), describing the interaction of the atmosphere with the underlying surface, the friction factor C_τ for a neutral stable stratification was over land or over an ice cover assumed to be equal to $0.02(1 + 3z_0/\delta)$, where δ is some scale of length. Over the ocean it was assumed to be 0.5×10^{-3} $U^{1/2}$ for a wind force $U = 15$m/s and 2.6×10^{-3} $U^{1/2}$ for $U > 15$ m/s. For an unstable stratification all these quantities have to be multiplied by $[1 + (T_s - T_h)^{1/2}\, U^{-1}]$. The coefficients of heat exchange and evaporation $c_H = c_E$ were specified similarly, the only difference being that for a neutral stable stratification over the ocean they are obtained from c_τ by multiplying by 5/6 for $U < 5$ m/s and by 10/7 for $U > 10$ m/s. The temperature T_h and the humidity q_h at the upper boundary of the surface layer of air were found from the condition that the flux of the pseudo-potential temperature Θ be continuous at this boundary:

$$-c_H \varrho_h U\,(\Theta_h - \Theta_s) = -\varrho_h \alpha U\, \frac{\Theta_1 - \Theta_h}{z_1 - z_h}, \tag{25.1}$$

where $\alpha \approx 1$ m, and a subscript 1 indicates the calculation level in the atmosphere closest to the Earth's surface; in addition, $q_h = r\, q_m(T_h)$. On the whole, this means of parametrizing the atmospheric boundary layer would appear to give a simplified approximation of the method of the Oceanology Institute described in §§9 and 23. Moreover, during the calculation of the surface temperature of the land or ice cover T_s from the heat-budget equation

(9.12), it was assumed that $H_s = 0$ on land and $H_s = \lambda(T_s - T_H)H^{-1}$ on an ice surface, where $H = 3$ m is its thickness, $\lambda = 0.005$ cal/(cm^2 s K) = 209.5 W/(m^2 K) is the thermal conductivity, and $T_H = 273.2$ K.

The albedo of the underlying surface was assumed to be 0.1 for the ocean, $0.2 + 0.4\bar{h}_s$ for the land, taking into account the water-equivalent thickness of the snow cover $\bar{h}_s$, and 0.6 for ice. If the value of T_s found from (9.10) for a region with snow or ice was greater than $T_{wi} = 273.2$ K, then it was replaced by T_{wi} and from the heat-budget equation for $T = T_{wi}$ the rate of melting of the snow or ice was calculated. The thickness $\bar{h}_s$ was found from equation (17.1) (neglecting evaporation). The relative air humidity r was set equal to 1 over the ocean, snow, ice, or damp soil (with a humidity of the upper meter layer $W > W_{cr} = 0.75W_s$, where $W_s = 15$ cm is the moisture capacity of the soil), while for $W \leqslant W_{cr}$ it was assumed that $r = W/W_{cr}$. Humidity W was calculated as indicated by Manabe and Bryan (see equation (23.3) and the explanation of it). The amount of nonconvective cloudiness was assumed to be $n = \alpha r + \beta$ (cf. (24.2)).

The "convective adjustment" was carried out similarly to (24.6), but with somewhat different formulas. For instance, the condition of appearance of dry convection and the conditions of matching had the form:

$$\frac{T_{k+1} - T_k}{p_{k+1} - p_k} > \frac{\gamma_a R}{g} \frac{T_{k+1} + T_k}{p_{k+1} + p_k}; \tag{25.2}$$

$$\frac{\hat{T}_{k+1} - \hat{T}_k}{p_{k+1} - p_k} > \frac{\gamma_m R}{g} \frac{\hat{T}_{k+1} + \hat{T}_k}{p_{k+1} + p_k};$$

$$(\hat{T}_{k+1} - \hat{T}_{k+1}) + (\hat{T}_k - T_k) = 0. \tag{25.3}$$

For wet convection these relations are more complicated:

$$\frac{T_{k+1} - T_k}{p_{k+1} - p_k} > \frac{\gamma_m R}{g} \frac{T_{k+1} + T_k}{p_{k+1} + p_k}; \qquad r_k, r_{k+1} > r_{cr} = 0.7; \tag{25.2a}$$

$$\gamma_m = \gamma_a \frac{1 - r}{1 - r_{cr}} + \gamma_{ma} \frac{r - r_{cr}}{1 - r_{cr}};$$

$$\frac{\hat{T}_{k+1} - \hat{T}_k}{p_{k+1} - p_k} = \frac{\gamma_m R}{g} \frac{\hat{T}_{k+1} + \hat{T}_k}{p_{k+1} + p_k};$$

$$(\hat{T}_{k+1} - T_{k+1}) + (\hat{T}_k - T_k) = \frac{\mathscr{L}}{c_p} m; \tag{25.3a}$$

$$(\hat{q}_{k+1} - q_{k+1}) + (\hat{q}_k - q_k) = m,$$

where γ_{ma} is the moist-adiabatic temperature gradient (lapse rate), m is the amount of condensed moisture, and $\hat{q}_i = r_{cr} q_m(\hat{T}_i)$. The large-scale condensation was described by equations (22.17), it being stipulated that the air in the clouds was saturated, that all cells were filled with clouds, that evaporation processes in the clouds were neglected, and that all condensed moisture departed as precipitation. The difference analog of equations (22.17) is

$$T^{n+1} - T^{n} = \frac{\mathcal{L}}{c_p}\,(q^{n+1} - q^{n}). \tag{25.4}$$

Horizontal mixing of momentum was described by formulas of the type

$$f_\theta = \frac{1}{a^2 \mathcal{P} \sin^2\theta}\left(\frac{\partial}{\partial\theta}\, K_{\mathrm{H}} \mathcal{P} \sin\theta\, \frac{\partial v_\theta}{\partial\theta} + \frac{\partial}{\partial\lambda}\, K_h\, \mathcal{P}\, \frac{\partial v_\theta}{\partial\lambda}\right);$$

$$f_\lambda = \frac{1}{a^2 \mathcal{P} \sin^2\theta}\left(\frac{\partial}{\partial\theta}\, K_h \mathcal{P} \sin^3\theta\, \frac{\partial}{\partial\theta}\, \frac{v_\lambda}{\sin\theta} + \frac{\partial}{\partial\lambda}\, K_h\, \mathcal{P}\, \frac{\partial v_\lambda}{\partial\lambda}\right), \tag{25.5}$$

ensuring conservation of the global angular momentum, and formulas like the one for v_θ were also used to describe the mixing of temperature and humidity. The short-wave fluxes of radiative energy were calculated separately for a cloudless atmosphere (for wavelengths less than 0.9 μm, absorption by ozone and Rayleigh scattering were taken into account) and a cloudy atmosphere (taking ozone and scattering in clouds into account but without Rayleigh scattering). For wavelengths greater than 0.9 μm the method of Katayama (1972) was used, and everywhere the procedure of layer doubling of Lacis and Hansen (1974) was employed as well. The fluxes of long-wave radiation were calculated with the aid of transmission functions, according to the calculated fields of T and q with linear approximation of the profiles inside the calculation layers.

In all the models constructed under the direction of Marchuk special attention was given to selecting the most precise and most economical methods of numerical solution of the equations. In the atmospheric block of the model of the Siberian Branch a vertical grid was employed, on which $\dot\sigma$ was shifted a half interval relative to the other unknown functions; on the horizontal grid this was done for v_θ and v_λ. Use was made of the method of splitting of the operators of transfer, diffusion, and adaptation. The transfer equation

$$\frac{\partial\psi}{\partial t} + \frac{1}{a\sin\theta}\left(\frac{1}{2}\,\frac{\partial\psi v_\theta \sin\theta}{\partial\theta} + \frac{1}{2}\, v_\theta \sin\theta\, \frac{\partial\psi}{\partial\theta} + \frac{1}{2}\,\frac{\partial\psi v_\lambda}{\partial\lambda} + \frac{1}{2}\, v_\lambda\, \frac{\partial\psi}{\partial\lambda}\right) + \frac{1}{2}\,\frac{\partial\psi\dot\sigma}{\partial\sigma} + \frac{1}{2}\,\dot\sigma\,\frac{\partial\psi}{\partial\sigma} = 0 \tag{25.6}$$

was split up into one-dimensional transfer equations in θ, λ, σ, written in terms of time according to the Krank–Nicholson method with skew-symmetric difference space operators, ensuring the preservation of quadratic integrals. These transfer equations are solved for λ using the method of cyclic elimination, for σ using scalar elimination, and for θ using cyclic elimination on meridians, shifted relative to one another by 180°, with a change in the sign of the vector quantities when passing through a pole. The adaptation equations were written in terms of time according to the Krank–Nicholson method and in terms of space according to the method of second-order accuracy. They were then solved using Richardson's iteration method, with the addition in polar regions of short-wave filtering of the gradients of pressure $\mathcal{P}$, geopotential, and zonal velocity v_λ.

The mean January circulation of the atmosphere was calculated on the basis of an individually considered atmospheric block, with climatically specified mean January values of both the temperature of the ocean surface $T_w(\theta, \lambda)$ and the distribution of sea ice and continental ice, as well as of the insolation in the absence of land topography. The calculation was carried out using a three-level model ($\sigma = \frac{1}{6}, \frac{1}{2}, \frac{5}{8}$), with horizontal grid intervals of 6° for θ and 10° for λ and with a time interval of 10 min. The initial conditions were $v_\lambda = v_\theta = 0$, $\mathcal{P} \equiv 1018$ mb, T (in K) $= 287 + 60\,(\sigma - 1)$, and $q = 0.8\, q_m(T)$. The calculation was carried out for 131 days, and the statistical data were processed for the period from the 100th to the 131st days.

The model gave a quite good representation of the level of the total kinetic energy of the atmosphere K, although the values obtained for the available potential energy A and the nonzonal parts K' and A' were a little low, because of the low spatial resolution. The spectra of the longitudinal wave numbers of the zonal and meridional components of K at the three levels, as well as the values of K' at 45° latitude, were close to reality (with an interval of inverse-cube law in the spectrum of K'). The zonal distributions of T and q obtained were close to the actual distributions, except that the subtropical jet stream in the distribution of v_λ for the Southern Hemisphere was too weak. However, the model gave a quite accurate picture of the tricellular meridional circulation in the winter hemisphere. The meridional fluxes of heat, moisture, and angular momentum showed a good qualitative fit with reality, but the quantitative fit was only fair. In the field of $\mathcal{P}$ the model gave a generally satisfactory description of stationary ultralong waves.

The oceanic block of the model of the Siberian Branch (Marchuk and Zalesnyi, 1974; Marchuk, Zalesnyi, and Kuzin, 1975; Marchuk, Kordadze, and Zalesnyi, 1979; and Zalesnyi, 1980) included a complete system of equations (see §12) of dynamics of an ocean of varying depth h_w with a realistic equation of state $\varrho_w(T, S)$. We will give here only the boundary conditions for temperature and salinity at the ocean surface $z = 0$ and at the shores Γ:

$$z = 0 : \alpha_T \frac{\partial T}{\partial z} + \beta_T T = Q_T; \qquad \alpha_s \frac{\partial S}{\partial z} + \beta_S S = 0; \tag{25.7}$$

$$\Gamma: \frac{\partial T}{\partial n} + \frac{z}{h_w}\frac{\partial h_w}{\partial n}\frac{\partial T}{\partial z} = \frac{\partial S}{\partial n} + \frac{z}{h_w}\frac{\partial h_w}{\partial n}\frac{\partial S}{\partial z} = 0. \tag{25.8}$$

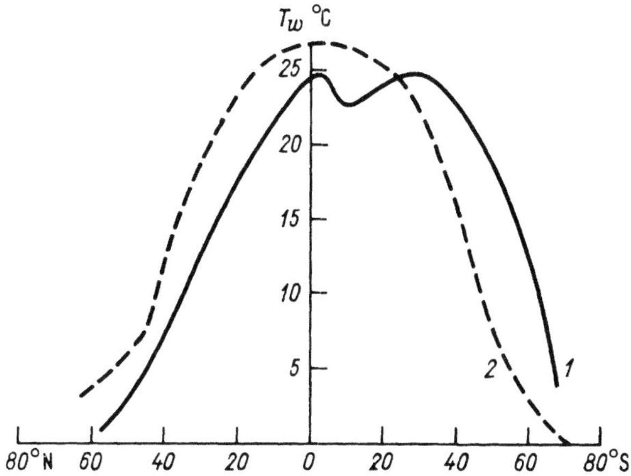

Fig. 25.1. Zonal T_w, according to January model of Siberian Branch (1) and actual values (2).

The horizontal mixing was described in the manner of (25.5). In the numerical solution of the equations, splitting of the operators of transfer + diffusion was carried out, for each of the three coordinates, using a variational method close to that of Galerkin. The splitting of the adaptation operator involved a semi-implicit method in time and a method of second-order accuracy in space. Calculations for a separately considered oceanic block were carried out for fixed $T_w(\theta, \lambda)$ over a period of 118 years (100 years of diffusion of heat and 18 years of interaction between the temperature and current fields).

Numerical experiments based on the complete model of the Siberian Branch were calculated for initial data obtained in the form of quasi-equilibrium states of the atmosphere and ocean for a fixed temperature of the ocean surface $T_w(\theta, \lambda)$ and fixed insolation. The computations pertained to 11 oceanic years $\approx$ 2 atmospheric months. Data on the characteristics of the atmospheric circulation over the last 20 days were statistically processed. It was found, first of all, that due to the ocean–atmosphere interaction the zonal values of T_w were reduced in the Northern Hemisphere and in the tropics and were increased in the Southern Hemisphere (Fig. 25.1). The evolution of the T_w field was found to be as follows: first negative anomalies of T_w appeared at the western shores of the Pacific and in the northwestern Indian Ocean (these gradually diminished), and the large gradients of T_w created very strong westward currents there, which enhanced the upwelling, and then also the meridional circulation in the ocean and the adaptation of the atmospheric circulation, with a breakup of cyclones and a weakening of the ocean-atmosphere interaction. It should be mentioned that, due to the absence in the model of a seasonal variation of the insolation and of heat sinks in the Antarctic, the calculations gave a positive temperature trend in the Southern Ocean, which remained even when T_w = 1°C was fixed on the coasts of

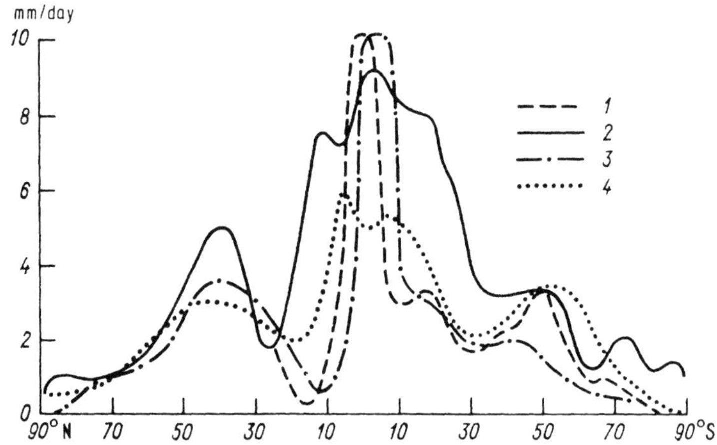

Fig. 25.2 Zonal distributions of precipitation according to models of Geophysical Fluid Dynamics Laboratory (1), Rand Corporation (2), Siberian Branch (3) and according to actual data (4).

Antarctica in an additional experiment (this led only to a cooling of the northern regions of the Atlantic and Pacific).

Inclusion of the ocean–atmosphere interaction reduced the meridional turbulent fluxes in the atmosphere, varied considerably the space spectra of kinetic energy (especially in the long-wave region of the upper troposphere), diminished the transport of heat by the mean meridional circulation, reduced the temperature of the middle troposphere in the tropics by about 5°C (for a slight variation of it outside the tropics), shifted the zone of tropical rain 10° southward, reduced somewhat the total kinetic energy of the atmosphere and the other elements of the energy cycle, weakened the subtropical jet stream of the Southern Hemisphere, while in the Northern Hemisphere weakening the Ferrel cell somewhat but enhancing the Hadley cell, lessened the baroclinic instability of the atmosphere, reduced the surface frictional stress of the wind in the Southern Hemisphere (especially at low latitudes), and, finally, deepened the centers of low pressure $\mathcal{P}$ over the continents of the Southern Hemisphere. Figure 25.2 compares the zonal distributions of precipitation according to the models of the Geophysical Fluid Dynamics Laboratory, the Rand Corporation, and Siberian Branch of the Soviet Academy with the actual distributions. The agreement between the results of the Siberian Branch and the Fluid Dynamics Laboratory is interesting. Taking the ocean-atmosphere interaction into account improves somewhat the distributions of precipitation and soil humidity, especially in South America and Eastern Australia. The next problem for the model of the Siberian Branch should be to calculate the seasonal fluctuations of the atmosphere-ocean-land system.

The further development of three-dimensional models of the climate system of the types described in §§23–25 will, in our opinion, open up new vistas for the theory of climate.

REFERENCES

Agafonova, E.G., Galerkin, L.I., Monin, A.S., 1972. 'On the origin of thermohaline circulation in the ocean'. *Okeanologiya*, **12** (6): 950–955 (in Russian).

Albrecht, F., 1949. 'Über die Wärme- und Wasserbilanz der Erde'. *Ann. Meteor.* **2** (5) 129–143.

Alexander R.C., Schlesinger M.E., 1975. 'An initial coupling experiment'. In: *A Review of Rand Climate Dynamics Research*. Chap. 6, 'Development of Global Ocean Models', Working note 9149, Rand Corp., pp. 186–191.

Andreev, O.A., Kagan, B.A., Oganesyan, L.A., 1976. 'On the global circulation in a two-layer ocean'. *Okeanologiya*, **16**(1) 5–14 (in Russian).

Angell, J.K., Korshover, J., 1976. 'Global analysis of recent total ozone fluctuations'. *Mon. Wea. Rev.* **104** (1) 63–75.

Avenhaus, R., Fenyi, S., Frick, H., 1980. 'Box models for the CO_2 cycle of the Earth'. *Environ. Int.* **2** (4–6) 379–385.

Barenblatt, G.I., 1978. 'Dynamics of turbulence spots and intrusion in a stably stratified fluid'. *Izv. Akad. Nauk SSSR, Fiz. Atmos. Okeana* **14** (2) 195–206 (in Russian).

Batchelor, G.K., 1969. 'Computation of the energy spectrum in homogeneous two-dimensional turbulence'. *Phys. Fluids*. **12**(11) 233–239.

Berger, A., 1975. In: *Proc. WMO/IAMAP Symposium on Long-Term Climatic Fluctuations*, pp. 65–74, WMO, No. 421, Geneva.

Berger, A.L., 1977. 'Support for the astronomical theory of climatic change'. *Nature*, **269** (5623), 44–45.

Bjerknes V., 1937. 'Application of line integral theorems to the hydrodynamics of terrestrial and cosmic vortices'. *Astrophys. Norw.* **2** 263–339.

Blinova, E.N., 1947. 'On the annual mean temperature distribution in the Earth's atmosphere taking into account continents and oceans'. *Izv. Akad. Nauk SSSR, Ser. Geogr., Geofiz.* **XI** (1), 3–13 (in Russian).

Bolin, B., 1976. 'Forandrer manniskan lufthavet och jordens klimat?' *Doc. Kgl. Vetenskapsakad.*

Bolin, B., and Keeling, C.D., 1963. 'Large-scale atmospheric mixing as deduced from the seasonal and meridional variations of carbon dioxide'. *J. Geophys. Res.* **68** (13), 3899.

Bowen, H.J.M., 1966. *Trace Elements in Biochemistry*. Acad. Press, 241 pp.

Bowen, R., 1966. *Palaeotemperature Analysis*. Elsevier, Amsterdam, 265 pp.

Brasseur, G., Nicolet, M., 1973. 'Chemospheric processes of nitric oxide in the mesosphere and stratosphere'. *Planet and Space Sci.* **21** (6) 939–961.

Brouwer, D., van Woerkom A.J. 1950. 'The secular variations of the orbital elements of the principal planets'. *Astronom. Papers. Am. Ephem.* **13**, Pt.2, 81.

Bryan, K., Manabe S., Pacanowski, R.C., 1975. 'A global ocean-atmospheric climate model. Pt. II. The oceanic circulation'. *J. Phys. Oceanogr.*. **5** (1), 30–46.

Bryan, K., Lewis, L.J., 1979. 'A water mass model of the World Ocean'. *J. Geophys. Res.* **84** (5).

Budyko, M.I. (ed.), 1963. *An Atlas of the Heat Balance of the Earth.* L. Gidrometeoizdat (in Russian).

Budyko, M.I., 1968. 'On the origin of ice ages'. *Meteorologiya i Gidrologiya*, No. 2, 3–12 (in Russian).

Chapman, S., 1930. 'A theory of upper-atmosphere ozone'. *Mem. R. Meteorol. Soc.* **3** 102.

Charney, J.G., 1960. 'Nonlinear theory of a wind driven homogeneous layer near the equator'. *Deep-Sea Res.* **6** (4) 303–310.

Charney, J.G., 1970, *J. Atmos. Sci.* **27** 376; 1971, *J. Atmos. Sci.* **28** 1087.

Charnock, H., 1955. 'Wind stress on a water surface'. *Quart. J. Roy. Meteorol. Soc.* **81** (350) 639–640.
Church, P.E., 1937. 'Temperatures of the western North Atlantic from thermograph records'. Assoc. d'Oceanogr. Phys., UFFI, Publ. Sci., No 4, Liverpool, p. 32.
Chýlek, P., Coakley J.A., 1975. 'The two-stream approximation in radiative transfer: including the angle of the incident radiation'. *J. Atmos. Sci.* **92** 409–19.
Coon, M.D., Maykut, G.A., Pritchard, R.S., Rothrock, D.A., Thorndike, A.S., 1974. 'Modeling the pack ice as an elastic-plastic material'. *Aidjex Bull.* **24** 1–105.
Cox, M.D., 1970. 'A mathematical model of the Indian Ocean'. *Deep-Sea Res.* **17** (1) 47–75.
Cox, M.D., 1975. 'A baroclinic numerical model of the World Ocean: preliminary results'. In: *Proc. of the Symposium on Numerical Models of Ocean Circulation* (Durham, New Hampshire, 17–20 October 1972). National Academy of Science, Washington, D.C., pp. 107–120.
Crutzen, P.J., 1971a. 'Energy conversions and mean vertical motions in the high-latitude summer mesosphere and lower thermosphere'. In: *Mesospher. Models and Relat. Experim., Proc. 4th ESRINESI. AB Symp.*, Frascati, 1970. Reidel, Dordrecht, pp. 78–88.
Crutzen, P.J., 1971b. 'Ozone production rates in an oxygen-hydrogen-nitrogen oxide atmosphere'. *J. Geophys. Res*, **76** (30) 7311–7327.
Crutzen, P.J., 1974a. 'Estimates of possible future ozone reduction from continued use of chlorofluoromethanes (CF_2Cl_2, $CFCl_3$)'. *Geophys. Res. Lett.* **1** 205–9.
Crutzen, P.J., 1974b. 'Estimates of possible variations in total ozone due to natural causes and human activities'. *Ambio* **3** 201.
Crutzen, P.J., 1974c. 'Photochemical reactions initiated by and influencing ozone in unpolluted tropospheric air'. *Tellus* **26** 47.
Dalton, J., 1793. *Meteorological Observations and Essays*. Manchester.
Defant, A., 1921. 'Die Zirkulation der Atmosphäre in den gemässigten Breiten der Erde'. *Geogr. Ann.* **3**, Stockholm.
Dolzhanskii, F.V., 1969. 'On calculation of the zonal atmospheric circulation'. *Izv. Akad. Nauk SSSR, Fiz. Atmos. Okeana* **5** (7) 659–671 (in Russian).
Dolzhanskii, F.V., 1971. 'Numerical experiments on modeling the Earth's zonal atmospheric circulation'. *Izv. Akad. Nauk SSSR, Fiz. Atmos. Okeana* **7** (1) 3–11 (in Russian).
Dove, H.W., 1837. *Meteorologische Untersuchungen*. Sanderische Buchhandlung, Berlin.
Drazin, P.G., Griffel, D.H., 1977, 'On the branching structure of diffusive climatological models', *J. Atmos. Sci.* **34** (2) 1696–1706.
Dymnikov, V.P., Perov, V.L., Lykosov, V.N., 1979. 'A hydrodynamic zonal model of the general atmospheric circulation'. *Izv. Akad. Nauk SSSR, Fiz. Atmos. Okeana* **15** (5) 484–497 (in Russian).
Dymnikov, V.P., Galin, V.A., Perov, V.L., 1980. 'Studies of climatic sensitivity to CO_2 doubling with the aid of a zonally averaged model of the general circulation'. In: *Mathematical Modeling of the Dynamics of the Atmosphere and Ocean. Part II, Mathematical Modelling of Atmospheric Motions*. Nauka, Novosibirsk, pp. 15–21 (in Russian).
Eddington, A., 1959, *Space, Time and Gravitation: An Outline of the General Relativity Theory*. Harper, New York, 213 pp.
Emden, R., 1913. 'Über Strahlungsgleichgewicht und atmosphärische Strahlung. Ein Beitrag zur Theorie der oberen Inversion'. Sitzungsberichte, Akademie der Wissenschaften, München, No. 1, 55–142.
Ertel, E., 1942. 'Ein neuer hydrodynamischer Wirbelsatz'. *Meteorol. Zeitsch.* **59** 277.
Faegre, A., 1972. 'An intransitive model of the earth-atmosphere-ocean system'. *J. Appl. Meteorol.* **11** 4–6.
Fedorov, E.E., Baranov, A.I., 1949. *The Climate of the Plain of the European U.S.S.R.* M.-L. Izd. Akad. Nauk SSSR, 412 pp. (in Russian).
Ferrel, W., 1856. 'An essay on the winds and the currents of the ocean'. Na. J. Medicine and Surgery **11** 287–301. Reprinted (1882) in:*Popular Essays on the Movements of the Atmosphere*. Prof. Pap. Signal Serv., No. 12, Washington, pp. 7–19.

Ferrel, W., 1858. 'The motions of fluids and solids relative to the Earth's surface'. *Math. Monthly* **1**, 140–147, 210–216, 300–307, 366–372, 379–406.
Fuglister, F., Worthington, L., 1951. 'Some results of a multiple ship survey of the Gulf Stream'. *Tellus* **3** (1) 1–14.
Galerkin, L.I., Gritsenko, A.M., 1980. 'Statistics of the circulation field of the Pacific Ocean'. *Dokl. Akad. Nauk SSSR* **251** (5) 1254–1257 (in Russian).
Galin, M.B., Kirichkov, S.E., 1979. 'A lumped-component model of the general atmospheric circulation with a variable Coriolis parameter'. *Izv. Akad. Nauk SSSR, Fiz. Atmos. Okeana* **15** 355–367 (in Russian).
Gavrilin, B.L., Monin, A.S., 1969. 'On primitive and balanced hydrodynamic equations of meteorological fields'. *Izv. Akad. Nauk SSSR, Fiz. Atmos. Okeana* **5** 763–774 (in Russian).
Gavrilin, B.L., Mirabel, A.P., Monin, A.S., 1972. 'On the energy spectrum of synoptic processes'. *Izv. Akad. Nauk SSSR, Fiz. Atmos. Okeana* **8** (5) 483–493 (in Russian).
Golitsyn, G.S., 1968. 'Estimates of some general circulation parameters in the planetary atmospheres of the terrestrial group'. *Izv. Akad. Nauk SSSR, Fiz. Atmos. Okeana* **4** (11) 1131–1138 (in Russian).
Golitsyn, G.S., 1970. 'Similarity theory for large-scale motions of planetary atmospheres'. *Dokl. Akad. Nauk SSSR* **190** (2) 323–326 (in Russian).
Golitsyn, G.S., 1973. *An Introduction to the dynamics of planetary atmospheres*. L. Gidrometeoizdat, 104 pp. (in Russian).
Golitsyn, G.S., Mokhov, I.I., 1978. 'Estimates of the sensitivity and role of cloudiness in simple climatic models'. *Izv. Akad. Nauk SSSR, Fiz. Atmos. Okeana* **14**(8) 803–814 (in Russian).
Golitsyn, G.S., Mokhov, I.I., 1978. 'On the stability and extremal properties of climatic models'. *Izv. Akad. Nauk SSSR, Fiz. Atmos. Okeana* **14**(4) 378–387 (in Russian).
Hadley, G., 1735. 'Concerning the cause of the general tradewinds'. *Phil. Trans. Roy. Soc. London* **39** 58.
Hall, R.T., Maycut, G.A., Rothrock, D.A., 1976. 'Sea ice modeling: Its testing with Landsat and potential use in FGGE'. In: *Proc. Symp. Met. Observ. from Space*, pp. 104–109. Committee on Space Research, ICSU, Paris.
Hamill, P., 1976. 'H_2SO_4 gas phase concentration in the stratosphere. Atmos. aerosols: opt. properties and eff. dig.' *Techn. Pap. Top. Molt.*, Williamsburg, Va, Washington, D.C., WA2/1-WA2.
Hays, J.D., Imbrie, J., Shackleton, M.J., 1976. 'Variations in the Earth's orbit: pacemaker of the ice ages'. *Science* **194** 1121–1132.
Helmholtz, von, H., 1888. *Über Atmosfärische Bewegungen*. Sitzber. Press. Akad. Wiss., Berlin, Math. Physik. KL., p. 647.
Hesstvedt, E., 1973. 'Effect of water vapor and nitrogen oxides upon the ozone layer'. *Water, Air and Soil Pollution*, **2** (1) 49–60.
Holland, W.R., Lin, L.B., 1975. 'On the generation of mesoscale eddies and their contribution to the oceanic circulation'. *J. Phys. Ocean.* **5** (4) 642–669.
Holton, J.R., Lindzen, R.S., 1968. 'A note on "Kelvin" waves in the atmosphere'. *Monthly Weather Rev.* **96** (6) 385–386.
Horn, L.H., Bryson, R.A., 1963. *J. Geophys. Res.*, **68**, 1059.
Houghton, H.G., 1954. 'On the annual heat balance of the northern hemisphere'. *J. Met.* **11**, (1) 1–9.
Humphreys, W., 1913. 'Volcanic dust and other factors in the production of climate changes and their possible relation to ice ages. *J. Frankl. Inst.* **176**.
Humphreys, W., 1929. *Physics of the Air*.
Hunt, B.G., and Manabe, S., 1968. 'Experiments with a stratospheric general circulation model: II. Large-scale diffusion of tracers in the stratosphere'. *Mon. Wea. Rev.*, **96** (8) 503–539.
Il'in, A.M., Kamenkovich, V.M., Zhugrina, T.G., Silkina, M.M., 1969. 'On the computation of total flows in the World Ocean (steady-state problem)'. *Izv. Akad. Nauk SSSR, Fiz. Atmos. Okeana* **5** (11) 1160–1172 (in Russian).

Jeffreys, H., 1926. 'On the dynamics of geostrophic winds'. *Q.J. Roy. Met. Soc.* **52**.

Julian, P.R., Washington, W.M., Hembree, L., Ridley, C., 1970. 'On the spectral distribution of large-scale atmospheric kinetic energy'. *J. Atmos. Sci.* **27** (3) 376–387.

Kagan, B.A., Laikhtman, D.P., Oganesyan, L.A., Pyaskovskii, R.V., 1972. 'A numerical experiment on the seasonal variability of the global circulation in a barotropic ocean'. *Izv. Akad. Nauk SSSR, Fiz. Atmos. Okeana* **8**(10) 1052–1071 (in Russian).

Kagan, B.A., Laikhtman, D.P., Oganesyan, L.A., Pyaskovskii, R.V., 1974. 'A two-dimensional thermal model of the World Ocean'. *Izv. Akad. Nauk SSSR, Fiz. Atmos. Okeana* **10** 1118–1122 (in Russian).

Kagan, B.A., Laikhtman, D.P., Oganesyan, L.A., Pyaskovskii, R.V., 1975. 'A numerical experiment on the general circulation of the ocean'. *Okeanologiya,* **15** (1) 5–11 (in Russian).

Kamenkovich, V.M., 1958. 'On some simplifications of the dynamic equations of steady currents in a baroclinic sea'. *Dokl. Akad. Nauk SSSR* **119** (6) 1134–1137 (in Russian).

Kamenkovich, V.M., 1967. 'On turbulent diffusion and viscosity in large-scale motions of the ocean and the atmosphere'. *Izv. Akad. Nauk SSSR, Fiz. Atmos. Okeana* **3** (12) 1326–1333 (in Russian).

Kant, I., 1756. 'Anmerkungen zur Erläuterung der Theorie der Winde'. Reference quoted from Hann, J. von. *Handbuch der Klimatologie.* J. Engelhorn, Stuttgart, 1908–1911.

Kao, S.K., Wendell, L.L., 1970. 'The kinetic energy of the large-scale atmospheric motion in wavenumber frequency space'. *J. Atmos. Sci.* **27** 359–375.

Kasahara, A., Washington, W.M.; 1971. 'General circulation experiments with six-layer NCAR model, including orography, cloudiness and surface temperature calculations'. *J. Atmos. Sci.* **28** 675–701.

Katayama, A.A., 1972. 'A simplified scheme for computing radiative transfer in the troposphere. Numerical simulation of weather and climate'. Techn. Rep., No. 6, 77 pp.

Kibel', I.A., 1943. 'The temperature distribution in the Earth's atmosphere'. *Dokl. Akad. Nauk SSSR* **39** (1) 18–22 (in Russian).

Kirichkov, S.E., 1978. 'Numerical experiments with a zonal atmospheric circulation model'. *Izv. Akad. Nauk SSSR, Fiz. Atmos. Okeana* **14** (7) 691–702 (in Russian).

Kitaigorodskii, S.A., 1970. *The Physics of the Interaction between the Atmosphere and Ocean.* L. Gidrometeoizdat, 284 pp. (in Russian).

Kochergin, V.P., Sukhorukov, V.A., 1975. 'A two-parameter model of developed turbulence'. In: *Numerical Methods of Continuum Mechanics.* Novosibirsk, Vol. 6, pp. 81–93 (in Russian).

Kochergin, V.P. Klimok, V.I., Sukhorukov, V.A., 1976. 'A turbulent model of the Ekman layer in the ocean'. In: *Numerical Methods of Continuum Mechanics.* Novosibirsk, Vol. 7, pp. 77–84 (in Russian).

Kochergin, V.P., Shcherbakov, A.V., 1977. 'Numerical modeling of a baroclinic World Ocean'. *Meteorologiya i Gidrologiya* **7** 56–63 (in Russian).

Kochin, N.E., (ed.), 1936. *Dynamic Meteorology.* Centr. Admin. United Hydrometeor. Service USSR, 602 pp.

Kolmogorov, A.N., 1942. 'Equations of turbulent motion of an "incompressible liquid"'. *Izv. Akad. Nauk SSSR, Ser. Fiz.* **6** (1–2) 56–58 (in Russian).

Koshlyakov, M.N., Grachev, Yu. M., 1974. 'Mesoscale currents within a hydrophysical test area in the tropical Atlantic'. In: *The Atlantic Hydrophysical Poligon-70 Experiment,* Nauka, pp. 163–180 (in Russian) (Also in *Deep-Sea Res.* **20** (6) 507–526, 1973).

Koshlyakov, M.N., Grachev, Yu. M., Enikeev, V. Kh., 1980. 'Kinematics of the synoptic-eddy field in the open ocean'. *Dokl. Akad. Nauk SSSR* **252** (3) 573–577 (in Russian).

Kraichnan R.H. 1967, Inertial ranges in two-dimensional turbulence. *Phys. Fluids* **10** (7) 1417–1423.

Kraus, E.B., Turner, J.S., 1967. 'A one-dimensional model of the seasonal thermocline. Part II. The general theory and its consequences'. *Tellus* **19** (1) 98–105.

Kurihara, Y., 1968. 'Note on finite-difference expressions for the hydrostatic relation and pressure gradient force'. *Mon. Wea. Rev.* **96** 654–656.

Kutzbach, J.E., Bryson, R.A., 1974. 'Variance spectrum of Holocene climatic fluctuations in the North Atlantic sector'. *J. Atmos. Sci.* **31** 1958–63.

Lacis, A.A., Hansen, J.E., 1974. 'Atmosphere of Venus: implications of Venera 8 sunlight measurements'. *Science* **184** (4140) 979–982.

Lai, D.Y., Richardson, P.L., 1974. 'Distribution and movements of Gulf Stream Rings'. *J. Phys. Oceanogr.* **7** (5) 670–683.

Laikhtman, D.L., Kagan, B.A., Oganesyan, L.A., Pyaskovskii, R.V., 1971. 'On the global circulation in a variable-depth barotropic ocean'. *Dokl. Akad. Nauk SSSR* **198** (2) 333–337 (in Russian).

Levitus, S., Oort, A.H., Ellis, J.S., vonder Haar, T.H., 1978. The annual variation in the global heat balance of the earth'. *J. Geophys. Res.* **C83** (4) 1958–1962.

Linden, P.F., 1975. 'The deepening of a mixed layer in a stratified fluid'. *J. Fluid. Mech.* **71** 385–405.

London, J., Kelley, J., 1974. 'Global trends in total atmospheric ozone'. *Science* **184** 988–989.

Longuet-Higgins, M.S., 1969. 'On wave breaking and the equilibrium spectrum of wind-generated waves'. *Proc. Roy. Soc. London* **A310** (1501) 151–159.

Lorenz, E.N., 1955. 'Available potential energy and the maintenance of the general circulation'. *Tellus* **7** (2) 157–167.

Lorenz, E.N., 1960. 'Maximum simplification of the dynamical equations'. *Tellus* **12** (3) 243–254.

Lorenz, E.N., 1967. *Nature and Theory of the General Circulation of the Atmosphere.* World Meteorol. Org., Geneva.

Luther, F.M., 1974a. 'Effect of stratospheric aerosol on solar heating rates'. Preprint, Lawrence Livermore Lab., Univ. Calif., June 18, 3 pp.

Luther, F.M., 1974b. 'Effect of increased stratospheric aerosol on planetary albedo'. Preprint, Lawrence Livermore Lab., Univ. Calif., UASG-74-14, June 27, 4 pp.

Luther, F.M., 1974c. 'Effect of stratospheric aerosol on long-wave radiative fluxes: variation with latitude and season'. Preprint, Lawrence Livermore Lab., Univ. Calif., UCID-16559, July 19, 6 pp.

Luther, F.M., 1975a. 'Relative influence of stratospheric aerosols on solar and long-wave fluxes for a tropical atmosphere'. Preprint, UCLR-75760 (Rev.1). Lawrence Livermore Lab., Univ. Calif., July, 18 pp.

Luther, F.M., 1975b. 'Comparison of one-wavelength and average aerosol property calculations with full-flux solar radiation calculations'. *Coll. Abstr., Second Conf. Atmospheric Radiation*, Arlington, Virginia, 29–31 Oct., pp. 39–40.

Manabe, S., Strickler, R., 1964. 'Thermal equilibrium of the atmosphere with a convective adjustment'. *J. Atmos. Sci.* **21** (4) 361–385.

Manabe, S., Smagorinsky, J., Strickler, R.F., 1965. 'Simulated climatology of a general circulation model with a hydrologic cycle'. *Mon. Wea. Rev.* **93** (12) 769–798.

Manabe, S., Bryan, K., 1969. 'Climate calculations with a combined ocean–atmosphere model'. *J. Atmos. Sci.* **26** 786–789.

Manabe, S., Smagorinsky, J., Holloway, J.L., Stone, H.M., 1970. 'Simulated climatology of a general circulation with a hydrologic cycle. III. Effects of increased horizontal computational resolution'. *Mon. Wea. Rev.* **98** (3) 175–212.

Manabe, S., Hahn, D.G., Holloway, J.L., 1974. 'The seasonal variation of the tropical circulation of the atmosphere'. *J. Atmos. Sci.* **31** 43.

Manabe, S., Holloway, J.L., 1975. 'The seasonal variation of the hydrologic cycle as simulated by a global model of the atmosphere'. *J. Geophys. Res.* **80** 1617–1649.

Manabe, S., Wetherald, R., 1975. 'The effects of doubling the CO_2 concentration on the climate of a general circulation model'. *J. Atmos. Sci.* **32** (1) 3–15.

Manabe, S., Bryan K., Spelman, M.J., 1975. 'A global ocean–atmosphere climate model: Part I. The atmospheric circulation'. *J. Phys. Oceanogr.* **5** (1) 3–29.

Manabe, S., Bryan, K., Spelman, M.J., 1979. 'A global ocean-atmosphere climate model with seasonal variation for future studies of climate sensitivity'. *Dynamics of Atm. and Ocean* **3** 393–420.

Marchuk, G.I., Zalesnyi, V.B., 1974. 'A numerical model of the large-scale circulation in the World Ocean'. In: *Numerical Methods of Calculating Ocean Currents*. Computation Centre of the Siberian Division of the USSR Academy of Sciences, Novosibirsk, pp. 3–20 (in Russian).

Marchuk, G.I., Zalesnyi, V.B., Kuzin, V.I., 1975. 'On the method of finite differences and finite elements in problems of global ocean circulation'. *Izv. Akad. Nauk SSSR, Fiz. Atmos. Okeana* **II** (12) 1294–1300 (in Russian).

Marchuk, G.I., Kordzadze, A.A., Zalesnyi, V.B., 1979. 'Problems of mathematical modeling of sea and ocean currents'. In: *Differential and Integral Equations. Boundary Value Problems*. Tbilisi, TGU, pp. 195–228 (in Russian).

Matsuno, T., 1971. 'A dynamical model of the stratospheric sudden warming'. *J. Atmos. Sci* **28** 1479–1494.

Maury, M.F., 1855. *The Physical Geography of the Sea*. New York, Harper, 287 pp.

Maykut, G.A., Untersteiner, N., 1971. 'Some results from a time dependent, thermodynamic model of sea ice'. *J. Geophys. Res.* **76** 1550–75.

Michkovitch, V.V., 1931. Glas Srpske Akad. Nauka, Odel. Prirodno-Matem. Nauka **143** (1) 70. Beograd.

Milankovitch, M., 1939. *Mathematical Climatology and the Astronomical Theory of Climatic Fluctuations*. Moscow (in Russian).

Monin, A.S., Obukhov, A.M., 1953. 'Dimensionless parameters of turbulence in the surface layer of the atmosphere'. *Dokl. Akad. Nauk SSSR* **93** (2) 223–226 (in Russian).

Monin, A.S., 1958. 'On precalculation of the zonal atmospheric circulation'. *Izv. Akad. Nauk SSSR, Ser. Geofiz.* **10** 1250–1253 (in Russian).

Monin, A.S., 1965. 'On the structure of the atmospheric boundary layer'. *Izv. Akad. Nauk SSSR, Fiz, Atmos. Okeana* **1** (3) 258–265 (in Russian).

Monin, A.S., Kolesnikova, V.N., 1965. 'On the oscillation spectra of meteorological fields'. *Izv. Akad. Nauk SSSR, Fiz. Atmos. Okeana* **1** (7) 653–669 (in Russian).

Monin, A.S., 1970. 'On turbulent mass flows in the ocean'. *Dokl. Akad. Nauk SSSR* **190** (2) 240–248 (in Russian).

Monin, A.S., 1972. *Earth's Rotation and Climate*. L. Gidrometeoizdat, 112 pp. (in Russian) (Engl. translation, 1976, Radok-Radhakrishna, Delhi, India).

Monin, A.S., 1974. 'On the equations of geophysical hydrodynamics'. *Izv. Akad. Nauk SSSR, Fiz. Atmos. Okeana* **10** (2) 119–126 (in Russian).

Monin, A.S., Shishkov, Yu. A., 1979. *History of Climate*. L. Gidrometeoizdat, 407 pp. (in Russian).

Monin, A.S., Simuni, L.M., 1980., 'A model of the differential rotation of the Sun. *Dokl. Akad. Nauk SSSR*, **250** (6) 1348–1351 (in Russian).

Munk, W.H., 1950. 'On the wind-driven ocean circulation'. *J. Meteorol.* **7** 79–83.

Needler, G.T., 1967. 'A model for thermohaline circulation in an ocean of finite depth'. *J. Marine. Res.* **25** (3) 329–343.

Needler, G.T., 1971. 'Thermocline models with arbitrary barotropic flow'. *Deep-Sea Res.* **18** (9) 895–904.

Newell, A.C., 1969. 'Rossby-wave packet interactions'. *J. Fluid. Mech.* **35** (2) 255–272.

North, G.R., 1975. *Numerical Models of Ocean Circulation*. Nation. Acad. Sci., Wash., No. 20418, 364 pp.

Novikov, E.A., 1975. 'Dynamics and statistics of the eddy system'. *J. Experim. and Theoret. Phys.* **68** (5) 1868–1882 (in Russian).

Obukhov, A.M., 1946. 'Turbulence in a temperature–inhomogeneous atmosphere'. *Trudy Instituta Teoret. Geofiz. Akad. Nauk SSSR* **24** (151) 3–42 (in Russian).

Obukhov, A.M., 1962. 'On the dynamics of a stratified fluid'. *Dokl. Akad. Nauk SSSR* **145** (6) 1239–1242 (in Russian).

Ogura, Y., 1952. 'The theory of turbulent diffusion in the atmosphere'. *J. Meteor. Soc., Japan* **30** (2) 53–58.

Ogura, J., 1964. 'Frictionally controlled, thermally driven circulations in a circular vortex with application to tropical cyclones'. *J. Atmos. Sci.* **21** (6).

Ohring, G., Adler, S., 1978. 'Some experiments with a zonally averaged climate model'. *J. Atmos. Sci.* **35** 186–205.

Oort, A.H., 1964a. 'On estimates of the atmospheric energy cycle'. *Mon. Wea. Rev* **92** (11) 483–493.

Oort, A.H., 1964b. 'On the energetics of the mean and eddy circulation in the lower stratosphere'. *Tellus* **16** (4).

Oort, A.H., Rasmusson, E.M., 1971. 'Atmospheric circulation statistics'. *NOAA Prof. Paper.* **5** 323 pp.

Oort, A.H., Peixoto, J.P., 1974. 'The annual cycle of the energetics of the atmosphere on a planetary scale'. *J. Geophys. Res.* **79** (18) 2705–2719.

Ozmidov, R.V., 1965. 'Some features of the energy spectrum of oceanic turbulence'. *Dokl. Akad. Nauk SSSR* **161** (4) 828–832 (in Russian).

Parkinson, C.L., Washington, W.M., 1979. 'A large-scale numerical model of sea ice'. *J. Geophys. Res.* **84** 311–337.

Penck, A., Brückner, E., 1909. *Die Alpen in Eiszeitalter*. Vols. 1–3, Leipzig, Tauchnitz, 1199 pp.

Perry, H., Landsberg, H.H., 1977. 'Projected world energy consumption'. *Energy and Climate* 35–51.

Phillips, N.A., 1956. 'The general circulation of the atmosphere: a numerical experiment'. *Quart. J. Roy. Met. Soc.* **82** (358) 123–164.

Prandtl, L., 1942. 'Bemerkungen zur Theorie des freien Turbulenz'. *Zeitschr. Angew. Math. und Mech.* **22** (5) 241.

Prinn, R.G., Alyea, F.N., Cunnold, D.M., 1975. 'Stratospheric distributions of odd nitrogen in a two-dimensional model' *J. Geophys. Res.* **80** (36) 4997–5004.

Pritchard, R.S., 1975. 'An elastic-plastic constitutive law for sea ice'. *J. Appl. Mech.* **42** 379–384. 1976. 'An estimate of the strength of arctic pack ice'. *Aidjex Bull.* **34** 94–113.

Resnyanskii, Yu. D., 1975. Parametrization of the integral dissipation of turbulent energy in the upper quasi-homogeneous ocean layer'. *Izv. Akad. Nauk SSSR, Fiz. Atmos. Okeana* **11** (7) 726–733 (in Russian).

Revelle, R., Munk, W., 1977. 'The carbon dioxide cycle and the biosphere'. *Energy and Climate* 140–156.

Rhines, P., 1975. 'Waves and turbulence on a beta-plane'. *J. Fluid Mech.* **69** (3) 417–444.

Robinson, G.D., 1977. 'Effluents of energy production: particulates'. *Energy and Climate*. 61–71.

Romankevich, E.A., 1977. *The Geochemistry of Organic Matter in the Ocean*. M. Nauka, 256 pp. (in Russian).

Rossby, C.-G., 1941. 'The scientific basis of modern meteorology. Climate and Man'. *Year-book of Agriculture*. Washington, U.S. Govt. Printing Off., pp. 599–655. Also: 1947. 'On the distribution of angular velocity in gaseous envelopes under the influence of large-scale horizontal mixing processes'. *Bull. Amer. Meteor. Soc* **28** 53–68.

Saltzman, B., Fleischer, A., 1962. 'Spectral statistics of the wind at 500 mb'. *J. Atmos. Sci* **19** (3) 195–206.

Saltzman, B., Vernekar, A.D., 1968. 'Parametrization of the large-scale transient eddy flux'. *Mon. Weath. Rev.* **96** (12) 854–857.

Saltzman, B., Vernekar, A.D., 1971a. 'Note on the effect of the earth orbital radiation variations on climate'. *J. Geophys Res.*, **76** 4195–4197.

Saltzman, B., Vernekar, A.D., 1971b. 'An equilibrium solution for the axially symmetric component of the earth macroclimate'. *J. Geophys. Res.*, **76** 1498–1524.

Saltzman, B., Vernekar, A.D., 1972. 'Global equilibrium solution for the zonally averaged macroclimate'. *J. Geophys. Res.* **77** 3936–3945.

Saltzman, B., Vernekar, A.D., 1975. 'A solution for the northern hemisphere climatic zonation during a glacial maximum'. *Quaternary Res.* **5** 307–320.

Sawyer, J.S., 1972. 'Numerical weather prediction within World Weather Watch–past, present and future'. *Tenth Anniversary of the World Weather Watch*, Geneva, WMO Publ. No. 342–44.

Schlesinger, M.E., Mintz, Y., 1979. 'Numerical simulation of ozone production, transport and distribution with a global atmospheric general circulation model'. *J. Atmos. Sci.* **36** (7) 1325–1330, 1358–1361.

Seidov, D.G., 1978. 'A numerical scheme for the study of synoptic eddies in the ocean'. *Izv. Akad. Nauk SSSR, Fiz. Atmos. Okeana*, **14** (7) 757–767 (in Russian).

Seidov, D.G., 1980. 'Synoptic eddies in the ocean. A numerical experiment'. *Izv. Akad. Nauk SSSR, Fiz. Atmos. Okeana*. **16** (1) 73–87 (in Russian).

Sela, J., Wiin-Nielsen. A., 1971. 'Simulation of the atmospheric annual energy cycle'. *Mon. Wea. Rev.* **99** 460–468.

Sellers, W., 1969. 'A global climatic model based on the energy balance of the Earth–atmosphere system'. *J. Appl. Met.* **8** (3) 392–400.

Semtner, A.J., Holland, W.R., 1980. 'Numerical simulation of equatorial ocean circulation'. *J. Phys. Oceanogr.* **10** 667–696.

Sergin, V. Ya., Sergin, S. Ya., 1978. *A Systematic Analysis of the Relation between Large Climatic Oscillations and the Earth's Glaciation*. L. Gidrometeoizdat, 297 pp. (in Russian).

Sharaf, Sh. G., Budnikova, N.A., 1968. 'Oscillations of the solar irradiation of the Earth caused by secular changes of the Earth's orbital elements'. *Dokl. Akad. Nauk SSSR* **182** (2) 291–293 (in Russian).

Shtokman, V.B., 1946. 'The vertical propagation of heat waves in the sea and indirect methods for determining the thermal conductivity'. *Trudy Instituta Okeanoligii Akad. Nauk SSSR* **1** 3–46 (in Russian).

Shtokman, V.B., 1948. *Equatorial Countercurrents in the Oceans*. L. Gidrometeoizdat, 156 pp. (in Russian).

Smagorinsky, J., Manabe, S., Holloway, J.L., 1965. 'Numerical results from a nine-level general circulation model of the atmosphere'. *Mon. Wea. Rev.* **93** (12) 727–768.

Stepanov, V.N., 1974. *The World Ocean*. n. Znaniye, 255 pp. (in Russian).

Stommel, H., 1948. 'The westward intensification of wind-driven ocean currents'. *Trans. Amer. Geophys. Union* **29** (2) 202–206.

Stone, P.H., 1972. 'A simplified radiative-dynamic model for the static stability of rotating atmospheres', *J. Atmos. Sci.* **29** (3) 405–418.

Stone, P.H., 1973. 'The effect of large-scale eddies on climatic change', *J. Atmos. Sci.* **30** (4) 521–529.

Stuiver, M., 1978. 'Atmospheric carbon dioxide and carbon reservoir changes'. *Science* **199** (4326) 253–258.

Sverdrup, H.U., 1947. 'Wind-driven currents in a baroclinic ocean, with application to the equatorial currents of the eastern Pacific'. *Proc. Nat. Acad. Sci. U.S.A.* **33** (11) 318–326.

Takano, K., 1969. 'General circulation in the global ocean'. *J. Oceanogr. Soc. Japan*. **25** (1) 48–51. (in Japanese).

Takano, K., 1975. 'Eddy viscosity and grid size in ocean model'. *J. Oceanogr. Soc. Japan*. **31** 105–108.

Takano, K., 1975. 'A numerical simulation of the world ocean: preliminary results'. In: *Pro. Symp.* on *Numerical Models of Ocean Circulation* (Durham, New Hampshire, 17–20 October, 1972). National Academy of Sciences, Washington, D.C., pp. 121–129.

Thomson, T., 1857. 'Grand currents of atmospheric circulation'. British Assoc. Meeting, Dublin.

Untersteiner, N., Maykut, G.A., 1969. 'Arctic sea ice'. *Naval Res.* **22** (5) 11–23.

Verbitskii, M.Ya., Chalikov, D.V., 1980. 'A thermodynamic model of a large ice sheet'. *Dokl. Akad. Nauk SSSR*, **250** (1) 59 (in Russian).

Verbitskii, M.Ya., 1981. 'Numerical modeling of the evolution of the cover glaciation'. *Dokl. Akad. Nauk SSSR*. **256** (6) 1333–1337 (in Russian).

Veronis, G., 1973a. 'Large-scale ocean circulation'. In: *Advances in Applied Mechanics*. Academic Press, New York, Vol. 13, pp. 1–92.

Veronis, G., 1973b. 'Model of world ocean circulation: I. Wind-driven, two-layer'. *J. Mar. Res.* **31** 228–288.

Veronis, G., 1976. 'Model of world ocean circulation: II. Thermally driven, two-layer'. *J. Mar. Res.* **34** (2) 199–216.

Wang, W.C., Yung, Y.L., Lacis, A.A., Mo.T., Hansen, J.E., 1976. 'Greenhouse effects due to man-made perturbations of trace gases', *Science* **194** 685–690.

Washington, W.M., Semtner, A.J., Meehl, G.A., Knight, D.J., Mayer, T.A., 1980. 'A general circulation experiment with a coupled atmosphere, ocean and sea ice model'. *J. Phys. Oceanogr.* **10** (12) 1887–1908.

Welander, P., 1959. 'On the vertically integrated mass transport in the oceans'. In: *The Atmosphere and the Sea in Motion*, B. Bolin (ed.), Rockefeller Inst. Press./Oxford Univ. Press, pp. 95–101.

Whittaker, R.H., Likens, G.E., 1972. 'Carbon in the biota'. *Brookhaven National Laboratory Symp. on Carbon and the Biosphere*, May 16–18, 1972.

Whittaker, R.H., Likens, G.E., 1975. 'The biosphere and man'. In: *Primary Productivity of the Biosphere*, H. Leith and R.H. Whittaker (eds.), Springer-Verlag.

Wiin-Nielsen, A., 1967. 'On the annual variation and spectral distribution of atmospheric energy'. *Tellus* **19** (4) 540–559.

Williams, G.P., Davis, D.R., 1965. 'A mean-motion model of the general circulation'. *Quart. J. Roy. Meteor. Soc.* **91** (390) 471–489.

Wong, C.S., 1978. 'Atmospheric input of carbon dioxide from burning wood'. *Science* **200** (4338) 197–200.

Woodwell, G.M., Whittaker, R.H., Reiners, W.A., Likens, G.E., Delwiche, C.C., Botkin, D.B., 1977. 'The biota and the world carbon budget'. *Science* **199** (4325) 141–146.

Woodwell, G.M., 1978. 'The carbon dioxide question'. *Sci. Amer.* **238** (1) 34–43.

Zalesnyi, V.B., 1980. 'A model of World-Ocean dynamics in an atmosphere–ocean calculation system'. In: *Mathematical Modeling of Ocean Dynamics*. Computation Center of the Siberian Division of the USSR Academy of Sciences, Novosibirsk, pp. 76–88 (in Russian).

Zilitinkevich, S.S., Monin, A.S., Turikov, V.G., Chalikov, D.V., 1976. 'Numerical modeling of the combined atmospheric and oceanic circulation'. *Dokl. Akad. Nauk SSSR* **230** (3) 556–559 (in Russian).

INDEX

GPSR Compliance

The European Union's (EU) General Product Safety Regulation (GPSR) is a set of rules that requires consumer products to be safe and our obligations to ensure this.

If you have any concerns about our products, you can contact us on ProductSafety@springernature.com

In case Publisher is established outside the EU, the EU authorized representative is:

Springer Nature Customer Service Center GmbH
Europaplatz 3
69115 Heidelberg, Germany

Batch number: 10607214

Printed by Printforce, the Netherlands